VOLUME FIFTY SIX

THE ENZYMES
Tyrosinase

Serial Editors

CLAUDIU T. SUPURAN
Neurofarba Department, Pharmaceutical and Nutraceutical Section, University of Florence, Sesto Fiorentino, Florence; Neurofarba Department, Sezione di Chimica Farmaceutica e Nutraceutica, Universita degli Studi di Firenze, Firenze, Italy

VOLUME FIFTY SIX

THE ENZYMES

Tyrosinase

Edited by

CLAUDIU T. SUPURAN

Neurofarba Department, Pharmaceutical and Nutraceutical Section, University of Florence, Sesto Fiorentino, Florence; Neurofarba Department, Sezione di Chimica Farmaceutica e Nutraceutica, Universita degli Studi di Firenze, Firenze, Italy

ACADEMIC PRESS

An imprint of Elsevier

ELSEVIER

Academic Press is an imprint of Elsevier
125 London Wall, London, EC2Y 5AS, United Kingdom
50 Hampshire Street, 5th Floor, Cambridge, MA 02139, United States
525 B Street, Suite 1650, San Diego, CA 92101, United States

First edition 2024

Notices
Knowledge and best practice in this field are constantly changing. As new research and experience broaden our understanding, changes in research methods, professional practices, or medical treatment may become necessary.

Practitioners and researchers must always rely on their own experience and knowledge in evaluating and using any information, methods, compounds, or experiments described herein. In using such information or methods they should be mindful of their own safety and the safety of others, including parties for whom they have a professional responsibility.

To the fullest extent of the law, neither the Publisher nor the authors, contributors, or editors, assume any liability for any injury and/or damage to persons or property as a matter of products liability, negligence or otherwise, or from any use or operation of any methods, products, instructions, or ideas contained in the material herein.

ISBN: 978-0-443-29520-1
ISSN: 1874-6047

For information on all Academic Press publications
visit our website at https://www.elsevier.com/books-and-journals

Publisher: Zoe Kruze
Acquisitions Editor: Leticia Lima
Editorial Project Manager: Palash Sharma
Production Project Manager: Abdulla Sait
Cover Designer: Gopalakrishnan Venkatraman

Typeset by MPS Limited, India

Working together
to grow libraries in
developing countries

www.elsevier.com • www.bookaid.org

Contents

Contributors

Suleyman Akocak
Department of Pharmaceutical Chemistry, Faculty of Pharmacy, Adıyaman University, Adıyaman, Türkiye

Yousef A. Bin Jardan
Department of Pharmaceutics, College of Pharmacy, King Saud University, Riyadh, Saudi Arabia

Alessandro Bonardi
NEUROFARBA Department, Pharmaceutical and Nutraceutical Section, Laboratory of Molecular Modeling Cheminformatics & QSAR, University of Florence, Sesto Fiorentino, Firenze, Italy

Cristina Campestre
Department of Pharmacy, G. d'Annunzio University of Chieti-Pescara, Chieti, Italy

Clemente Capasso
Department of Biology, Agriculture and Food Sciences, Institute of Biosciences and Bioresources, CNR, Napoli, Italy

Simone Carradori
Department of Pharmacy, G. d'Annunzio University of Chieti-Pescara, Chieti, Italy

Luigi Franklin Di Costanzo
Department of Agriculture, Department of Excellence, University of Naples Federico II, Palace of Portici, Piazza Carlo di Borbone, Portici NA, Italy

Aslınur Doğan
Department of Pharmaceutical Chemistry, Faculty of Pharmacy, Adıyaman University, Adıyaman, Türkiye

Fosca Errante
Interdepartmental Laboratory of Peptide and Protein Chemistry and Biology; Department of Neurosciences, Psychology, Drug Research and Child Health, University of Florence, Sesto Fiorentino, Florence, Italy

Arianna Granese
Department of Drug Chemistry and Technology, "Sapienza" University of Rome, Rome, Italy

Paola Gratteri
NEUROFARBA Department, Pharmaceutical and Nutraceutical Section, Laboratory of Molecular Modeling Cheminformatics & QSAR, University of Florence, Sesto Fiorentino, Firenze, Italy

Huma Hameed
Faculty of Pharmaceutical Sciences, University of Central Punjab (UCP), Lahore, Pakistan

Ali Irfan
Department of Chemistry, Government College University Faisalabad, Faisalabad, Pakistan

Francesco Melfi
Department of Pharmacy, G. d'Annunzio University of Chieti-Pescara, Chieti, Italy

Amar Osmanović
Faculty of Pharmacy, University of Sarajevo, Sarajevo, Bosnia and Herzegovina

Anna Maria Papini
Interdepartmental Laboratory of Peptide and Protein Chemistry and Biology; Department of Chemistry "Ugo Schiff", University of Florence, Sesto Fiorentino, Florence, Italy

Luigi Pisano
Section of Dermatology, Health Sciences Department, University of Florence, Florence, Italy

Paolo Rovero
Interdepartmental Laboratory of Peptide and Protein Chemistry and Biology; Department of Neurosciences, Psychology, Drug Research and Child Health, University of Florence, Sesto Fiorentino, Florence, Italy

Laila Rubab
Department of Chemistry, Sargodha Campus, The University of Lahore, Sargodha, Pakistan

Ali Akbar Saboury
Institute of Biochemistry and Biophysics, University of Tehran, Tehran, Iran

Lucrezia Sforzi
Interdepartmental Laboratory of Peptide and Protein Chemistry and Biology; Department of Neurosciences, Psychology, Drug Research and Child Health, University of Florence, Sesto Fiorentino, Florence, Italy

Claudiu T. Supuran
Neurofarba Department, Pharmaceutical and Nutraceutical Section; Department of Neurosciences, Psychology, Drug Research and Child Health; Department of NEUROFARBA—Section of Pharmaceutical and Nutraceutical Sciences, University of Florence, Sesto Fiorentino, Florence, Italy

Martina Turco
Health Sciences Department (DSS), University of Florence, Florence, Italy

Ameer Fawad Zahoor
Department of Chemistry, Government College University Faisalabad, Faisalabad, Pakistan

Samaneh Zolghadri
Department of Biology, Jahrom Branch, Islamic Azad University, Jahrom, Iran

Preface

Volume 56 of *The Enzymes* presents the copper enzyme tyrosinase. This enzyme is extensively investigated due to its particular di-copper active site center, with various different states of the metal ions/enzyme being known in detail nowadays due to the relevant advances in structural biology achieved in the last decade. In addition, ultimately, tyrosinase is gaining an increasing attention due to its biomedical and biotechnological applications. Indeed, being involved in the biosynthesis of the skin pigment melanin (in vertebrates and other organisms), tyrosinase and the related enzymes tyrosinase-related protein 1 (TYRP1) and tyrosinase -related protein 2 started to be considered as drug targets for sking pigmentation diseases (both hypo- and hyperpigmentation problems) as well as in melanoma, a skin tumor difficult to treat with the presently available armamentarium of anticancer drugs and with an increasing incidence due to exposure to UV radiation. Furthermore, tyrosinase is present in many diverse organisms, from bacteria, plants, fungi, arthropods and mollusks, to vertebrates, since melanin is important in all of them for various functions related to the synthesis of secondary metabolites (in bacteria, plants and fungi), protection from UV radiation (mollusks and vertebrates) and so on.

The present volume is divided in 9 chapters which address various topics related to tyrosinase. Chapter 1 by Capasso and Supuran presents an overview of these enzymes present in bacteria, plants, fungi, arthropodes and vertebrates, with a general presentation of their molecular biology, phylogenetic relationship, catalyzed reactions (also comparing them with related enzymes such as the catecholoxidases and laccases), homology and functional divergence with the phylogenetically related hemocyanins, di-copper proteins involved in oxygen transport in invertebrates. A brief discussion of the potential of these enzymes as antibacterial drug targets is also provided.

Chapter 2, by Zolghadri and Saboury presents in detail the catalytic mechanism of tyrosinase, which involves electron transfer reactions between the copper(II) sites, leading to hydroxylation of monophenolic compounds to diphenols as well as the subsequent oxidation of these diphenols to corresponding dopaquinones. The fact that two different reactions are catalyzed by the same enzyme (present in four diverse oxidation states, with or without O_2 bound) leads to a rather complex kinetic behavior, which is analyzed in detail in the chapter. Not only the kinetics but also the inhibition mechanism is presented in detail, also considering the fact that there is much confusion in the literature between compounds acting only as tyrosinase inhibitors or those which can be both substrates and inhibitors of the enzyme, mainly due to the unfortunate erroneous papers from just one author.

Chapter 3, by Di Costanzo, discuss the structural biology of these enzymes. The first X-ray crystal structure of a tyrosinase has been reported only in 2006 from *Streptomyces castaneoglobisporus*, but thereafter many other such enzymes have been characterized, crystallized and analyzed in many other organisms. However, up to now only 122 crystal structures of such proteins (alone or with bound inhibitors are available in the Protein Database). The chapter presents in detail the main features of these proteins as determined by this important technique, highligthing the substrates/ inhibitors binding modes too.

Chapter 4, by Doğan and Akocak, presents the natural compounds (products) acting as tyrosinase inhibitors. Indeed, many bacterial, plant and fungal metabolites act as effective inhibitors of this enzyme, presumably due to its involvement in various physiological functions dependent on its oxiziding properties. Flavonoids, flavones, flavonoles, isoflavones, hydro-quinones, coumarins, kojic acid derivatives, chalcones and many other classes of natural products are examined in detail. I stress again here that there are several confusing papers in the literature denying the inhibitory effects of some of these compounds due to erroneous statements and poorly controlled experiments performed by one researcher which should be discarded as scientifically not sound.

Chapter 5, by Melfi et al. presents a detailed review of the drug design of tyrosinase inhibitors. Heterocyclic, aromatic and amino acid-based tyrosinase inhibitors are discussed based on the different design strategies used to obtain them together with the biological data for the inhibition of the enzyme from different sources (mostly the fungal enzyme has been used in initial work, but later the human enzyme was also available, although data on its inhibition are still rather scarce).

Chapter 6, by Errante et al., examines the peptide and peptidomimetic tyrosinase inhibitors. Amino acids and (oligo)peptides of natural and synthetic origin are reviewed, together with the design strategies which led to compounds incorporating peptide conjugated with other functionalities, which afforded highly effective tyrosinase inhibitors.

Chapter 7 by Bonardi and Gratteri discusses the computational studies of tyrosinase inhibitors. Both ligand-based and structure-based drug design strategies of tyrosinase inhibitors are available in the literature, based both on the crystal structures of crystallized enzymes or on homology models developed for enzymes not yet crystallized. Virtual screening, docking of inhibitors inside the tyrosinase active site as well as molecular dynamics studies are also available, with various degrees of reliability, depending on the used constraints/critical reasoning. As with all computational studies, the experimental validation is crucial for understanding whether a certain model is valid or not, and the chapter presents very well this situation discussing the critical issues to be considered when adopting computational procedures for drug design of enzyme inhibitors.

Chapter 8, by Irfan et al., reviews the bacterial tyrosinases and their inhibition. Five different types of bacterial tyrosinases are known to date, which differ by their biochemical and structural features. Apart *Streptomyces castaneoglobisporus* tyrosinase structure, recently the X-ray crystal structure of the enzyme from *Bacillus megaterium* has also been reported. Most of the natural product and synthetic bacterial tyrosinase inhibitors are thus discussed also considering future design strategies based on these X-ray data.

Chapter 9 by Pisano et al. discusse the biomedical applications of tyrosinases and of their inhibitors. Both the biomedical and the biotechnological applications are considered. In the first case, various hypopigmentation and depigmentation conditions (vitiligo, idiopathic guttate hypomelanosis, pityriasis versicolor, pityriasis alba) but also hyperpigmentations (melasma, lentigines, post-inflammatory and periorbital hyperpigmentation, cervical idiopathic poikiloderma and acanthosis nigricans) are presented together with the modalities to intervene pharmacologically, which are however rather limited. Indeed, hydroquinone, azelaic acid and tretinoin (all-*trans*-retinoic acid) are clinically used in the management of some hyperpigmentations, whereas many novel chemotypes acting as tyrosinase inhibitors with potential antimelanoma action are being investigated. Thiamidol, a synthetic heterocyclic derivative incorporating hydroquinone was recently approved as a new tyrosinase inhibitor for the treatment of melasma. Kojic acid and vitamin C are used for avoiding vegetable/food oxidative browning due to the tyrosinase-catalyzed reactions, whereas some bacterial tyrosinases have biotechnological applications for obtaining L-DOPA, an anti-Parkinson's disease drug.

Overall, the book presents updated information on many well-studied tyrosinases present in organisms all over the phylogenetic tree, a field in expansion in the last years.

I am grateful to all authors who contributed to this book, providing insightful chapters. I also wish to thank Leticia Lima from Elsevier Brasil and the desk editor Palash Sharma as well as the production editor Abdulla Sait for their constant help throughout the project.

CLAUDIU T. SUPURAN

CHAPTER ONE

Overview on tyrosinases: Genetics, molecular biology, phylogenetic relationship

Clemente Capasso[a,*] and Claudiu T. Supuran[b]

[a]Department of Biology, Agriculture and Food Sciences, Institute of Biosciences and Bioresources, CNR, Napoli, Italy
[b]Neurofarba Department, Pharmaceutical and Nutraceutical Section, University of Florence, Sesto Fiorentino, Florence, Italy
*Corresponding author. e-mail address: clemente.capasso@ibbr.cnr.it

Contents

Abstract

Tyrosinases (TYRs) are enzymes found in various organisms that are crucial for melanin biosynthesis, coloration, and UV protection. They play vital roles in insect cuticle sclerotization, mollusk shell formation, fungal and bacterial pigmentation, biofilm formation, and virulence. Structurally, TYRs feature copper-binding sites that are essential for catalytic activity, facilitating substrate oxidation via interactions with conserved histidine residues. TYRs exhibit diversity across animals, plants, fungi, mollusks, and bacteria, reflecting their roles and function. Eukaryotic TYRs undergo post-translational modifications, such as glycosylation, which affect protein folding and activity. Bacterial TYRs are categorized into five types based on their structural variation, domain organization and enzymatic properties, showing versatility across bacterial species. Moreover, bacterial

ISSN 1874-6047, https://doi.org/10.1016/bs.enz.2024.05.010

1

TYRs, akin to fungal TYRs, have been implicated in the synthesis of secondary metabolites with antimicrobial properties. TYRs share significant sequence homology with hemocyanins, oxygen-carrier proteins in mollusks and arthropods, highlighting their evolutionary relationships. The evolution of TYRs underscores the dynamic nature of these enzymes and reflects adaptive strategies across diverse taxa.

1. Introduction

Tyrosinases (TYRs, EC 1.14.18.1) represent a class of enzymes that are ubiquitous across diverse biological kingdoms, including animals, plants, insect, fungi, and bacteria [1–7]. These enzymes use molecular oxygen to catalyze the hydroxylation of monophenols (molecules with one phenol group) to o-diphenols (molecules with two phenol groups) and the subsequent oxidation of o-diphenols to o-quinones (molecules with two oxygen atoms attached to the benzene ring in place of the hydroxyl groups), pivotal reactions in the biosynthesis of melanin, the pigment responsible for coloration in various organisms (Fig. 1) [8–11]. After these

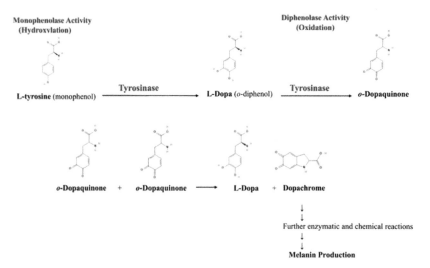

Fig. 1 Schematic representation of the reaction catalyzed by TYR leading to melanin formation. Exerting monophenolase activity, TYR hydroxylates monophenols, such as L-tyrosine, to o-diphenols, exemplified by L-Dopa. For the diphenolase activity, TYR catalyzes the oxidation of o-diphenols to o-quinones, specifically o-dopaquinone. Simultaneously, concerted chemical reactions occur wherein two molecules of o-dopaquinone undergo a self-reaction, yielding an o-diphenol molecule (L-Dopa) and dopachrome which is the "monomer" for the formation of melanin through a cycloaddition type chain reaction.

enzymatic reactions, two o-quinone molecules, o-dopaquinone, undergo a reaction yielding dopachrome and regenerating a molecule of o-diphenol.

After enzymatic oxidation of substrates by TYR, reactive quinone intermediates are generated. These quinones possess high reactivity and propensity to undergo non-enzymatic polymerization reactions [12]. This polymerization process involves spontaneous binding of quinone molecules to each other, forming larger macromolecular structures known as melanins, which are heterogeneous, insoluble pigments with diverse structures and properties. The non-enzymatic polymerization of quinones to melanins is a crucial step in melanogenesis, as it allows for the stabilization and accumulation of pigments within the melanocytes [13–15]. Melanins serve important biological functions, including protection against UV radiation, regulation of skin pigmentation, and scavenging of free radicals [15,16]. The spontaneous polymerization of quinones contributes to the structural and functional diversity of melanin in various biological systems [14,15,17]. This intricate process underscores the multifaceted nature of TYR catalytic abilities, contributing significantly to melanogenesis. TYR is the initial and rate-limiting enzyme of the melanin biosynthetic pathway. However, melanin production is not solely dependent on TYR; it is a complex process involving multiple enzymes and regulatory factors [1,8,14–16,18]. For instance, downstream enzymes, such as TYR-related protein 1 (TYRP1) and TYR-related protein 2 (TYRP2 and known as dopachrome tautomerase (DCT) or dopachrome isomerase (DCI)), play distinct roles in melanin synthesis. TYRP1 is involved in the conversion of dopachrome to 5,6-dihydroxyindole-2-carboxylic acid (DHICA), a critical step in the synthesis of eumelanin, a dark-brown to black pigment [19,20]. TYRP1 is localized within melanosomes, specialized organelles where melanin synthesis occurs, and it interacts with other melanogenic enzymes to regulate melanin polymerization and maturation [21,22]. DCT catalyzes the conversion of dopachrome to 5,6-dihydroxyindole (DHI), a precursor in both the eumelanin and pheomelanin (yellow to reddish-brown pigment) synthesis pathways [23–25]. DCT is predominantly expressed in melanocytes, where it participates in the regulation of melanin pigment production and contributes to the determination of skin and hair color. Genetic variations in genes encoding melanogenic enzymes, such as TYR, TYRP1, and DCT, can influence melanin production and skin pigmentation [26–28]. Moreover, melanin production is tightly regulated by various factors, including hormonal signals, cytokines, UV radiation, and genetic factors. Hormones, such as melanocyte-stimulating hormone (MSH), adrenocorticotropic hormone (ACTH), and estrogen play crucial roles in regulating melanin synthesis [29,30].

1.1 Exploring the catalytic spectrum of phenoloxidases

In the literature, the term "phenoloxidase" is frequently employed to encompass a group of enzymes, including TYRs, catecholoxidases, and laccases [31–34]. These enzymes share similarities in their catalytic mechanisms and substrate specificity (Table 1).

TYRs primarily catalyze the oxidation of both monophenols and o-diphenols, whereas whereas catecholoxidases specifically act on o-diphenols to produce o-quinones [8–11]. On the other hand, laccases can oxidize a broad range of substrates, including phenolic and non-phenolic compounds, through different catalytic mechanisms. This versatility makes laccases valuable enzymes in various biotechnological applications, including lignin degradation, pollutant detoxification, and more [35]. TYRs, catecholoxidases, and laccases may exhibit overlapping substrate specificities, but they are not identical in their enzymatic activities [36,37].

Table 1 Characteristics of TYRs, catecholoxidases, and laccases.

Enzyme	Substrate specificity	Structural features	Biological roles
TYR	Monophenols and o-diphenols	Contains copper ions in active sites, typically organized into two copper centers	Involved in melanin synthesis, wound healing, and defense mechanisms
Catecholoxidase	o-Diphenols	Contains copper ions in active sites, typically organized into three copper centers	Participates in various oxidation reactions, including melanin formation
Laccase	Broad range including phenolic and non-phenolic compounds	Contains multiple copper ions and a distinctive protein structure. Typically organized into four copper centers	Participates in lignin degradation, pollutant detoxification, and more

Some phenoloxidases can perform both catecholase and cresolase activities but having one does not necessarily mean having the other because of differences in enzyme structure and substrate preferences [38]. Catecholase activity refers specifically to the enzymatic conversion of catechol to ortho-quinone, whereas cresolase activity involves the conversion of cresols to methylquinones [39]. A positive test for catecholase activity does not necessarily indicate the presence of cresolase activity [3–7,40]. Different structural configurations of phenoloxidases can lead to unique preferences for substrate binding and catalytic abilities [41,42]. Enzymes utilize specific mechanisms to facilitate substrate oxidation, and differences in these mechanisms can influence their ability to catalyze different reactions [3–7]. Enzymes may have evolved to perform specific functions in different biological contexts, resulting in variations in their activities [41,42]. Therefore, when characterizing phenoloxidases, it is essential to consider their specific enzymatic activities and substrate preferences rather than assume uniformity across the entire group [3–7]. This distinction ensures accurate interpretation of the experimental results and a deeper under-standing of the functional diversity within the phenoloxidase family.

1.2 TYRs in biological systems

TYRs exert a significant influence on critical processes essential for orga-nismal survival and functionality across diverse taxa [43]. One of their pri-mary functions is the synthesis of melanin, the primary pigment responsible for coloration in organisms ranging from vertebrates to invertebrates [44].

In humans, melanin synthesis, facilitated by TYR activity is indis-pensable because it confers protective benefits against harmful ultraviolet radiation and plays a significant role in physiological processes such as thermoregulation and camouflage [45,46].

TYR is indispensable in the process of insect cuticle sclerotization, a crucial step in the development of insect exoskeletons [47]. Through their enzymatic action, TYRs catalyze the cross-linking of cuticular proteins, imparting strength and rigidity to insect exoskeleton [47]. This process is essential for providing structural support, facilitating locomotion, and protecting insects from environmental stresses [47].

In mollusks, particularly in bivalves TYR have expanded significantly [48,49]. Species such as *Pinctada fucata martensii* (pearl oyster cultured as seafood) exhibit a remarkable abundance of TYRs genes compared with other organisms, highlighting their importance in shell formation and other physiological processes [48,49]. These enzymes are not only integral to melanin production but also contribute to the formation of periostracum, biogenesis of initial non-calcified shell structures, and coloration of shells. Moreover, studies have underscored their involvement in the mollusk innate immune system, indicating their multifaceted role beyond melanin biosynthesis [48]. Notably, they are highly expressed in calcified tissues, where they directly participate in the synthesis of shell matrix proteins (SMPs). The silencing of tyrosinase in certain species leads to disordered growth of the nacreous layer, underscoring their direct involvement in shell formation. Thus, the expansion of mollusk tyrosinases not only broadens their catalytic functions, but also underscores their direct participation in the intricate process of calcified shell formation.

In fungi, TYR is involved in various biological processes, including pigmentation, pathogenesis, and formation of structural components such as cell walls and spore coats [2]. Melanin production by fungal TYRs enhances resistance to environmental stresses such as UV radiation and host immune responses, contributing to fungal virulence and persistence [50]. TYRs participate in the synthesis of secondary metabolites with diverse biological activities, including toxins and antibiotics, highlighting their importance in fungal physiology and ecology [51].

TYR is implicated in pigment production, biofilm formation, and virulence in bacteria [52]. Bacterial TYR contributes to pigmentation by catalyzing the production of melanin-like compounds, which provide protection against oxidative stress and host defense [53]. Furthermore, TYR plays a role in the formation and maintenance of bacterial biofilms (multicellular communities that confer resistance to antimicrobial agents) and host immune responses [54]. The bacterial TYRs, as the fungal TYRs, are associated with the synthesis of secondary metabolites with antimicrobial properties, suggesting their potential as targets for novel antibiotic development [55,56].

In addition to their functions in animal biology, fungi, and bacteria, TYR is crucial in plant pigmentation and plays a significant role in regulating the synthesis of pigments related to coloration, photoprotection, and signaling [57,58]. Through the enzymatic conversion of phenolic compounds, TYR contributes to the production of pigments, such as anthocyanins, flavonoids, and lignins, which play diverse roles in plant physiology, including attraction of pollinators, defense against herbivores, and adaptation to environmental cues [59,60].

The multifaceted functions of TYR in melanin synthesis underscore their significance in shaping phenotypic traits, ecological interactions, and adaptive strategies across biological systems. Elucidating the molecular mechanisms governing TYR activity has profound implications for understanding the dynamics of organismal development, evolution, and ecological resilience.

1.3 Description of the genes encoding TYRs in various organisms

The genetic basis underlying TYR function exhibits notable diversity across taxa, reflecting the evolutionary divergence and specialization of this enzyme across biological kingdoms. In humans, the gene encoding TYR is located on chromosome 11q14-q21 and spans approximately 65 kb [61]. *TYR* gene consists of five exons and four introns, with exon 1 containing the translation initiation site and exons 2–5 encoding the structural domains crucial for enzymatic activity [62]. The exon–intron organization of *TYR* plays a crucial role in the regulation of gene expression, mRNA processing, and protein synthesis. Exon 1 contains a transcription start site and encodes the 5′ untranslated region (UTR) of the mRNA. Exons 2 encodes the protein-coding regions of the TYR, including the conserved copper-binding domains essential for enzymatic activity [62]. These exons are interspersed with intronic regions that are removed during mRNA splicing, leading to the formation of mature mRNA transcript for translation into functional TYR protein of 529 amino acids (Fig. 2).

```
tyrosinase [Homo sapiens]
GenBank: AAB60319.1

    1 mllavlycll wsfqtsaghf pracvssknl mekeccppws gdrspcgqls grgscqnill
   61 snaplgpqfp ftgvddresw psvfynrtcq csgnfmgfnc gnckfgfwgp ncterrllvr
  121 rnifdlsape kdkffayltl akhtissdyv ipigtygqmk ngstpmfndi niydlfvwmh
  181 yyvsmdallg gseiwrdidf aheapaflpw hrlfllrweq eiqkltgden ftipywdwrd
  241 aekcdictde ymggqhptnp nllspasffs swqivcsrle eynshqslcn gtpegplrrn
  301 pgnhdksrtp rlpssadvef clsltqyesg smdkaanfsf rntlegfasp ltgiadasqs
  361 smhnalhiym ngtmsqvqgs andpifllhh afvdsifeqw lqrhrplqev ypeanapigh
  421 nresymvpfi plyrngdffi sskdlgydys ylqdsdpdsf qdyiksyleq asriwswllg
  481 aamvgavlta llaglvsllc rhkrkqlpee kqpllmeked yhslyqshl
```

Signal Peptide: 1-17 aa

Tyrosinase CuA-binding region signature: 202-219 aa

Tyrosinase and hemocyanins CuB-binding region signature: 383-394 aa

Fig. 2 The amino acid sequence of human TYR with residues involved in CuA- and CuB-binding are highlighted in yellow and green, respectively. CuA- and CuB-binding sites were identified using PROSITE Motif Identification, a bioinformatic tool for identifying conserved motifs and functional domains in protein sequences. The SignalP program (version SignalP 5.0) was used to predict the presence of signal peptides and the location of cleavage sites in human TYR.

The N-terminal signal peptide shown in Fig. 2 serves as a signal for targeting the protein to a specific cellular location. In this case, it likely targets the protein to melanosomes, which are specialized compartments within melanocytes where melanin synthesis occurs [63]. The copper-binding sites allow the catalytic activity of tyrosinase. Copper ions serve as cofactors, directly participating in the enzymatic reaction by accepting and donating electrons [64]. Transcriptional regulation of TYR harbors binding sites for transcription factors involved in the spatial and temporal control of TYR expression, thereby modulating melanin synthesis in response to developmental, hormonal, and environmental cues [65–67]. Overall, the regulation of TYR gene expression involves a complex interplay of transcriptional and post-transcriptional mechanisms that fine-tune the abundance, stability, and translational efficiency of TYR mRNA molecules in response to intrinsic and extrinsic signals [65–67]. In mammals, birds, reptiles, and amphibians, TYR genes exhibit considerable conservation in their structural organization and functional domains [34,43,44]. Orthologs of the human *TYR*'s gene have been identified in various animal species, highlighting the evolutionary conservation of TYR function in mediating pigmentation and other physiological processes [68]. However, interspecies variations in TYR gene sequences and regulatory elements contribute to diverse phenotypic manifestations of pigmentation across animal taxa, reflecting adaptations to ecological niches and selective pressures.

In animals, plants, mollusks, and fungi *TYR* genes may exhibit variations in exon-intron organization while retaining structural and functional conservation [69]. For example, animal *TYR* genes often consist of a similar number of exons encoding conserved catalytic domains. However, intron lengths and positions may vary between species [69]. Similarly, plant and fungal TYR genes may display differences in exon-intron structure compared to their animal counterparts, reflecting the unique evolutionary history and physiological requirements of these organisms [70]. The genetic architecture of *TYR* genes in plants is complex, with multiple gene family members encoding functionally diverse isoforms involved in distinct aspects of pigment biosynthesis and cellular differentiation [70]. Moreover, the expression of plant TYR genes is tightly regulated by developmental and environmental stimuli, and hormonal signaling pathways, reflecting the dynamic integration of endogenous and exogenous factors in modulating plant pigmentation [31,32,36,37,40,51]. Fungal TYR genes display structural similarities with their counterparts in animals and plants, comprising conserved catalytic domains and regulatory elements involved in transcriptional control [2,32,39,50,58]. Furthermore, fungal TYR genes often exhibit functional diversification, with distinct isoforms specialized in melanin synthesis in different cell types or under specific environmental conditions, thereby enhancing the adaptability and virulence of fungal pathogens [71].

In bacteria, TYRs are encoded by genes typically found within the bacterial genome and their genetic organization varies across bacterial species [53,72]. Although the presence of *TYR* genes is widespread among diverse bacterial taxa, their sequence conservation and structural features can differ significantly, reflecting the evolutionary divergence and functional specialization of these enzymes in different ecological niches [53,72]. The genetic architecture of bacterial TYR genes often consists of a single open reading frame (ORF) encoding the TYR protein. These genes may be located in the bacterial genome or carried within plasmids, mobile genetic elements that facilitate horizontal gene transfer and genetic exchange between bacterial populations [73]. The organization of bacterial TYR genes may also include regulatory elements such as promoter regions, transcription factor binding sites, and ribosome binding sites, which govern gene expression in response to environmental or developmental signals [73]. Bacterial TYRs are enzymes responsible for melanin synthesis and play crucial roles in pigment production. Studies show they help bacteria withstand environmental stresses like UV radiation, oxidative damage, and antimicrobial agents [52,55,72–74]. By producing melanins with antioxidant properties, TYRs help scavenge harmful reactive oxygen species, reduce oxidative stress, and boost bacterial fitness and stress tolerance. Moreover, TYRs have been implicated in the development of infectious diseases caused by bacterial pathogens, highlighting their multifaceted importance in both bacterial survival and virulence. Certain pathogenic bacteria utilize TYRs to produce melanin-like pigments within host tissues, facilitating immune evasion, host colonization, and tissue invasion [52,55,72–74]. The production of melanin by bacterial pathogens can also confer resistance to antimicrobial agents and host defense mechanisms, thereby exacerbating the severity and persistence of bacterial infections [75]. Finally, bacterial TYRs have garnered attention as biocatalysts for biotechnological applications, including the synthesis of melanin-based materials, enzymatic transformations, and environmental bioremediation [74]. The enzymatic activity of bacterial TYRs enables the production of melanin nanoparticles with unique physicochemical properties suitable for diverse biomedical and industrial applications [76,77]. Bacterial TYRs have been exploited for their ability to degrade aromatic pollutants and xenobiotic compounds, thereby offering potential solutions for environmental remediation and pollution control [78]. Overall, genetic characterization of bacterial TYR genes provides insights into their functional versatility, ecological significance, and biotechnological potential. By elucidating the genetic basis of TYR-mediated processes in bacteria, researchers can unravel the fundamental principles governing microbial adaptation, host-pathogen interactions, and environmental resilience in bacterial ecosystems.

2. Structural homology and functional divergence in TYRs and hemocyanins

2.1 TYRs and hemocyanins

TYR enzymes exhibit intricate structural features that are pivotal for their catalytic functions. At the heart of their catalytic process lies the coordination of copper ions within a distinct active site [79,80]. The copper binding sites found in TYRs exhibit a remarkable degree of sequence similarity with those present in hemocyanins, which are oxygen carrier proteins predominantly found in mollusks and arthropods [81,82]. This noteworthy homology suggests a shared evolutionary origin and implies a functional relationship between the two protein families [81,83]. Table 2 shows the distinguishing features of tyrosinases and hemocyanins.

Specifically, the copper binding sites in both TYRs and hemocyanins play pivotal roles in facilitating their respective biological functions. Hemocyanins serve as oxygen transporters in the circulatory systems of mollusks and arthropods, and function analogously to hemoglobin in vertebrates [81,83]. These proteins contain copper ions coordinated within their active sites, which enable the reversible binding and release of oxygen molecules. The highly conserved sequence motifs surrounding the copper binding sites in hemocyanins reflect the critical role of these residues in maintaining the structural integrity and oxygen-binding capacity of the protein [84]. In contrast, TYRs are enzymes involved in melanin biosynthesis and other metabolic processes [43]. Despite their enzymatic function, the structural resemblance between TYRs and hemocyanins, particularly in the copper-binding sites, suggests a potential evolutionary link between these protein families [80,85].

Table 2 Comparative features of TYRs and hemocyanins.

Feature	Tyrosinases	Hemocyanins
Biological Function	Enzymes involved in melanin synthesis and other processes	Respiratory proteins for oxygen transport
Active Site	Binuclear copper active site	Copper-containing active site
Catalytic Activity	Catalyzes enzymatic reactions	Primarily involved in oxygen binding/release
Oxygen Transport	No	Yes
Structural Composition	Single polypeptide chain with copper site	Multi-subunit protein with copper-containing sites
Distribution	Found in various organisms (bacteria, fungi, plants, animals)	Primarily found in mollusks and arthropods

3. Evolutionary dynamics and functional diversification of type-3 copper proteins

In the realm of molecular biology, understanding the evolutionary dynamics of protein families illuminates the intricate pathways that have shaped life on Earth. Among these, type-3 copper proteins are crucial players with diverse functions across taxa. Aguilera et al. [80] delved deeply into the origins and evolutionary trajectories of these proteins, shedding light on their intricate evolutionary history and functional diversification. It has been proposed that during the course of evolution, there may have been a functional transition within these protein families from enzymatic oxygen detoxification, as seen in TYRs, towards oxygen transport, as observed in hemocyanins [80,85]. This proposed functional shift underscores the dynamic nature of protein evolution and the adaptive strategies employed by organisms to meet the changing physiological demands. Although TYRs and hemocyanins have distinct biological roles, their shared ancestry and structural similarities highlight the intricate interplay between enzymatic function and oxygen transport mechanisms across diverse taxa [85]. By meticulously examining sequenced genomes across different metazoan phyla, Aguilera et al. unveiled the ubiquity of type-3 copper proteins and elucidated their evolutionary relationships. The classification of type-3 copper proteins into three subclasses is predicated on their domain architecture and the presence of conserved residues within copper-binding sites. A systematic examination of the sequenced genomes across the three domains of cellular life revealed the ubiquity of type-3 copper proteins, with representatives found in each domain [80]. All type-3 copper exhibited a conserved pair of copper-binding sites, denoted as Cu(A) and Cu(B). However, these can be further classified into subclasses based on additional conserved domains or motifs. The α-subclass typically features an N-terminal signal peptide, suggesting secretion or vesicular localization, while the β-subclass lacks this domain and is presumably localized to the cytosol. The γ-subclass, on the other hand, possesses an N-terminal signal peptide, a cysteine-rich region potentially implicated in protein–protein interactions or dimerization, and a transmembrane domain indicative of membrane-bound localization [80]. Among the kingdoms, Metazoa is unique in possessing all three subclasses of type-3 copper proteins. Non-holozoan unikonts (e.g., amoebozoans and fungi) lack γ-subclass members, with only the α-subclass present in non-unikont eukaryotes, bacteria, and archaeal genomes surveyed. This distribution aligns with the notion that the secreted α-subclass may be ancestral, potentially existing in the last universal common ancestor of all cellular life. Subsequent duplications and divergence along the unikont stem likely gave rise to the cytosol-localized β-subclass. The membrane-bound γ form, exclusive to metazoan genomes, appears to have evolved from a second duplication event in an α-subclass type-3 copper protein gene, as evidenced by its close relationship with α-subclass proteins. This inference is supported by the absence of type-3 copper proteins in the holozoans *Monosiga brevicolis* and *Capsaspora owczarzaki* [80]. The three subclasses of type-3 copper proteins delineated earlier were supported by phylogenetic analyses, with the β-subclass demonstrating the highest degree of divergence [80]. Specifically, phylogenetic examination situates the γ-subclass as a clade within unresolved α-subclass polytomy. This form of membrane-bound type-3 copper proteins encompasses only tyrosinases and tyrosinase related-proteins. Previous structural and enzymatic investigations have established that the α-subclass encompasses tyrosinases, catechol oxidases, mollusks and urochordate hemocyanins. Phylogenetic analyses revealed the absence of α-subclass in various metazoan phyla and subphyla, including poriferans, placozoans, platyhelminths, arthropods, hemichordates, echinoderms, cephalochordates, and chordates. Molluscan and urochordate hemocyanins, residing within the α-subclass, form a robustly supported monophyletic group, with the exception of *Lottia gigantea* hemocyanin. This observation suggests two plausible scenarios: either the last common ancestor of extant bilaterians possessed an α-subclass hemocyanin, which was subsequently lost in most phyla, or convergent evolution of this gene occurred in molluscs and urochordates. Within each subclass, lineage-specific gene expansion was observed. Notable examples include soybeans (α-subclass), nematodes (α-subclass), mosquitoes (β-subclass), and amphioxus (γ-subclass). Generally, these expansions entail the proliferation of a particular functional class of type-3 copper proteins, such as the presence of 18 tyrosinases in amphioxus. In arthropods, duplication of β-subclass type-3 copper proteins gave rise to contemporary arthropod hemocyanin and tyrosinase proteins. Examination of the genomic architecture of type-3 copper protein genes revealed clusters of two-seven linked genes in the genomes of soybean (Glycine max, α-subclass), brown algae (Ectocarpus siliculosus, α-subclass), mosquito (Anopheles gambiae, β-subclass), and amphioxus (Branchiostoma floridae, γ-subclass), consistent with subclass expansions primarily resulting from tandem gene duplication events. Thus, the phylogenetic analysis suggests that cytosolic (β) and membrane-bound (γ) forms evolved from an ancestral type-3 copper protein, which likely originated from predominantly secreted precursor (α). This evolutionary transition suggests a dynamic process wherein the ancestral protein, likely involved in extracellular functions, undergoes structural modifications and adaptations to fulfill intracellular or membrane-associated roles. Such evolutionary shifts are often driven by changes in gene expression, protein localization, and interactions with other biomolecules, ultimately leading to functional divergence and adaptation to new cellular environments.

4. Structural characteristics of TYR proteins
4.1 Structural characteristics of human TYR

Currently, there is no existing crystallographic structure of human TYR. However, researchers have addressed this gap by constructing a model that relies on the crystal structure of TYRP1 [86]. This structural blueprint was sourced from the SWISS-MODEL template library, an extensive repository housing experimentally determined protein structures extracted from the Protein Data Bank (Figs. 3 and 4) [86].

The structural sequence alignment of human TYR is presented in Fig. 3, employing the SWISS-MODEL template library to identify and align the structural templates with the target sequence of human TYR.

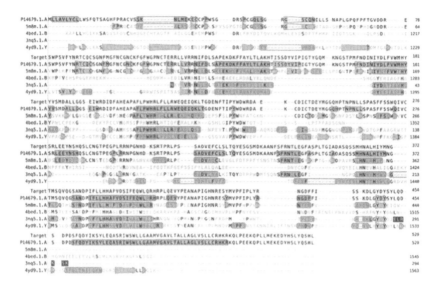

Fig. 3 Structural sequence alignment of the human TYR. Structural templates were identified and aligned with the target sequence of human TYR using the SWISS-MODEL template library. Legend: P14679.1.A, Human TYR sequence; 5m8m.1.A, tyrosinase related protein 1 in complex with kojic acid; 4bed.1.B, Keyhole limpet hemocyanin (KLH); 3nq5.1.A, Tyrosinase from *Bacillus megaterium* R209H mutant; 4yd9.1.Y, squid hemocyanin. Violet indicates secondary structure color for α-helices; green indicates secondary structure color for β-strands.

Fig. 4 shows a model of the 3D structure of the human TYR. This figure offers valuable insights into the structural characteristics of the protein. Notably, it highlighted the α-helical transmembrane domain of TYR situated within the cell membrane, which anchors the protein to the melanosome membrane. As shown in Fig. 4, a small flexible cytoplasmic and a large intra-melanosomal domains are visible. The small region at the C-terminus of the protein extends into the cytoplasm of melanocytes, while the large intra-melanosomal domain is located within the melanosome, where the main activity of the protein, such as catalyzing chemical reactions involved in melanin synthesis, takes place.

Typically, characteristic shared among TYRs is the presence of a "type 3 copper center", harboring two essential copper-binding sites (Fig. 5) [1,79].

In this configuration, two copper atoms, typically denoted as CuA and CuB, are coordinated by six conserved histidine residues (H_{A1}, H_{A2}, H_{A3}, H_{B1}, H_{B2}, H_{B3}) [87]. These copper ions can bind to one oxygen molecule and serve as electron acceptors, facilitating the oxidation of substrates

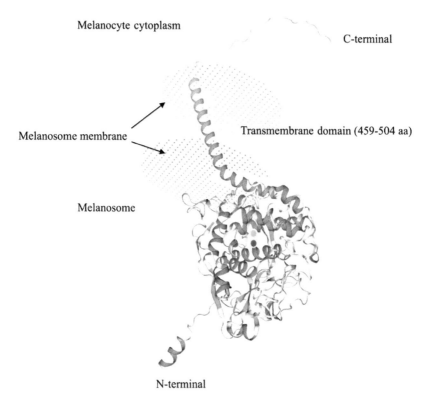

Melanocyte cytoplasm

C-terminal

Melanosome membrane

Transmembrane domain (459-504 aa)

Melanosome

N-terminal

Fig. 4 Model of the 3D structures of human TYR. The figure reveals the transmembrane domain, which is positioned within the cell membrane (black dots). Moreover, the N- and C-termini are illustrated, which serve as the starting and concluding points of the protein sequence, respectively. Additionally, the figure shows copper ions (A and B), represented by red and orange spheres, which are indispensable for the proper functioning of TYR. The β-strands and α-helix are portrayed in blue and green, respectively.

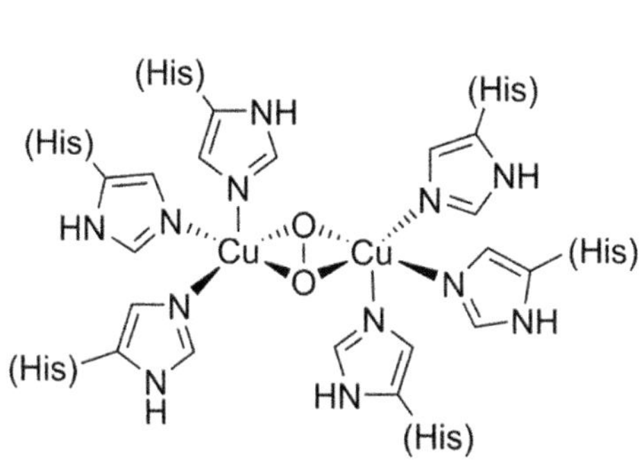

Fig. 5 Schematic representation of a "type 3 copper center".

during enzymatic reactions. Histidine residues, integral components of the active site, engage in coordinated interactions with copper ions, thereby fortifying their binding and facilitating electron transfer during substrate oxidation [1]. This coordination forms a distinct structural motif, often described as a 'four α-helix bundle,' forming a hydrophobic pocket near the protein surface and indicating the arrangement of protein helices that surround the copper atoms (Fig. 4) [88]. In most TYR proteins, four or five of the six essential histidine residues are located within these helical fragments. For CuA, one histidine ligand (H_{A1}) was positioned before a pair of histidines (H_{A2}, H_{A3}), with H_{A2} likely located in a coil fragment. The arrangement of these histidines differs from that of hemocyanins, in which the three histidine ligands are situated within α-helical fragments. Similarly, for CuB, two histidines (H_{B1} and H_{B2}) are separated by three residues, with the third ligand (H_{B3}) located after this pair. These histidines are distributed across different α-helix fragments. Notably, H_{B3} formed a His-His arrangement [88]. Recent studies have demonstrated that H_{B3} serves as the second ligand for CuB, although the first histidine is crucial for the stereospecificity of o-diphenolic substrates. Mutations affecting the second histidine have been associated with human albinism, highlighting their importance in enzyme activity and substrate specificity. Thus, histidine residues are instrumental in preserving the structural integrity and catalytic activity of TYR [85,89].

Additional residues beyond the conserved catalytic core have been identified as being crucial for catalysis and substrate specificity. Among these are a "gatekeeper" amino acid, positioned near CuA, and a "waterkeeper" residue located at the entrance of the active site. The gatekeeper amino acid is proposed to have multiple functions, including influencing copper ion incorporation, stabilizing substrate binding within the active site, and regulating enzymatic activity to prevent undesirable oxidation of phenolic compounds [85,89]. By controlling access to the active site, the gatekeeper ensures that only the appropriate substrates are oxidized. The waterkeeper residue, situated at the entrance of the active site, is believed to play a role in the deprotonation of monophenolic substrates, facilitating their interaction with copper ions and subsequent oxidation [90,91]. Together, these additional residues contribute to the fine-tuning of enzymatic activity and substrate specificity of TYRs and related enzymes.

Moreover, TYR proteins possess a carboxyl C-terminal tail that plays a pivotal role in protein sorting and in targeting specific cellular compartments, such as melanosomes in melanocytes [92,93]. Experimental evidence suggests that the C-terminal tail contains essential signals, including the dileucine motif (LL) and tyrosine-based motif (YXXB), which interact with adaptor proteins to facilitate protein transport [94]. Glycosphingolipids are also implicated in the processing in the Golgi complex, further highlighting the complexity of protein trafficking mechanisms [95,96]. In eukaryotic TYRs, C-terminal extension regulates enzyme activity by obstructing substrate access to the active site. This extension may also play a role in copper ion incorporation [94].

4.2 TYRs in plants, fungi, and mollusks

Plant TYRs represent a diverse group of enzymes distributed among various plant species that play pivotal roles in pigment biosynthesis, defense mechanisms, and environmental adaptation [31,32,36,37,51,97,98]. These enzymes exhibit structural characteristics that distinguish them from their mammalian counterparts, reflecting their diverse functions in plant cells. In terms of size, plant TYRs vary widely in length, with primary structures encompassing varying numbers of amino acid residues, depending on the specific enzyme and plant species [31,32,36,37,51,97,98]. This variability in sequence length contributes to the diversity of TYR isoforms found across different plants. Structurally, plant TYRs display notable differences in their primary sequence and domain organization compared to mammalian TYRs [32,97,98]. While sharing certain conserved catalytic residues with human TYR, plant TYRs often exhibit distinct domain arrangements and amino acid composition. These structural variances likely underlie the differences in enzymatic properties and substrate specificity between plant and mammalian TYRs [99]. One significant structural difference is the absence of transmembrane domains in the plant TYRs. Unlike mammalian TYRs, which are often membrane-bound and localized within cellular compartments, such as melanosomes, plant TYRs may lack transmembrane domains [99,100]. Instead, they are commonly found within plastids or other cellular compartments, where they participate in pigment bio-synthesis and other metabolic processes. Despite these structural disparities, plant TYRs have functional similarities with their mammalian counter-parts. Both enzyme groups catalyze the oxidation of phenolic compounds, leading to pigment production and other physiological responses. However, the unique structural features of plant TYRs likely confer specific enzymatic properties tailored to the requirements of plant metabolism and environmental adaptation [100].

Fungal TYRs play essential roles in pigment production, pathogenesis, and maintaining structural integrity within fungal organisms. Structurally, these enzymes exhibit notable variations compared with their human counterparts [97]. In terms of structural features, fungal TYRs may display differences in primary sequence, domain organization, and copper-binding motifs compared with human TYR. These structural disparities likely underlie the differences in enzymatic properties and substrate specificity between fungal and human TYRs. One salient feature of fungal TYRs is the presence of transmembrane domains that anchor enzyme within the cellular membranes [97]. These domains play a crucial role in facilitating melanin synthesis and other cellular processes by localizing the enzyme to specific subcellular compartments, where pigment production occurs. Additionally, the glycosylation patterns in fungal TYRs may differ from those in humans, influencing protein folding, stability, and enzymatic activity [97]. Glycosylation of fungal TYRs is essential for proper protein folding and stability as well as for modulating enzymatic activity in response to environmental cues.

Fungal and plant TYRs exhibit notable differences in their biochemical characteristics and enzymatic activities [97]. Fungal TYRs, such as those from white rot fungus *Pycnoporus sanguineus* (PsT) and filamentous fungus *Trichoderma reesei* (TrT), often demonstrate broader substrate specificity compared to plant TYRs like those from apple and potato [97]. Fungal TYRs may exhibit higher activity on various phenolic compounds, including monophenols and diphenols, whereas plant TYRs may show lower activity, particularly on monophenols [97]. On the other hand, plant TYRs may exhibit slower kinetics or lower activity on certain substrates. Furthermore, fungal TYRs may demonstrate different stereo-specificity patterns compared to plant TYRs, influencing their overall catalytic efficiency and substrate preferences.

Genome sequence analysis revealed that the TYR family expanded in mollusks, such as the Pacific oyster (*Crassostrea gigas*) [101]. The TYR gene sequences from the *C. gigas* genome and the *Pinctada fucata martensii* transcriptome were verified. Phylogenetic analysis of these TYRs with functionally known TYRs from other mollusk species identified eight subgroups [101]. Structural data and surface pockets of the binuclear copper center in the eight subgroups of molluscan TYR were obtained using the latest versions of online prediction servers. Structural comparison with other TYR proteins from the protein databank revealed functionally important residues and their location within these protein structures [101]. The structural and chemical features of these pockets, which may be related to substrate binding, showed considerable variability among mollusks, which undoubtedly defines TYR substrate binding. Based on these observations, it can be concluded that the TYR family rapidly evolved as a consequence of substrate adaptation in mollusks.

4.3 Other features of eukaryotic TYR

Eukaryotic TYR proteins exhibit notable structural features related to glycosylation that play crucial roles in protein sorting, stability, and enzymatic activity, particularly in mammalian TYRs and TYR-related proteins (TYRP1 and TYRP2) [79]. The presence of N-glycan sites, which are exclusive to animal TYRs, has been extensively studied for their effect on protein processing and trafficking. N-glycosylation sites are strategically positioned within TYR proteins, particularly in the regions preceding the transmembrane fragment [95,102]. These N-glycans, added to the endoplasmic reticulum (ER), serve as important mediators during the secretory pathway, facilitating correct protein folding by promoting the binding of lectin chaperones [102]. N-glycan processing during transit through the Golgi network may contribute to the differential transport of TYR and TYR-related proteins to premelanosomes, influencing their sorting and transport mechanisms at different maturation stages. The efficiency and occupancy of N-glycans at specific sequons (specific amino acid motifs within protein sequences that serve as recognition sites for the attachment of carbohydrate molecules) within TYR proteins have significant implications for protein maturation and activity [95,102]. The occupancy of N-glycans at certain sites such as the N86-N371 pair is crucial for promoting proper folding and enzymatic activity [93]. The glycosylation profile of TYR proteins varies between species and paralogs, with differences observed in the composition and distribution of N-glycan forms [103]. Mutations affecting N-glycosylation sites have been linked to oculocutaneous albinism (OCA)1, highlighting the importance of proper N-glycan processing for TYR function and pigmentation [104]. Furthermore, TYR proteins can be targeted for ubiquitination, a process in which ubiquitin molecules are covalently attached to lysine residues in the protein sequence. It has been identified a membrane-associated ubiquitin ligase, the RNF152, as a critical regulator of melanogenesis by targeting tyrosinase [105]. RNF152 attaches ubiquitin molecules to tyrosinase through ubiquitination, marking it for lysosomal degradation. This process controls the levels of tyrosinase, thereby influencing melanin production in melanocytes. The co-localization of RNF152 and tyrosinase in the trans-Golgi network underscores the specificity of this regulatory mechanism [105]. The molecular mechanisms of Tyr post-translational modifications enhance our understanding of pigmentation disorders, skin biology, and melanin-related diseases. Moreover, studies have demonstrated the critical role of endoplasmic reticulum (ER) α-glucosidases in TYR processing, since inhibition of these enzymes leads to impaired folding and copper acquisition [106].

5. Structural features of bacterial TYRs

Bacterial TYRs exhibit diverse structural characteristics and enzymatic properties, and can be categorized into five types based on domain organization and the requirement for auxiliary proteins [107]. Type I TYRs, such as those found in *Streptomyces casstaneoglobisporus* and *Streptomyces antibioticus*, typically require a secondary helper protein (caddie protein) for secretion, proper folding, copper atom assembly, and enzymatic activity [108] (Fig. 6).

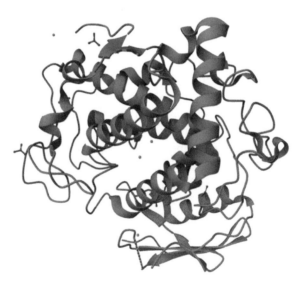

Fig. 6 Crystal structure (PDB entry, 5Z0D) of the deoxy form of tyrosinase (colored in green) from *S. castaneoglobisporus* in complex with the caddie protein (colored in brown). Copper ions are clearly visible at the center of the molecule. Four nitrate ions are shown in the figure.

In Streptomyces TYRs, the bicistronic operons (*melC1* and *melC2*) contain two genes. *melC2* encodes an inactive apotyrosinase (30–35 kDa), necessitating *melC1* (13–19 kDa) for secretion and activation [108]. Type II TYRs are similar to Streptomyces tyrosinases but do not require a caddie protein during secretion (e.g., *Bacillus megaterium* tyrosinase) [109]. Type III TYRs are exemplified by tyrosinase from *Verrucomicrobium spinosum* (Fig. 7).

They are small monomeric enzymes such as TYPE II TRYs, which contain only the catalytic domain and do not rely on additional helper proteins, potentially allowing for secretion. These TYRs might be capable of independent secretion. Type III TYRs found in *V. spinosum* resemble fungal tyrosinases and possess a C-terminal domain, the removal of which significantly influences enzyme activity, suggesting a regulatory role for C-terminal extension [110]. This observation aligns with the C-terminal function proposed for plant and fungal TYRs, which maintains the enzyme in an inactive state within the cell [98]. Type IV tyrosinases are among the smallest (14 kDa) tyrosinases reported in bacteria and are only active in homodimer forms (e.g., *Bacillus thuringiensis*). However, there is ongoing debate regarding whether these proteins exhibit true TYR activity owing to their size and structural characteristics [107]. Type V TYRs include

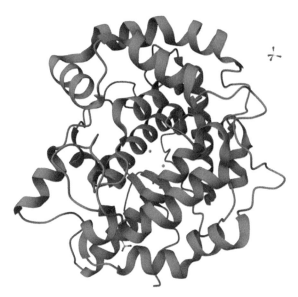

Fig. 7 Three-dimensional structure of active tyrosinase from the bacterium *Verrucomicrobium spinosum*. The copper ions, located at the center of the molecule, are clearly visible. In addition, two sulfate ions are present.

enzymes that lack the sequence features typical of TYRs but instead exhibit characteristics resembling laccases. These enzymes may have marginal activity on tyrosine but are more proficient on substrates like ABTS (2,2′-azino-bis(3-ethylbenzothiazoline-6-sulfonic acid)), which is a chemical compound commonly used in biochemical assays as a substrate to test the activity of certain enzymes, particularly those that have peroxidase-like activity [107]. Examples include membrane-bound TYRs from *Marinomonas mediterranea*, which display classical laccase substrate specificity and are activated by SDS [111]. The classification of Types IV and V as tyrosinases is debated because of their structural features, since they lack the binuclear type-3 copper center characteristic of tyrosinases. These diverse bacterial TYRs underscore the versatility and adaptability of these enzymes across different bacterial species. Their structural variations and enzymatic properties contribute to a wide range of biological functions and potential industrial applications [111].

6. Structural role of cysteine residues in TYR proteins: implications for stability and function across organisms

In addition to the catalytic center and C-terminal tail, TYR proteins contain cysteine (Cys) residues and disulfide bridges that are critical for their structural integrity and enzymatic activity. The presence and distribution of Cys residues in TYR proteins vary across different organisms, impacting their structural stability and function [79]. Specific clusters of Cys residues, particularly in mammalian TYRs, contribute to proper protein folding and maintenance. Mutations in these Cys residues can lead to misfolding and inactivation of TYR, resulting in albinism. TYR from gram–positive bacteria, such as Streptomyces, lacks Cys residues, while gram–negative bacteria and fungi TYRs do contain them [79]. However, the thermostability, proteolytic resistance, and copper affinity of Streptomyces TYRs are lower than those of the gram–negative bacteria. Mammalian TYRs typically contain 17 Cys residues grouped into three clusters [91]. The first two clusters are located at the N-terminus of the CuA site, whereas the third cluster is located between CuA and CuB [91]. Plant catechol oxidases only possess the first cluster [112]. The first cluster, characterized by consecutive CC pairs, is crucial for the correct folding and maintenance of the N-terminal domain, as supported by crystallographic data from plant catecholoxidase studies [113]. The second cluster, known as the epidermal growth factor (EGF)-like region, contained five Cys residues. This region shares homology with EGF-like motifs and laminin-LE motifs, suggesting a role in protein-protein interactions. The third cluster, with four to five Cys residues, lacks homology with EGF-like motifs, but contains a CxxC element.

7. Conclusions

TYRs are versatile enzymes found across the three domains of cellular life. They catalyze key reactions in melanin biosynthesis that are crucial for coloration and UV protection in organisms. Enzymatic oxidation of substrates by TYR leads to the generation of reactive quinone intermediates, which spontaneously polymerize to form melanins, providing structural and functional diversity. Melanins play essential roles in UV protection, skin pigmentation, and antioxidant defense mechanisms. Genetic variations in the genes encoding melanogenic enzymes can influence melanin production and skin pigmentation. Furthermore, melanin production is tightly regulated by various factors, including hormonal signals, cytokines, UV radiation, and genetic factors. TYRs belong to the broader category of phenoloxidases, which also includes catecholoxidases and laccases. Although these enzymes share similarities in their catalytic mechanisms and substrate specificities, they exhibit distinct enzymatic activities and structural features. TYRs play diverse roles in biological systems, including insect cuticle sclerotization, mollusk shell formation, fungal pigmentation, bacterial biofilm formation, and microbial virulence. They contribute to the adaptation and survival strategies across diverse taxa, highlighting their significance in ecological interactions and evolutionary processes. Classification of bacterial enzymes into different types based on their domain organization offers avenues for specific inhibition strategies tailored to each type. Bacterial and fungal TYRs are promising candidates for drug targeting owing to their enzymatic properties. The inhibition of bacterial tyrosinases can disrupt melanin synthesis in pathogenic bacteria, offering a novel approach to combat infections. Moreover, targeting bacterial TYRs may have implications in other conditions in which melanin production plays a role, such as hyperpigmentation disorders.

Genetic characterization of *TYR* genes provides insights into their functional versatility, ecological significance, and biotechnological potential.

Investigation of the evolutionary dynamics and functional diversification of type-3 copper proteins revealed a complex landscape shaped by gene duplication, divergence, and adaptation across diverse organisms. Type-3 copper proteins, with representatives found in each domain, share a conserved pair of copper-binding sites, but exhibit further subclassification based on additional domains or motifs. Among the subclasses, the α-subclass typically features an N-terminal signal peptide, suggesting secretion or vesicular localization, while the β-subclass lacks this domain and is presumed to localize to the cytosol. The γ-subclass possesses an N-terminal signal peptide, a cysteine-rich region potentially involved in protein-protein interactions or dimerization, and a transmembrane domain indicative of membrane-bound localization. The findings from the phylogenetic analysis imply that both the cytosolic (β) and membrane-bound (γ) forms likely descended from an ancient type-3 copper protein, which probably had a predominant secretion (α) pattern. This evolutionary transformation suggests a dynamic process, wherein the original protein, presumably engaged in extracellular functions, undergoes structural alterations and adaptations to serve intracellular or membrane-associated roles.

References

[1] M. Kanteev, M. Goldfeder, A. Fishman, Structure-function correlations in tyrosinases, Protein Sci. 24 (9) (2015) 1360–1369.

[2] S. Halaouli, M. Asther, J.C. Sigoillot, M. Hamdi, A. Lomascolo, Fungal tyrosinases: new prospects in molecular characteristics, bioengineering and biotechnological applications, J. Appl. Microbiol. 100 (2) (2006) 219–232.

[3] K. Yu, W. He, X. Ma, Q. Zhang, C. Chen, P. Li, et al., Purification and biochemical characterization of polyphenol oxidase extracted from wheat bran (Wan grano), Molecules 29 (6) (2024).

[4] R. Sanchez, L. Arroyo, P. Luaces, C. Sanz, A.G. Perez, Olive polyphenol oxidase gene family, Int. J. Mol. Sci. 24 (4) (2023).

[5] O. Samuel Ilesanmi, O. Funke Adedugbe, D. Adeniran Oyegoke, R. Folake Adebayo, O. Emmanuel Agboola, Biochemical properties of purified polyphenol oxidase from bitter leaf (Vernoniaamygdalina), Heliyon 9 (6) (2023) e17365.

[6] L. Ruckthong, M. Pretzler, I. Kampatsikas, A. Rompel, Biochemical characterization of Dimocarpus longan polyphenol oxidase provides insights into its catalytic efficiency, Sci. Rep. 12 (1) (2022) 20322.

[7] M. Moeini Alishah, S. Yildiz, C. Bilen, E. Karakus, Purification and characterization of avocado (Persea americana) polyphenol oxidase by affinity chromatography, Prep. Biochem. Biotechnol. 53 (1) (2023) 40–53.

[8] G.M. Casanola-Martin, H. Le-Thi-Thu, Y. Marrero-Ponce, J.A. Castillo-Garit, F. Torrens, A. Rescigno, et al., Tyrosinase enzyme: 1. An overview on a pharmacological target, Curr. Top. Med. Chem. 14 (12) (2014) 1494–1501.

[9] A. De, S. Mandal, R. Mukherjee, Modeling tyrosinase activity. Effect of ligand topology on aromatic ring hydroxylation: an overview, J. Inorg. Biochem. 102 (5-6) (2008) 1170–1189.

[10] Y. Song, S. Chen, L. Li, Y. Zeng, X. Hu, The hypopigmentation mechanism of tyrosinase inhibitory peptides derived from food proteins: an overview, Molecules 27 (9) (2022).

[11] S. Zolghadri, A. Bahrami, M.T. Hassan Khan, J. Munoz-Munoz, F. Garcia-Molina, F. Garcia-Canovas, et al., A comprehensive review on tyrosinase inhibitors, J. Enzyme Inhib. Med. Chem. 34 (1) (2019) 279–309.

[12] R. Sarangarajan, S.P. Apte, The polymerization of melanin: a poorly understood phenomenon with egregious biological implications, Melanoma Res. 16 (1) (2006) 3–10.

[13] J.P. Ebanks, R.R. Wickett, R.E. Boissy, Mechanisms regulating skin pigmentation: the rise and fall of complexion coloration, Int. J. Mol. Sci. 10 (9) (2009) 4066–4087.

[14] R.M. Slominski, T. Sarna, P.M. Plonka, C. Raman, A.A. Brozyna, A.T. Slominski, Melanoma, melanin, and melanogenesis: the Yin and Yang relationship, Front. Oncol. 12 (2022) 842496.

[15] A.B. Mostert, Melanin, the what, the why and the how: an introductory review for materials scientists interested in flexible and versatile polymers, Polymers-Basel 13 (10) (2021).

[16] M. Brenner, V.J. Hearing, The protective role of melanin against UV damage in human skin, Photochem. Photobiol. 84 (3) (2008) 539–549.

[17] M. Sugumaran, Reactivities of quinone methides versus o-quinones in catecholamine metabolism and eumelanin biosynthesis, Int. J. Mol. Sci. 17 (9) (2016).

[18] M. Snyman, R.E. Walsdorf, S.N. Wix, J.G. Gill, The metabolism of melanin synthesis—from melanocytes to melanoma, Pigment. Cell Melanoma Res. (2024).

[19] T. Hida, T. Kamiya, A. Kawakami, J. Ogino, H. Sohma, H. Uhara, et al., Elucidation of melanogenesis cascade for identifying pathophysiology and therapeutic approach of pigmentary disorders and melanoma, Int. J. Mol. Sci. 21 (17) (2020).

[20] J. Zhang, R. Yu, X. Guo, Y. Zou, S. Chen, K. Zhou, et al., Identification of TYR, TYRP1, DCT and LARP7 as related biomarkers and immune infiltration characteristics of vitiligo via comprehensive strategies, Bioengineered 12 (1) (2021) 2214–2227.

[21] G. Raposo, M.S. Marks, Melanosomes-dark organelles enlighten endosomal membrane transport, Nat. Rev. Mol. Cell Bio 8 (10) (2007) 786–797.

[22] L. Le, J. Sirés-Campos, G. Raposo, C. Delevoye, M.S. Marks, Melanosome biogenesis in the pigmentation of mammalian skin, Integr. Comp. Biol. 61 (4) (2021) 1517–1545.

[23] C. Salinas, J.C. Garcia-Borron, F. Solano, J.A. Lozano, Dopachrome tautomerase decreases the binding of indolic melanogenesis intermediates to proteins, Biochim. Biophys. Acta 1204 (1) (1994) 53–60.

[24] J. Wan, X.M. Liu, T.C. Lei, S.Z. Xu, Effects of mutation in dopachrome tautomerase on melanosome maturation and anti-oxidative potential in cultured melanocytes, Zhonghua Yi Xue Za Zhi 89 (24) (2009) 1707–1710.

[25] A. Tingaud-Sequeira, E. Mercier, V. Michaud, B. Pinson, I. Gazova, E. Gontier, et al., The Dct(-/-) mouse model to unravel retinogenesis misregulation in patients with albinism, Genes (Basel) 13 (7) (2022).

[26] M. GEB, L. Dosh, H. Haidar, A. Gerges, S. Baassiri, A. Leone, et al., Nerve growth factor and burn wound healing: update of molecular interactions with skin cells, Burns 49 (5) (2023) 989–1002.

[27] T. Kondo, V.J. Hearing, Update on the regulation of mammalian melanocyte function and skin pigmentation, Expert. Rev. Dermatol. 6 (1) (2011) 97–108.

[28] S.A. Ali, I. Naaz, K.U. Zaidi, A.S. Ali, Recent updates in melanocyte function: the use of promising bioactive compounds for the treatment of hypopigmentary disorders, Mini Rev. Med. Chem. 17 (9) (2017) 785–798.

[29] A.C.C. Esposito, D.P. Cassiano, C.N. da Silva, P.B. Lima, J.A.F. Dias, K. Hassun, et al., Update on melasma—Part I: Pathogenesis, Dermatol. Ther. (Heidelb.) 12 (9) (2022) 1967–1988.

[30] D.P. Cassiano, A.C.C. Esposito, C.N. da Silva, P.B. Lima, J.A.F. Dias, K. Hassun, et al., Update on melasma—Part II: Treatment, Dermatol. Ther. (Heidelb.) 12 (9) (2022) 1989–2012.

[31] L. Blaschek, E. Pesquet, Phenoloxidases in plants-how structural diversity enables functional specificity, Front. Plant. Sci. 12 (2021) 754601.

[32] C.M. Marusek, N.M. Trobaugh, W.H. Flurkey, J.K. Inlow, Comparative analysis of polyphenol oxidase from plant and fungal species, J. Inorg. Biochem. 100 (1) (2006) 108–123.

[33] N. Durán, M.A. Rosa, A. D'Annibale, L. Gianfreda, Applications of laccases and tyrosinases (phenoloxidases) immobilized on different supports: a review, Enzyme Microb. Tech. 31 (7) (2002) 907–931.

[34] G. Janusz, A. Pawlik, U. Swiderska-Burek, J. Polak, J. Sulej, A. Jarosz-Wilkolazka, et al., Laccase properties, physiological functions, and evolution, Int. J. Mol. Sci. 21 (3) (2020).

[35] P.A. Pinto, I. Fraga, R.M.F. Bezerra, A.A. Dias, Phenolic and non-phenolic substrates oxidation by laccase at variable oxygen concentrations: selection of bisubstrate kinetic models from polarographic data, Biochem. Eng. J. 153 (2020).

[36] A. Todaro, R. Cavallaro, S. Argento, F. Branca, G. Spagna, Study and characterization of polyphenol oxidase from eggplant (Solanum melongena L.), J. Agric. Food Chem. 59 (20) (2011) 11244–11248.

[37] M. Perez-Gilabert, F. Garcia Carmona, Characterization of catecholase and cresolase activities of eggplant polyphenol oxidase, J. Agric. Food Chem. 48 (3) (2000) 695–700.

[38] A. Biundo, V. Braunschmid, M. Pretzler, I. Kampatsikas, B. Darnhofer, R. Birner-Gruenberger, et al., Polyphenol oxidases exhibit promiscuous proteolytic activity, Commun. Chem. 3 (1) (2020).

[39] E. Nikolaivits, M. Dimarogona, I. Karagiannaki, A. Chalima, A. Fishman, E. Topakas, Versatile fungal polyphenol oxidase with chlorophenol bioremediation potential: characterization and protein engineering, Appl. Environ. Microb. 84 (23) (2018).

[40] J.G. Liao, X.M. Wei, K.L. Tao, G. Deng, J. Shu, Q. Qiao, et al., Phenoloxidases: catechol oxidase—the temporary employer and laccase—the rising star of vascular plants, Hortic. Res-England 10 (7) (2023).

[41] M.A. McLarin, I.K.H. Leung, Substrate specificity of polyphenol oxidase, Crit. Rev. Biochem. Mol. 55 (3) (2020) 274–308.

[42] J.F. Li, Z.Y. Deng, H.H. Dong, R. Tsao, X.R. Liu, Substrate specificity of polyphenol oxidase and its selectivity towards polyphenols: unlocking the browning mechanism of fresh lotus root (Nelumbo nucifera Gaertn.), Food Chem. 424 (2023).

[43] R. Esposito, S. D'Aniello, P. Squarzoni, M.R. Pezzotti, F. Ristorate, A. Spagnuolo, New insights into the evolution of metazoan tyrosinase gene family, PLoS One 7 (2012) 4.

[44] M.E. McNamara, V. Rossi, T.S. Slater, C.S. Rogers, A.L. Ducrest, S. Dubey, et al., Decoding the evolution of melanin in vertebrates, Trends Ecol. Evol. 36 (5) (2021) 430–443.

[45] L. Guo, W. Li, Z. Gu, L. Wang, L. Guo, S. Ma, et al., Recent advances and progress on melanin: from source to application, Int. J. Mol. Sci. 24 (5) (2023).

[46] O. Badejo, O. Skaldina, A. Gilev, J. Sorvari, Benefits of insect colours: a review from social insect studies, Oecologia 194 (1–2) (2020) 27–40.

[47] M.J. Gorman, Y. Arakane, Tyrosine hydroxylase is required for cuticle sclerotization and pigmentation in. Insect Biochem. Molec. 40 (3) (2010) 267–273.

[48] X.W. Xiong, Y.F. Cao, Z.X. Li, R.L. Huang, Y. Jiao, L.Q. Zhao, et al., Insights from tyrosinase into the impacts of modified morphology of calcium carbonate on the nacre formation of pearl oysters, Front. Mar. Sci. 9 (2022).

[49] X.D. Du, G.Y. Fan, Y. Jiao, H. Zhang, X.M. Guo, R.L. Huang, et al., The pearl oyster Pinctada fucata martensii genome and multi-omic analyses provide insights into biomineralization, Gigascience 6 (8) (2017).

[50] H.C. Eisenman, A. Casadevall, Synthesis and assembly of fungal melanin, Appl. Microbiol. Biotechnol. 93 (3) (2012) 931–940.

[51] J.M. Al-Khayri, R. Rashmi, V. Toppo, P.B. Chole, A. Banadka, W.N. Sudheer, et al., Plant secondary metabolites: the weapons for biotic stress management, Metabolites 13 (6) (2023).

[52] H. Agarwal, S. Bajpai, A. Mishra, I. Kohli, A. Varma, M. Fouillaud, et al., Bacterial pigments and their multifaceted roles in contemporary biotechnology and pharmacological applications, Microorganisms 11 (3) (2023).

[53] K.U. Zaidi, A.S. Ali, S.A. Ali, I. Naaz, Microbial tyrosinases: promising enzymes for pharmaceutical, food bioprocessing, and environmental industry, Biochem. Res. Int. 2014 (2014) 854687.

[54] N.E. El-Naggar, W.I.A. Saber, Natural melanin: current trends, and future approaches, with especial reference to microbial source, Polymers (Basel) 14 (7) (2022).

[55] Y. Yuan, W.L. Jin, Y. Nazir, C. Fercher, M.A.T. Blaskovich, M.A. Cooper, et al., Tyrosinase inhibitors as potential antibacterial agents, Eur. J. Med. Chem. 187 (2020).

[56] P.K. Mantravadi, K.A. Kalesh, R.C.J. Dobson, A.O. Hudson, A. Parthasarathy, The quest for novel antimicrobial compounds: emerging trends in research, development, and technologies, Antibiotics (Basel) 8 (1) (2019).

[57] J.V.O. Barreto, L.M. Casanova, A.N. Junior, M. Reis-Mansur, A.B. Vermelho, Microbial pigments: major groups and industrial applications, Microorganisms 11 (12) (2023).

[58] L. Lin, J.P. Xu, Fungal pigments and their roles associated with human health, J. Fungi 6 (4) (2020).

[59] M. Riebel, A. Sabel, H. Claus, P. Fronk, N. Xia, H. Li, et al., Influence of laccase and tyrosinase on the antioxidant capacity of selected phenolic compounds on human cell lines, Molecules 20 (9) (2015) 17194–17207.

[60] K. Min, G.W. Park, Y.J. Yoo, J.S. Lee, A perspective on the biotechnological applications of the versatile tyrosinase, Bioresour. Technol. 289 (2019) 121730.

[61] Y.H. Kim, S.J. Park, S.H. Choe, J.R. Lee, H.M. Cho, S.U. Kim, et al., Identification and characterization of the tyrosinase gene (TYR) and its transcript variants (TYR_1 and TYR_2) in the crab-eating macaque (Macaca fascicularis), Gene 630 (2017) 21–27.

[62] S. Ponnazhagan, L. Hou, B.S. Kwon, Structural organization of the human tyrosinase gene and sequence analysis and characterization of its promoter region, J. Invest. Dermatol. 102 (5) (1994) 744–748.

[63] N.J. Kus, M.B. Dolinska, K.L. Young, E.K. Dimitriadis, P.T. Wingfield, Y.V. Sergeev, Membrane-associated human tyrosinase is an enzymatically active monomeric glycoprotein, PLoS One 13 (6) (2018).

[64] E. Akyilmaz, E. Yorganci, E. Asav, Do copper ions activate tyrosinase enzyme? A biosensor model for the solution, Bioelectrochemistry 78 (2) (2010) 155–160.

[65] M. Reinisalo, J. Putula, E. Mannermaa, A. Urtti, P. Honkakoski, Regulation of the human tyrosinase gene in retinal pigment epithelium cells: the significance of transcription factor orthodenticle homeobox 2 and its polymorphic binding site, Mol. Vis. 18 (4–6) (2012) 38–54.

[66] C.A. Ferguson, S.H. Kidson, The regulation of tyrosinase gene transcription, Pigm Cell Res. 10 (3) (1997) 127–138.

[67] D. Fang, Y. Tsuji, V. Setaluri, Selective down-regulation of tyrosinase family gene TYRP1 by inhibition of the activity of melanocyte transcription factor, MITF, Nucleic Acids Res. 30 (14) (2002) 3096–3106.

[68] J. Elkin, A. Martin, V. Courtier-Orgogozo, M.E. Santos, Analysis of the genetic loci of pigment pattern evolution in vertebrates, Biol. Rev. Camb. Philos. Soc. 98 (4) (2023) 1250–1277.

[69] A. Camacho-Hubner, C. Richard, F. Beermann, Genomic structure and evolutionary conservation of the tyrosinase gene family from Fugu, Gene 285 (1–2) (2002) 59–68.

[70] F. Murisier, F. Beermann, Genetics of pigment cells: lessons from the tyrosinase gene family, Histol. Histopathol. 21 (4-6) (2006) 567–578.

[71] X.N. Fan, P.H. Zhang, W. Batool, C. Liu, Y. Hu, Y. Wei, et al., Contribution of the tyrosinase (MoTyr) to melanin synthesis, conidiogenesis, appressorium development, and pathogenicity in Magnaporthe oryzae, J. Fungi 9 (3) (2023).

[72] H. Claus, H. Decker, Bacterial tyrosinases, Syst. Appl. Microbiol. 29 (1) (2006) 3–14.

[73] I. Bervoets, D. Charlier, Diversity, versatility and complexity of bacterial gene regulation mechanisms: opportunities and drawbacks for applications in synthetic biology, FEMS Microbiol. Rev. 43 (3) (2019) 304–339.

[74] P. Manivasagan, J. Venkatesan, K. Sivakumar, S.K. Kim, Actinobacterial melanins: current status and perspective for the future, World J. Microbiol. Biotechnol. 29 (10) (2013) 1737–1750.

[75] J.D. Nosanchuk, A. Casadevall, Impact of melanin on microbial virulence and clinical resistance to antimicrobial compounds, Antimicrob. Agents Chemother. 50 (11) (2006) 3519–3528.

[76] G. Faccio, K. Kruus, M. Saloheimo, L. Thöny-Meyer, Bacterial tyrosinases and their applications, Process. Biochem. 47 (12) (2012) 1749–1760.

[77] A. Mavridi-Printezi, M. Guernelli, A. Menichetti, M. Montalti, Bio-applications of multifunctional melanin nanoparticles: from nanomedicine to nanocosmetics, Nanomaterials-Basel 10 (11) (2020).

[78] A.M. McMahon, E.M. Doyle, S. Brooks, K.E. O'Connor, Biochemical characterisation of the coexisting tyrosinase and laccase in the soil bacterium Pseudomonas putida F6, Enzyme Microb. Tech. 40 (5) (2007) 1435–1441.

[79] X.L. Lai, H.J. Wichers, M. Soler-Lopez, B.W. Dijkstra, Structure and function of human tyrosinase and tyrosinase-related proteins, Chem. Eur. J. 24 (1) (2018) 47–55.

[80] F. Aguilera, C. McDougall, B.M. Degnan, Origin, evolution and classification of type-3 copper proteins: lineage-specific gene expansions and losses across the Metazoa, BMC Evol. Biol. 13 (2013).

[81] B. Lieb, B. Altenhein, J. Markl, A. Vincent, E. van Olden, K.E. van Holde, et al., Structures of two molluscan hemocyanin genes: significance for gene evolution, Proc. Natl. Acad. Sci. U. S. A. 98 (8) (2001) 4546–4551.

[82] T. Burmester, Molecular evolution of the arthropod hemocyanin superfamily, Mol. Biol. Evol. 18 (2) (2001) 184–195.

[83] J. Markl, Evolution of molluscan hemocyanin structures, BBA-Proteins Proteom. 1834 (9) (2013) 1840–1852.

[84] C.J. Coates, E.M. Costa-Paiva, Multifunctional roles of hemocyanins, Subcell. Biochem. 94 (2020) 233–250.

[85] H. Decker, F. Tuczek, Tyrosinase/catecholoxidase activity of hemocyanins: structural basis and molecular mechanism, Trends Biochem. Sci. 25 (8) (2000) 392–397.

[86] T. Schwede, J. Kopp, N. Guex, M.C. Peitsch, SWISS-MODEL: an automated protein homology-modeling server, Nucleic Acids Res. 31 (13) (2003) 3381–3385.

[87] H. Noh, S.J. Lee, H.J. Jo, H.W. Choi, S. Hong, K.H. Kong, Histidine residues at the copper-binding site in human tyrosinase are essential for its catalytic activities, J. Enzym. Inhib. Med. Ch. 35 (1) (2020) 726–732.

[88] N. Fujieda, S. Yabuta, T. Ikeda, T. Oyama, N. Muraki, G. Kurisu, et al., Crystal structures of copper-depleted and copper-bound fungal pro-tyrosinase: Insights into endogenous cysteine-dependent copper incorporation, J. Biol. Chem. 288 (30) (2013) 22128–22140.

[89] H. Decker, T. Schweikardt, F. Tuczek, The first crystal structure of tyrosinase: all questions answered? Angew. Chem. Int. Ed. 45 (28) (2006) 4546–4550.

[90] L. Kampatsikas, A. Rompel, Similar but still different: which amino acid residues are responsible for varying activities in type-III copper enzymes? Chembiochem 22 (7) (2021) 1161–1175.

[91] F. Solano, On the metal cofactor in the tyrosinase family, Int. J. Mol. Sci. 19 (2) (2018).

[92] T.F. Liu, G. Kandala, V. Setaluri, PDZ domain protein GIPC interacts with the cytoplasmic tail of melanosomal membrane protein gp75 (tyrosinase-related protein-1), J. Biol. Chem. 276 (38) (2001) 35768–35777.

[93] D. Cioaca, S. Ghenea, L.N. Spiridon, M. Marin, A.J. Petrescu, S.M. Petrescu, C-terminus glycans with critical functional role in the maturation of secretory glyco-proteins, PLoS One 6 (5) (2011).

[94] J.C. García-Borrón, F. Solano, Molecular anatomy of tyrosinase and its related proteins: beyond the histidine-bound metal catalytic center, Pigm. Cell Res. 15 (3) (2002) 162–173.

[95] N. Branza-Nichita, G. Negroiu, A.J. Petrescu, E.F. Garman, F.M. Platt, M.R. Wormald, et al., Mutations at critical N-glycosylation sites reduce tyrosinase activity by altering folding and quality control, J. Biol. Chem. 275 (11) (2000) 8169–8175.

[96] H. Sprong, S. Degroote, T. Claessens, J. van Drunen, V. Oorschot, B.H.C. Westerink, et al., Glycosphingolipids are required for sorting melanosomal proteins in the Golgi complex, J. Cell Biol. 155 (3) (2001) 369–379.

[97] E. Selinheimo, D. NiEidhin, C. Steffensen, J. Nielsen, A. Lomascolo, S. Halaouli, et al., Comparison of the characteristics of fungal and plant tyrosinases, J. Biotechnol. 130 (4) (2007) 471–480.

[98] W.H. Flurkey, J.K. Inlow, Proteolytic processing of polyphenol oxidase from plants and fungi, J. Inorg. Biochem. 102 (12) (2008) 2160–2170.

[99] A. Bijelic, M. Pretzler, C. Molitor, F. Zekiri, A. Rompel, The structure of a plant tyrosinase from walnut leaves reveals the importance of "substrate-guiding residues" for enzymatic specificity, Angew. Chem. Int. Ed. 54 (49) (2015) 14677–14680.

[100] C.W.G. VanGelder, W.H. Flurkey, H.J. Wichers, Sequence and structural features of plant and fungal tyrosinases, Phytochemistry 45 (7) (1997) 1309–1323.

[101] R.L. Huang, L. Li, G.F. Zhang, Structure-based function prediction of the expanding mollusk tyrosinase family, Chin. J. Oceanol. Limn. 35 (6) (2017) 1454–1464.

[102] G. Gupta, S. Sinha, N. Mitra, A. Surolia, Probing into the role of conserved N-glycosylation sites in the Tyrosinase glycoprotein family, Glycoconj. J. 26 (6) (2009) 691–695.

[103] G. Negroiu, N. Branza-Nichita, A.J. Petrescu, R.A. Dwek, S.M. Petrescu, Protein specific N-glycosylation of tyrosinase and tyrosinase-related protein-1 in B16 mouse melanoma cells, Biochem. J. 344 (1999) 659–665.

[104] M. Patel, C. Kassouf, Y.V. Sergeev, Functional in silico analysis of human tyrosinase and OCA1 associated mutations, Invest. Ophth Vis. Sci. 61 (7) (2020).

[105] R. Ueda, R. Hashimoto, Y. Fujii, J. Menezes, H. Takahashi, H. Takeda, et al., Membrane-associated ubiquitin ligase RING finger protein 152 orchestrates melanogenesis via tyrosinase ubiquitination, Membranes (Basel) 14 (2) (2024).

[106] N. Branza-Nichita, A.J. Petrescu, R.A. Dwek, M.R. Wormald, F.M. Platt, S.M. Petrescu, Tyrosinase folding and copper loading in vivo: a crucial role for calnexin and alpha-glucosidase II, Biochem. Biophys. Res. Commun. 261 (3) (1999) 720–725.

[107] M. Fairhead, L. Thöny-Meyer, Bacterial tyrosinases: old enzymes with new relevance to biotechnology, N. Biotechnol. 29 (2) (2012) 183–191.

[108] Y. Matoba, S. Kihara, N. Bando, H. Yoshitsu, M. Sakaguchi, K. Kayama, et al., Catalytic mechanism of tyrosinase implied from the quinone formation on the Tyr98 residue of the caddie protein, PLoS Biol. 16 (12) (2018).

[109] M. Sendovski, M. Kanteev, V.S. Ben-Yosef, N. Adir, A. Fishman, First structures of an active bacterial tyrosinase reveal copper plasticity, J. Mol. Biol. 405 (1) (2011) 227–237.

[110] M. Fekry, K.K. Dave, D. Badgujar, E. Hamnevik, O. Aurelius, D. Dobritzsch, et al., The crystal structure of tyrosinase from Verrucomicrobium spinosum reveals it to be an atypical bacterial tyrosinase, Biomolecules 13 (9) (2023).

[111] D. López-Serrano, F. Solano, A. Sanchez-Amat, Involvement of a novel copper chaperone in tyrosinase activity and melanin synthesis in, Microbiology-Sgm 153 (2007) 2241–2249.

[112] C. Eicken, B. Krebs, J.C. Sacchettini, Catechol oxidase—structure and activity, Curr. Opin. Struc. Biol. 9 (6) (1999) 677–683.

[113] S.M. Prexler, M. Frassek, B.M. Moerschbacher, M.E. Dirks-Hofmeister, Catechol oxidase versus tyrosinase classification revisited by site-directed mutagenesis studies, Angew. Chem. Int. Ed. 58 (26) (2019) 8757–8761.

CHAPTER TWO

Catalytic mechanism of tyrosinases

Samaneh Zolghadri[a],* and Ali Akbar Saboury[b],*

[a]Department of Biology, Jahrom Branch, Islamic Azad University, Jahrom, Iran
[b]Institute of Biochemistry and Biophysics, University of Tehran, Tehran, Iran
*Corresponding authors. e-mail address: szjahromi@yahoo.com; saboury@ut.ac.ir

Contents

Abstract

Tyrosinases (TYR) play a key role in melanin biosynthesis by catalyzing two reactions: monophenolase and diphenolase activities. Despite low amino acid sequence homology, TYRs from various organisms (from bacteria to humans) have similar active site architectures and catalytic mechanisms. The active site of the TYRs contains two copper ions coordinated by histidine (His) residues. The catalytic mechanism of TYRs involves electron transfer between copper sites, leading to the hydroxylation of monophenolic compounds to diphenols and the subsequent oxidation of these to corresponding dopaquinones. Although extensive studies have been conducted on the structure, catalytic mechanism, and enzymatic capabilities of TYRs, some mechanistic aspects are still debated. This chapter will delve into the structure of the active site, catalytic function, and inhibition mechanism of TYRs. The goal is to improve our understanding of the molecular mechanisms underlying TYR activity. This knowledge can help in developing new strategies to modulate TYR function and potentially treat diseases linked to melanin dysregulation.

The Enzymes, Volume 56
ISSN 1874-6047, https://doi.org/10.1016/bs.enz.2024.05.001

1. Introduction

The catalytic mechanism of tyrosinases (TYRs) is a complex and important area of research that has been extensively studied. The catalytic mechanism of TYRs is highly regulated by several factors, including the concentration of substrates and cofactors, temperature, and pH. The optimal temperature for TYR activity is around 37 °C and the optimal pH is around 6.5–7.5 [1–5]. Generally, TYRs perform two successive enzymatic reactions using molecular oxygen (O_2): the *ortho*-hydroxylation of a monophenol (M) (cresolase or monophenolase activity) and the oxidation of an *o*-diphenol (D) (catecholase or diphenolase activity) to the corresponding quinone (Q) [6,7]. Then, the reactive Qs are polymerized spontaneously into melanins [8]. Due to their ability to oxidize tyrosine residues and small phenolic molecules, TYRs have various biotechnological applications, such as biopolymer cross-linking, dye production, and bioremediation [9]. TYRs play a vital role in the enzymatic browning of vegetables and fruits and depigmentation disorders in humans. Therefore, understanding the catalytic mechanism of TYRs is crucial for developing therapies for skin pigmentation disorders in humans or preventing enzymatic browning in vegetables. As researchers continue to explore the underlying mechanisms of TYRs, they gain new insights into their regulation and potential applications, which could have valuable implications for medicine, pharmaceuticals, biotechnology, and beyond.

TYRs have a general structure consisting of a short N-terminal signal peptide, a C-terminal segment, and a central domain. The N-terminal signal peptide is involved in intracellular trafficking and processing and is proteolytically removed [4]. In plants, this segment of TYR handles the enzyme transfer to the chloroplast [10], while in humans and animals, it may play an essential role in melanosome transfer [11]. In fungi, TYRs are cytoplasmic or associated with the cell wall and do not contain a transit peptide [12]. In bacteria, TYRs consist of a TAT signal peptide (identified at the N-terminal region) in some species and are responsible for protein secretion [13–15]. The central domain of TYRs is composed of a conserved active site consisting of two copper ions. These copper ions are coordinated by imidazole groups of three histidine (His) residues and a water molecule, making a type-3 copper protein active site. The oxidation state of these ions and their binding ability to O_2 affects the TYR catalytic cycle [16–18]. Fekry et al. determined the three-dimensional structure of TYR (from *Verrucomicrobium spinosum*), and they identified the structural features of the central domain and its importance for substrate specificity and TYR activity [19]. Finally, the C-terminal segment protects the entrance of the catalytic site. The latent state of TYR, i.e. (pro-TYR), consists of the C-terminal and the central domain. *In-vivo* activation of pro-TYR happens by proteolytic breakdown within the C-terminal segment, which increases the active site accessibility and enzymatic activity [20].

Animal TYRs exhibit several remarkable structural features, including cysteine (Cys) clusters and *N*-glycosylation sites [4]. Previous studies have suggested that *N*-glycosylation is necessary for the activity and stability of human TYR in vivo [17,21,22]. However, several studies have reported that recombinant human TYR from *E. coli* could display activity without post-translational modifications such as glycosylation [23]. Furthermore, Cys clusters are essential for correct TYR folding [24].

In the last decade, research on the structure of TYRs from different species [18,25,26] has opened a new perspective into the mechanism of TYRs catalytic cycle. This chapter will summarize and discuss recent advances and current knowledge in the structure, catalytic mechanism, and inhibition of TYRs.

2. The overall architecture of the TYR active site

As mentioned, the type-3 copper proteins, including TYR, have a conserved active site with six His residues located in α-helical fragments [27,28]. The structure of TYR is maintained among these fragments by electrostatic and cation-π interactions. The imidazole group of His residues coordinated with two copper ions (CuA and CuB) facilitates the transfer of electrons between the substrates during the catalytic activity [29,30]. Fig. 1 depicts the active site of *Agaricus bisporus* TYR.

In fungi, five His residues are located in a stable α-helical fragment, and one His coordinating CuA motif is placed in a flexible loop region stabilized by an unusual thioether bond with an adjacent Cys residue [30]. However, this feature is not found in all sources, such as bacterial TYRs [32].

Despite the similar spectroscopic properties of copper ions in this family, there are differences in the activities of various members attributed to the active site architecture and substrate accessibility [9,33,34]. Some crystallographic investigations on *Aspergillus. oryzae* TYR (*Ao*TYR) [35], and *Streptomyces. castaneoglobisporus* TYR (*Sc*TYR) [3] revealed the flexibility of two copper ions, suggesting different positions within the active

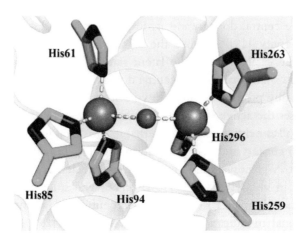

Fig. 1 The active site of *Agaricus bisporus* TYR (PDB ID 2Y9W). The side chains of His residues are presented as sticks. The copper ions, bridging oxygen, carbon, and nitrogen atoms are colored brown, red, green, and blue, respectively [31]. *Represented with permission from C. Zou, W. Huang, G. Zhao, X. Wan, X. Hu, Y. Jin, et al., Determination of the bridging ligand in the active site of tyrosinase, Molecules 22 (2017) 1836.*

site. This flexibility permits the conserved His to lose their interactions with the Cu ions and facilitates the binding of M substrates. Studies on TYRs with known crystal structures and site-directed mutagenesis experiments have provided deeper insight into the conformational stability, enzymatic activity, and mechanistic details of TYR active sites [36–40]. Several investigations have highlighted the essential role of His residues in the copper binding and catalytic activity of TYRs from different species [41–43]. Understanding the role of His residues could lead to new therapies for conditions such as hyperpigmentation and melanoma.

Spritz et al. investigated the mutations of three conserved His residues (His-363, -367, and -390) to alanine (Ala) at the CuB position. They found that the His-363 and -367 mutations completely lost TYR activity, probably due to the loss of copper ions. In contrast, the copper-binding of the His390Ala mutant compared to the wild-type enhanced approximately 2.5-fold [44].

Jackman et al. reported that the replacement of His-193 and -215 by glutamine (Glu) reduced diphenolase activity by 6250- and 2778-fold compared to the wild type of *Streptomyces glaucescens* TYR (SgTYR). However, mutations of His-37 and -53 to Gln and His-62 and -189 to asparagine (Asn) were almost inactive on L-dopa [45]. Furthermore, Matoba et al. demonstrated the abolishment of the monophenolase activity of *Sc*TYR by replacing His-63 with Phe [3].

Nakamura et al. [16] found that single mutations of conserved His residues (His-63, -84, -93, -290, -294, and -333) in *Ao*TYR largely abolished copper binding and catalysis. Also, the site-directed mutation of Cys-82 to Ala showed that Cys-82 is the essential residue for TYR activity. Similarly, by the three His residues (of CuB) substitution to Ala, Kaintz et al. [46] confirmed that His residues are essential for copper binding of a *Coreopsis grandiflora* polyphenol oxidase.

Recently, Noh et al. [47] predicted a 3D model of TYR and analyzed the catalytic activity of recombinant wild-type and seven His mutants in the active site of the human TYR enzyme overexpressed in E. coli BL21 (DE3) to determine the role of His residues in copper binding and catalysis of the enzyme. They found at catalytic activity was lost by replacing the His residues (His-180, -202, -211, -363, -367, and -390) with Ala at the copper-binding site. They mentioned that the loss of hydroxylase activity of the CuA site mutants was about 50% lower than the wild-type TYR, without significant changes in the dopa oxidation activity. By contrast, both catalytic activities were decreased upon mutations at CuB, confirming that the catalytic sites are at least partially distinct for these two activities. In this study, they mentioned that His residues play a role in TYR activity regulation by acting as acid–base catalysts in addition to their role in copper binding. These residues can maintain the optimal pH for TYR activity by accepting or donating protons depending on the pH environment changes.

According to the mutagenesis experiments, the conserved His residues are chemically and structurally critical to TYR activity. Thus, most mutants concerning the conserved His residues result in an inactive or partially active TYR [48].

The contribution of other amino acid residues to the catalytic activity of TYR, such as tyrosine (Tyr), glutamate (Glu), valine (Val), and Asn, has been investigated by several researchers. Fujieda et al. found that Tyr acts as a proton donor during the catalytic activity [35]. Goldfeder et al. showed that the monophenolase activity of *Bacillus. megaterium* TYR (*Bm*TYR) increased by exchanging Val-218 with phenylalanine (Phe). They also mentioned the functional role of Asn and Glu in the deprotonation process [49]. Decker et al. (2018) reviewed the role of Asn and Glu in the activity of type-3 copper proteins through site-directed mutagenesis experiments. As demonstrated by Decker et al., all TYRs exhibit a conserved water molecule that is hydrogen-bound to Asn and Glu and is necessary to deprotonate M substrates [36].

3. Oxidation states of the TYR active site

The properties of TYR vary depending on the oxidation states of the two copper ions [11,50,51]. Ramsden and Riley [52] reported four oxidation states of the TYR active site:

1. The oxidized met state (Em; [Cu(II)–Cu(II)]), which is inactive towards M.
2. The reduced deoxy state (Ed; [Cu(I)–Cu(I)]). Two copper ions in the active site of Ed bind to dioxygen and generate Eox.
3. The oxygenated state (Eox; [Cu(II)–O_2^{2-}–Cu(II)]). In Eox, dioxygen binds in a μ-η^2:η^2 side-on bridging mode as a peroxide ion. Eox catalyzes the conversion of the catechol and phenol substrates to Qs.
4. Deactivated form ([Cu(II)-Cu(0)], suicide inactivation) (Scheme 1).

4. TYR mechanism

The catalytic mechanism of TYRs involves a complex process of oxidation and reduction reactions. As mentioned, TYRs catalyzes two

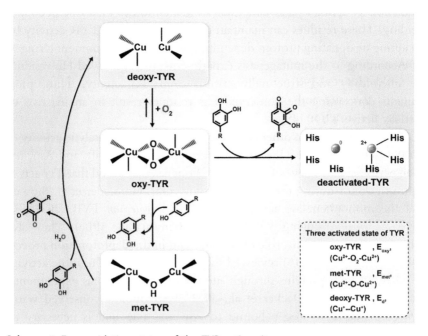

Scheme 1 Four oxidation states of the TYR active site.

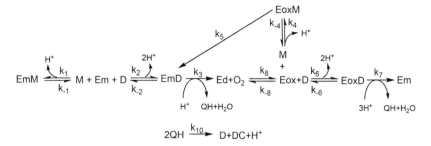

Scheme 2 The mechanism of two activities of TYR (monophenolase and diphenolase activities). Em, met-TYR; M, monophenol (such as L-tyrosine); D, o-diphenol (such as L-dopa); Ed, deoxy-TYR; Eox, oxy-TYR; QH, protonated o-dopaquinone; and DC, dopachrome [56]. *Represented with permission from S. Zolghadri, M. Beygi, T.F. Mohammad, M. Alijanianzadeh, T. Pillaiyar, P. Garcia-Molina, et al., Targeting tyrosinase in hyperpigmentation: current status, limitations and future promises, Biochem. Pharmacol. (2023) 115574.*

different processes: monophenolase or cresolase activity (the hydroxylation of M to D) and diphenolase or catechol oxidase activity (the oxidation of D to Q) [53–55] (Scheme 2).

The monophenolase reaction involves transferring one electron from M substrates to the reduced copper ion (Cu(I)), and transferring another electron and a proton to O_2. The resulting product (D) can be further oxidized to its corresponding Q by the diphenolase activity. Q molecules spontaneously polymerize to form melanin [57].

Goldfeder et al. [49] mentioned that despite binding M and D substrates at the active site and orienting identically, just the M substrate rotates during the reaction. Several structural features, including a thioether binding restriction on the active site and a bulky residue (e.g., a bulky Phe residue) above the active site, prevent the substrate's rearrangement necessary for the hydroxylation of M substrates. They found that these features allow the oxidation of D substrates, resulting in the differentiation between enzymes with both monophenolase and diphenolase activities and those with only diphenolase reactions [49]. However, recent biochemical and structural studies of TYRs have raised doubts about these findings. Bijelic et al. [58], who presented the first three-dimensional structure of TYR from a plant called *Juglans regia*, claimed that the lack of monophenolase activity in plant polyphenol oxidase is not related to the existence of a bulky residue above CuA and the degree of restriction around the CuA site, as assumed previously. They mentioned that the distinction between monophenolase and diphenolase activities depends on the electrostatic environment, conformation, and type of second-shell residues at the entrance of the site where activity occurs [58].

Unlike diphenolase activity, it is not possible to study of mono-phenolase activity independently because the chemical reactions of this activity must occur simultaneously with diphenolase activity [59]. In the following section, we will discuss these two activities in more detail.

5. Monophenolase activity of TYR

Despite intensive efforts, some details of the mechanism of the monophenolase activity of TYR are still under debate among researchers. According to previous studies, the peroxide intermediate in the oxy-form mediates the hydroxylation of M substrates. However, a dead-end complex (EmM) is formed by binding M to Em forms that cannot bind to oxygen. Therefore, for monophenolase activity, Em must reduce to Ed, and Ed binds to oxygen to form Eox. In this regard, copper ions are bound to dioxygen to originate a $\mu:\eta^2:\eta^2$-peroxide dicopper (II) intermediate (Eox). This intermediate performs the regioselective monooxygenation of *para*-substituted M substrates to catechols through an electrophilic aromatic substitution mechanism [60,61].

Furthermore, in monophenolase activity, deprotonation of M substrates is an obligatory step involving the participation of a conserved water molecule and conserved Glu and Asn residues [9]. Substrates may adjust the suitable orientation for the deprotonation or oxidation step [62], indicating that the enzyme active site must be flexible to allow substrate rotation and reorganization [63,64]. It has been suggested that this reorganization occurs through the rotation of three His residues coordinating CuA at the catalytic sites, followed by butterfly twisting of the Cu_2O_2 and substrate reor-ientation [65,66]. Spectroscopic analyses and crystal structures of the dioxygen-bound TYRs have revealed that the substrate binding induces a movement of the copper ion with the rotation of the peroxide ligand, weakening the dioxygen bond and providing one of the peroxide oxygen atoms with access to the ϵ carbon of the phenolic substrate [35].

Recently, Kipouros et al. [1] investigated this critical reaction and provided new insights into fundamental O_2 activation mechanisms through coupled binuclear copper active sites. They characterized the $TYR/O_2/M$ ternary intermediate under single-turnover conditions by spectroscopic, kinetic, and computational methods. Their study revealed that the M substrate is docked in the fully protonated active-site pocket of Eox without cleavage of the O-O bond (μ:η^2:η^2-peroxide) or coordination to a copper. The construction of this intermediate includes displacing water molecules of active site and replacing their hydrogen bonds to the μ:η^2:η^2-peroxide with a single hydrogen bond from the substrate hydroxyl group. This exchange enables the mono-oxygenation mechanism, where the cleavage of dioxygen O-O bond occurs to receive the proton of the phenolic group, followed by coordination of phenolate to a copper site attendant with ortho-hydroxylation of its aromatic group by the unprotonated μ-oxo [67].

In 2022, García-Molina et al. studied the monophenolase activity on four types of M substrates (A–D). They emphasized that each substrate shows unique characteristics due to the chemical nature of its hydroxylated and Q derivatives [68]. Here, we will summarize their work and the proposed catalytic mechanism of TYR on four types of M substrates.

5.1 Type A

These substrates produce Q, whose chemical evolution creates D in the medium, such as, the physiological M substrate of hTYR, L-tyrosine, and related compounds (tyramine, sinephrine, and L-tyrosine methyl ester). Also, the substrates require the addition of a secondary reagent to generate D, such as L-serine or 3-methyl-2-benzothiazolinone hydrazone (MBTH) [68]. In this type, the system needs a lag period (τ) to accumulate a quantity of D in the reaction medium until it reaches the final steady state. The steady-state rate and τ are affected by the substrate and enzyme concentrations and pH. Furthermore, the catalytic concentrations of D substrates affected τ but not the steady-state rate [69]. The mechanism of this type of substrate has been investigated in different studies in detail and presented in Scheme 2. In this regard, kinetic variations of the TYR mechanism can occur in the other three types of M substrates (B, C, and D) regarding the lack of D accumulation in the medium. We will consider these types of substrates in the following.

5.2 Type B

These substrates cannot accumulate D due to easy oxidation and require hydrogen peroxide (H_2O_2) for TYR activity (Scheme 3), such as hydroquinone (HQ). HQ was formerly known as a depigmenting agent, but it is now recognized as an alternative substrate of TYR in the presence of ascorbic acid and H_2O_2. TYR exhibits activity on HQ through the stages handled by K_5, k_3, and k_{-2}. HHQ is unstable and rapidly oxidized to HPB

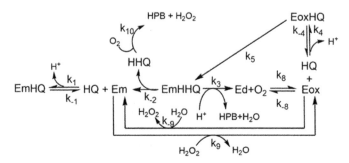

Scheme 3 Monophenolase activity of TYR on HQ. Em, met-TYR; HQ, hydroquinone; HPB, hydroxyparabenzoquinone; HHQ, hydroxyhydroquinone; Ed, deoxy-TYR; and Eox, oxy-TYR. *Represented with permission from P. García-Molina, F. García-Molina, J.A. Teruel-Puche, J.N. Rodríguez-López, F. García-Cánovas, J.L. Muñoz-Muñoz, Considerations about the kinetic mechanism of tyrosinase in its action on monophenols: a review, Mol. Catal. 518 (2022) 112072.*

by the molecular oxygen. This process continues until all the enzyme is converted to Em (an inactive form of TYR on HQ), causing the system to stop and the catalytic cycle to remain incomplete. Due to this reason, HQ was not previously mentioned as a substrate. However, further investigations have shown that adding H_2O_2 to the medium can enable HQ to be an alternative substrate for TYR. Indeed, H_2O_2 closes the catalytic cycle by transforming Em into Eox, resulting in TYR activity on HQ and other M substrates. Thus, a deviation from the kinetic mechanism of the monophenolase reaction on HQ is observed in this case [68].

5.3 Type C

These substrates (such as 4–tert-butylphenol (4–TBF) and related compounds) release D. Then, TYR oxidizes it to produce stable Q, thus preventing further D production [68]. Based on previous studies on the catalytic behavior of TYR on 4–TBF, there is an inverse dependency of the pseudo-steady state rate to the substrate concentration and a direct dependency of the rate to the square of TYR concentration [68,70].

As shown in Scheme 4, from the activity of TYR on 4–TBF, 4–tert-butylcatechol (TBC) accumulates and is oxidized to tert-butylquinone (o-TBQ). As TBC remains stable over the time scale of measurement and does not accumulate into TBC, the system gets a pseudo-steady state [68,70,71]. In the presence of H_2O_2, the conversion from Em to Eox occurs, the velocity increases, the reliances on the substrate and enzyme concentrations vanish, and the TBC accumulation gives rise to a delay in the product accumulation [68].

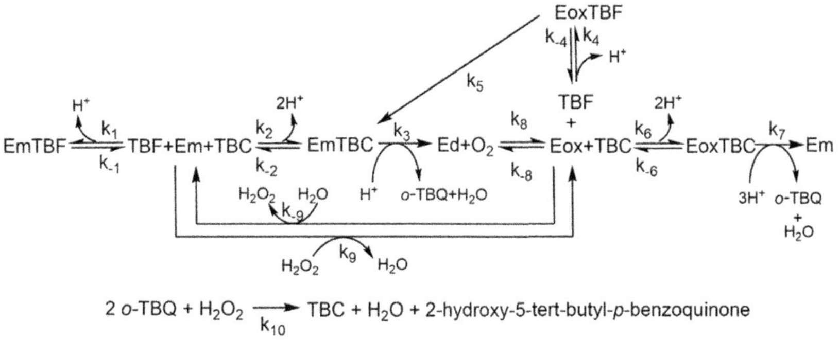

Scheme 4 Monophenolase activity of TYR on 4-TBF. Em, met-TYR; 4-TBF, 4-tertbutylphenol; TBC, 4-tertbutylcatechol; Ed, deoxy-TYR; Eox, oxy-TYR; and TBQ, tertbutylquinone. *Represented with permission from P. García-Molina, F. García-Molina, J.A. Teruel-Puche, J.N. Rodríguez-López, F. García-Cánovas, J.L. Muñoz-Muñoz, Considerations about the kinetic mechanism of tyrosinase in its action on monophenols: a review, Mol. Catal. 518 (2022) 112072.*

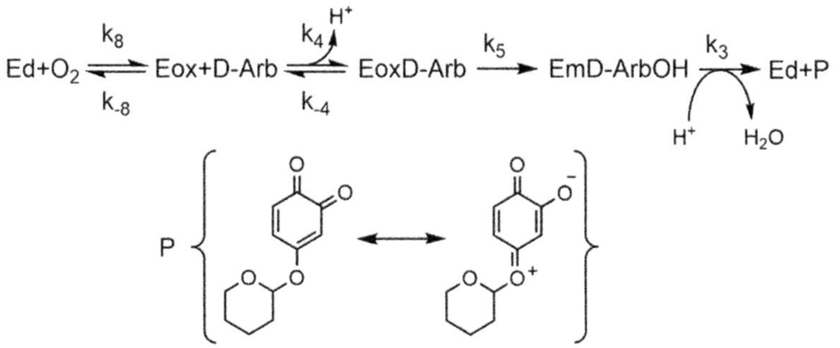

Scheme 5 Monophenolase activity of TYR on D-Arb. Ed, deoxy-TYR; Eox, oxy-TYR; D-Arb, deoxyarbutin; Em, met-TYR; and ArbOH, hydroxydeoxyarbutin. *Represented with permission from P. García-Molina, F. García-Molina, J.A. Teruel-Puche, J.N. Rodríguez-López, F. García-Cánovas, J.L. Muñoz-Muñoz, Considerations about the kinetic mechanism of tyrosinase in its action on monophenols: a review, Mol. Catal. 518 (2022) 112072.*

5.4 Type D

These substrates (such as deoxyarbutin (D-Arb), a monophenol derived from HQ) do not release D into the medium or form it to generate a stable Q [68]. As you see in Scheme 5, the proposed kinetic mechanism of TYR on D–Arb is radically separate from that predicted on other M substrates [68].

6. Diphenolase activity of TYR

For diphenolase activity, Ed is bound to O_2 with a high affinity and converted to Eox. Then, Eox is bound to another D and forms EoxD. Subsequently, D is oxidized to QH, and Eox is converted to Em, completing the catalytic cycle. Finally, as you see in Scheme 6, a dopachrome molecule (DC) and a substrate D are produced by the reaction between two QH molecules [56,72] (Scheme 6).

The Michaelis constant (a measure of the enzyme's affinity for the substrate) for O_2, K_m^{O2}, in the presence of D substrates, has been widely studied. Ingraham (1957) reported a variation in the K_m^{O2} according to the hydrogen donor structure and concluded that TYR combines with O_2 before complexing with D [73]. Duckworth and Coleman (1970) observed the same variation but proposed that the reaction might follow the opposite order, with oxygen not binding first and two binding sites for TYR [74]. The kinetic analysis of TYR (from *Neurospora Crassa*) by Gutteridge and Robb (1975) showed random binding of D and O_2 [75]. Rodríguez-López et al. (1993) mentioned that K_m^{O2} depends on the nature of M and D substrates, and it is always lower in the presence of M substrates than D substrates [76].

To gain new information into the mechanism of D substrate oxidation, Rodríguez-López et al. (2000) conducted a study on the reaction of TYR (from *Agaricus bisporus*) with O_2 in the presence of various D substrates using transient and steady-state kinetics transient-phase and steady-state kinetics. The determination of individual rate constants for several partial reactions comprising the catalytic cycle was presented for the first time by

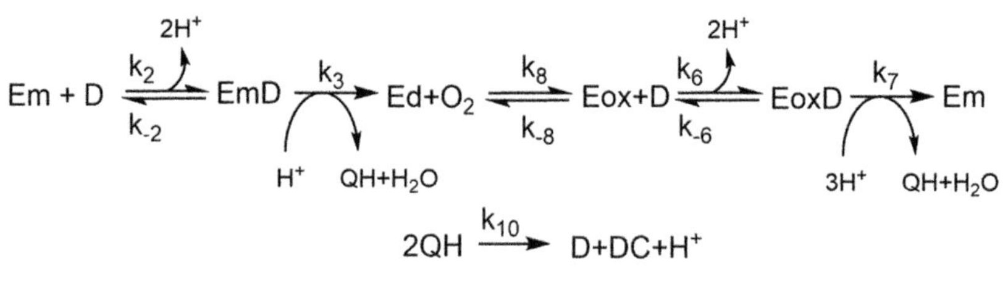

Scheme 6 Kinetic Mechanism for Diphenolase activity. D, o-diphenol; Em, met-TYR; Ed, deoxy-TYR; Eox, oxy-TYR; QH, protonated o-dopaquinone; and DC, dopachrome. *Represented with permission from S. Zolghadri, M. Beygi, T.F. Mohammad, M. Alijanianzadeh, T. Pillaiyar, P. Garcia-Molina, et al., Targeting tyrosinase in hyperpigmentation: current status, limitations and future promises, Biochem. Pharmacol. (2023) 115574.*

kinetic analysis. Their study revealed that TYR binds to O_2 very fast ($k_{+8} = 2.3 \times 10^7$ M/s), similar to hemocyanins (($k_{+8} = 1.3 - 5.7) \times 10^6$ M/s). Ed binds O_2 reversibly at the active site with $K_D^{O2} = 46.6 \, \mu M$, similar to the dissociation constant for deoxyhemocyanin from *Octopus vulgaris* ($K_D^{O2} = 90 \, \mu M$). Transient-phase and steady-state kinetics demonstrated that D substrates react with Em significantly faster ($k_{+2} = 9.02 \times 10^6$ M/s) than with Eox ($k_{+6} = 5.4 \times 10^5$ M/s). This disparity originates from differential polar and steric elements modulating the access of D substrates to the active site for Em and Eox. The k_{cat} values for several D substrates are also compatible with polar and steric aspects governing the orientation, mobility, and substrate reactivity at the active site of TYR [55].

7. TYR inhibition

Scientists have made significant efforts to study the mechanism of monophenolase and diphenolase inhibition of TYR [77–82]. Scheme 7 depicts the generalized inhibition mechanism of TYR monophenolase and diphenolase activities.

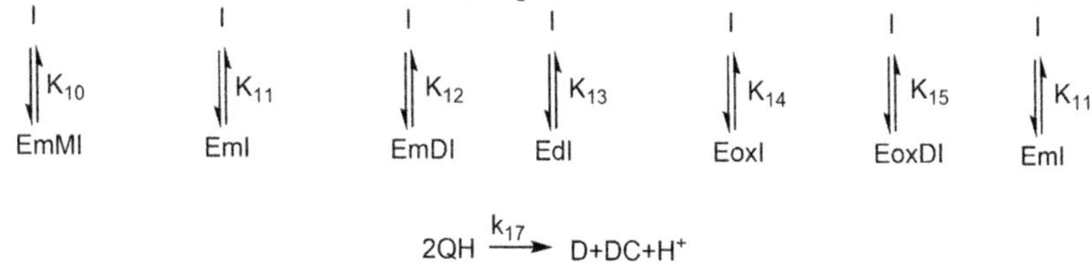

Scheme 7 Generalized inhibition of two activities of TYR (monophenolase and diphenolase activities). M, monophenol; D, o-diphenol; I, inhibitor; Em, met-TYR; Ed, deoxy-TYR; Eox, oxy-TYR; QH, protonated o-dopaquinone; and DC, dopachrome. *Represented with permission from S. Zolghadri, M. Beygi, T.F. Mohammad, M. Alijanianzadeh, T. Pillaiyar, P. Garcia-Molina, et al., Targeting tyrosinase in hyperpigmentation: current status, limitations and future promises, Biochem. Pharmacol. (2023) 115574.*

The lag period of monophenolase activity before reaching a steady state is one of the challenges of studying monophenolase inhibition, making the study difficult. In order to overcome this challenge and achieve a steady state, it is necessary to add a certain amount of L-dopa to the reaction mixture, and TYR behaves as a Michaelis-Menten enzyme. The Michaelis-Menten model can describe the TYR activity, and its kinetic parameters (such as Km and Vmax) can provide valuable insights into the mechanism of inhibition and help identify potential inhibitors for further development [83–85].

Diphenolase activity also follows the Michaelis-Menten model. In the study of the inhibition of the diphenolase activity of TYR, it is essential to determine the inhibition strength based on the inhibition constants and the inhibition types (competitive, uncompetitive, non-competitive, and mixed) [83]. One of the main parameters used in enzyme inhibition studies is the IC_{50} (half maximal inhibitory concentration). The inhibition degree of diphenolase activity (iD), varying the inhibitor concentration to a fixed amount of D substrate, can be used for this purpose. By non-linear regression adjustment of iD concerning the initial inhibitor concentration $[I]_0$, IC_{50} is determined. The analytical expression of the IC_{50} for the different types of inhibition is related to the apparent inhibition constant (K_I^{app}) [83].

Generally, TYR inhibitors can be categorized into three group:
a. True inhibitors
b. Specific TYR inactivators
c. Alternative substrates [86,87].

Scientists have proposed several strategies for data analysis related to iD, including the dependence of $1/IC_{50}$ vs. V_0/V_{max}, which can successfully discriminate between mechanisms by considering a mono-substrate reaction in rapid equilibrium and the representation of IC_{50} vs. $[S]_0/K_m$ as a new tool for studying the kinetics of inhibitors and characterizing the strength of the inhibitor and the type of inhibition [83].

7.1 True inhibitors

True inhibitors interact directly with the free TYR or TYR/substrate complex, creating four types of inhibition (competitive, uncompetitive, non-competitive, and mixed) [88]. Under the fast equilibrium, the relationship between IC_{50} and K_I^{app} for competitive, uncompetitive, and non-competitive inhibitors of a mono-substrate reaction reveals the dependence of iD on the substrate concentration used in the experiments, except for non-competitive inhibition. Furthermore, the relationship between IC_{50} and substrate concentration ($[D]_0$) varies depending on the type of inhibition and allows us to distinguish between them. The characteristic of non-competitive inhibition is the lack of dependence of iD on $[D]_0$, and the value of K_I^{app} is equal to the IC_{50}. On the other hand, in the case of the competitive and type 1 mixed inhibition mechanisms, there is an ambiguity between them due to the dependence of iD on $[D]_0$ of both. The plot of iD vs. n, with $n = [D]_0/K_m$, permits us to resolve this issue. A competitive inhibitor binds preferentially to Em and Eox (in such a way that $K_{I2}^{app} \to \infty$), and K_{I1}^{app} is calculated from the IC_{50} expression. A non-competitive inhibitor binds to Em, Eox, EmD, and EoxD. In this case, K_{I1}^{app} and K_{I2}^{app} are approximately equal. Uncompetitive inhibitors bind preferentially to the EmD and EoxD (K_{I1}^{app} is effectively infinite, $K_{I1}^{app} \to \infty$). In this case, the plot of iD vs. n is hyperbolic and passes through the coordinate origin. The calculation of K_{I2}^{app} is immediate from the IC_{50} value. For the mixed inhibitors that exhibit both competitive and non-competitive inhibitions, the values of the apparent inhibition constant of K_{I1}^{app}, Eox, Em and those constants of K_{I2}^{app}, EoxD, and EmD are extracted from the dependence of iD vs. n. The obtained results must correspond to the IC_{50} value, and K_{I1}^{app} is not equal to K_{I2}^{app} [83]. Note that the inhibitor does not bind to Ed at the saturation concentration of oxygen ($K_{13} \to \infty$) due to the high affinity of TYR for oxygen [89].

7.1.1 Specific TYR inactivators

Specific TYR inactivators are mechanism-based inhibitors or suicidal TYR substrates. These compounds can be o-diphenols, triphenols, or reducing agents such as ascorbic acid, tetrahydrofolic acid, tetrahydropterin, and nicotinamide adenine dinucleotide (NADH) [90–92]. Interestingly, TYR inactivation is stereoselective in the substrate binding but not in the suicide inactivation constant. For example, D-dopa and L-ascorbic acid showed a high affinity for TYR binding compared to L-dopa and D-ascorbic acid, while the maximum inactivation rate was the same [93]. M substrates induce no suicidal inactivation of TYR [68,94,95].

Although the diversity in the chemical structure of the different substrates, they all possess an oxidation/reduction stage. A proton transfer to the peroxide moiety on the active site of Eox results in copper (0), H_2O_2, and a Q product, thus inactivating TYR [91,96]. Due to the high affinity of oxygen for TYR, it protects the enzyme from irreversible inhibition. Therefore, the suicide inactivation process proceeds slowly under aerobic

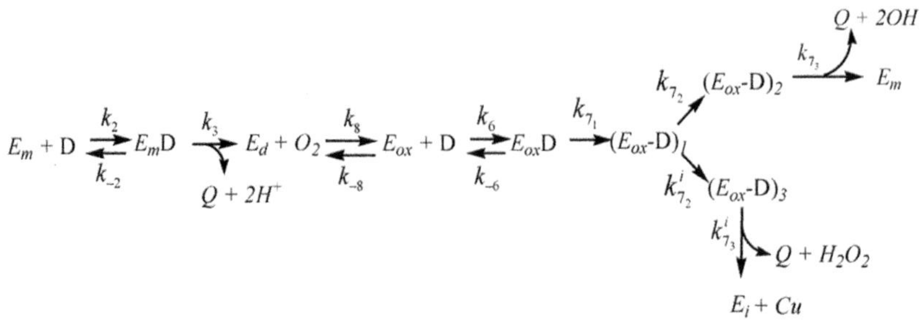

$$Q + 2OH$$

$$E_m + D \underset{k_{-2}}{\overset{k_2}{\rightleftharpoons}} E_mD \overset{k_3}{\underset{Q + 2H^+}{\searrow}} E_d + O_2 \underset{k_{-8}}{\overset{k_8}{\rightleftharpoons}} E_{ox} + D \underset{k_{-6}}{\overset{k_6}{\rightleftharpoons}} E_{ox}D \overset{k_{7_1}}{\rightarrow} (E_{ox}\text{-}D)$$

$$k_{7_2}(E_{ox}\text{-}D)_2 \overset{k_{7_3}}{\rightarrow} E_m$$

$$k_{7_2}^i(E_{ox}\text{-}D)_3$$

$$k_{7_3}^i \searrow Q + H_2O_2$$

$$E_i + Cu$$

Scheme 8 Proposed mechanism of suicide inactivation of TYR acting on D substrates. D, o-diphenol; Em, met-TYR; Ed, deoxy-TYR; Eox, oxy-TYR; E$_i$, inactive enzyme; and Q, o-quinone. *Represented with permission from J. Munoz-Munoz, J. Acosta-Motos, F. Garcia-Molina, R. Varon, P. Garcia-Ruíz, J. Tudela, et al., Tyrosinase inactivation in its action on dopa, Biochim. Biophys. Acta Proteins Proteom. 1804 (7) (2010) 1467–1475.*

conditions, while the enzyme inactivation occurs quickly in the presence of o-diphenols under anaerobic conditions [89,97].

The suicide substrates must have at least two OH groups in the benzene ring in the ortho position of their structures. For example, experimental results showed that trihydroxybenzene (e.g., pyrogallol) is the most potent suicide substrate of mushroom TYR because of autooxidation, not due to the action of TYR [90].

The parameters of λ_{max} (maximum apparent inactivation constant), K_m for the substrate, and r (a partition ratio between the catalytic and the inactivation pathways) characterized the kinetics of suicide substrates [92]. In a review, Muñoz–Muñoz et al. (2010) explained this process and proposed its mechanism [92]. Scheme 8 depicts the kinetic mechanism of the suicide inactivation of TYR acting on D substrates proposed by them.

7.1.2 Alternative substrates

Some phenolic compounds have been known as TYR inhibitors in the literature, while they may be alternative substrates of TYR [85,87]. Alternative substrates, such as compounds with a free hydroxyl in an aromatic ring at the meta and ortho positions, are not TYR substrates by themselves. However, they can become substrates by a reducing agent, which initiates the conversion of Em to Ed. Once bound to O_2, Ed conversion to Eox occurs, which can hydroxylate these compounds to D and then to Q, changing the melanogenesis pathway. Additionally, in the presence of H_2O_2, the transition from Em to Eox (active on these monophenols) occurs [98,99]. Thus, it is essential to discriminate between inhibitors and alternative substrates of TYR. For this purpose, Ortiz-Ruiz et al. (2015) proposed an experiment with several stages. First, the degree of inhibition of monophenolase (iM) and diphenolase activities (iD) in the presence of the molecule are determined. If they are equal, it is a TYR inhibitor; otherwise, the molecule can be an inhibitor or a substrate of TYR. They also suggested two additional experiments to resolve the ambiguity that the molecules were inhibitors or alternative substrates. The first step involved measuring product formation resulting from the oxidation of a D substrate (such as TBC and L-dopa) by TYR until total oxygen consumption. TBC produces TBQ, a very stable Q, while L-dopa produces DC, a relatively stable product. By measuring the accumulation of TBQ or DC, researchers can determine when all the oxygen has been consumed. If the concentration of the product remains constant, the target molecule functions as an inhibitor. Otherwise, if the product concentration decreases, it is related to oxygen consumption by the target, and the phenols are substrates. In this condition, the stability of Qs needs to be tested to ensure that any changes in absorbance are related to the different coefficients of absorptivity of the Qs from the target compound and not due to the interaction of this compound with the Qs. When Arbutin was studied, they observed that the absorbance decreased by increasing the concentration of Arbutin. As there was no change in absorbance over time, they concluded that Arbutin acts as an alternative substrate for TYR. While the initial step was effective for studying alternative substrates such as Arbutin. However, due to an interaction of the target molecule with the Qs of TBC or L-dopa substrates, it was impossible to do the first step correctly for them. Therefore, they applied an alternative step (step 2) to assay the target phenol with TYR and H_2O_2. They observed a variation in the absorbance of the target phenols in the presence of TYR and H_2O_2. After verifying that this deviation was not due to the H_2O_2 attack on the molecules, they concluded that these molecules are alternative substrates for TYR. They applied this step for various phenols (including TBP, iso-eugenol, arbutin, eugenol, guaiacol, and carvacrol) and measured the increase of the absorbance of the generated Qs. They observed that all molecules are alternative substrates of TYR. Interestingly, Arbutin fulfilled Step 1 and Step 2. These findings confirm that TYR can interact with various phenols as a versatile enzyme and provide insights into the mechanisms underlying phenol oxidation by TYR [100].

8. Conclusion

In this chapter, we summarized and discussed the current insights into the architecture of the active site, catalytic function, and inhibition mechanism of TYR. Understanding these features of TYRs is crucial in developing strategies to prevent enzymatic browning in vegetables, fruits, food, and seafood, as well as in developing treatments for pigmentation disorders. As described, the TYR active site contains six His residues coordinating two copper ions connected by a bridging a hydroxide anion or a water molecule. The implications for the TYR reaction vary depending on the oxidation states of the two copper ions in the active site. Moreover, the kinetic mechanism of the monophenolase activity of TYR is very complex due to overlapping by diphenolase activity in time. Given these findings, researchers have been working on distinguishing between monophenolase and diphenolase activities for decades. However, recent biochemical and structural studies have raised doubts about the mono-phenolase activity of TYR. As mentioned in this chapter, a complexity in monophenolase activity leads to the dual behavior of some monophenolic compounds as TYR inhibitors or alternative substrates. Thus, due to this aspect, the application of these compounds as TYR inhibitors must be controlled and verified. All this information opens new insights into the individual stages of the TYR reaction and will be helpful in the design and development of a new class of TYR inhibitors.

References

[1] I. Kipouros, E.I. Solomon, New mechanistic insights into coupled binuclear copper monooxygenases from the recent elucidation of the ternary intermediate of tyrosinase, FEBS Lett. 597 (1) (2023) 65–78.
[2] N. Manh Khoa, N. Viet Phong, S.Y. Yang, B.S. Min, J.A. Kim, Spectroscopic ana-lysis, kinetic mechanism, computational docking, and molecular dynamics of active metabolites from the aerial parts of Astragalus membranaceusBunge as tyrosinase inhibitors, Bioorg Chem. 134 (2023) 106464.
[3] Y. Matoba, S. Kihara, N. Bando, H. Yoshitsu, M. Sakaguchi, K. Kayama, et al., Catalytic mechanism of the tyrosinase reaction toward the Tyr98 residue in the caddie protein, PLoS Biol. 16 (12) (2018) e3000077.
[4] C. Olivares, F. Solano, New insights into the active site structure and catalytic mechanism of tyrosinase and its related proteins, Pigment. Cell Melanoma Res. 22 (6) (2009) 750–760.
[5] M. Sendovski, M. Kanteev, V.S. Ben-Yosef, N. Adir, A. Fishman, First structures of an active bacterial tyrosinase reveal copper plasticity, J. Mol. Biol. 405 (1) (2011) 227–237.
[6] A. Mouadili, D. Mazouzi, R. Touzani, Towards efficient catalysts via biomimetic chemistry for diphenols and aminophenols aerobic oxidation, Iran. J. 42 (4) (2023) 1111–1125.

[7] E. Selvarajan, R. Veena, N. Manoj Kumar, Polyphenol oxidase, beyond enzyme browning, Microbial Bioprospecting for Sustainable Development, Springer Singapore, 2018, pp. 203–222.

[8] L. Guo, W. Li, Z. Gu, L. Wang, L. Guo, S. Ma, et al., Recent advances and progress on melanin: from source to application, Int. J. Mol. Sci. 24 (5) (2023) 4360.

[9] M. Kanteev, M. Goldfeder, A. Fishman, Structure-function correlations in tyrosinases, Protein Sci. 24 (9) (2015) 1360–1369.

[10] F. Panis, A. Rompel, Identification of the amino acid position controlling the different enzymatic activities in walnut tyrosinase isoenzymes (jrPPO1 and jrPPO2), Sci. Rep. 10 (1) (2020) 10813.

[11] S. Carradori, F. Melfi, J. Rešetar, R. Şimşek, Tyrosinase enzyme and its inhibitors: an update of the literature, Metalloenzymes (2024) 533–546.

[12] C.W. Van Gelder, W.H. Flurkey, H.J. Wichers, Sequence and structural features of plant and fungal tyrosinases, Phytochemistry 45 (7) (1997) 1309–1323.

[13] H. Claus, H. Decker, Bacterial tyrosinases, Syst. Appl. Microbiol. 29 (1) (2006) 3–14.

[14] M. Fairhead, L. Thöny-Meyer, Bacterial tyrosinases: old enzymes with new relevance to biotechnology, N. Biotechnol. 29 (2) (2012) 183–191.

[15] K. Schaerlaekens, L. Van Mellaert, E. Lammertyn, N. Geukens, J. Anne, The importance of the Tat-dependent protein secretion pathway in Streptomyces as revealed by phenotypic changes in tat deletion mutants and genome analysis, Microbiology 150 (1) (2004) 21–31.

[16] M. Nakamura, T. Nakajima, Y. Ohba, S. Yamauchi, B.R. Lee, E. Ichishima, Identification of copper ligands in Aspergillus oryzae tyrosinase by site-directed mutagenesis, Biochem. J. 350 (2) (2000) 537–545.

[17] F. Solano, On the metal cofactor in the tyrosinase family, Int. J. Mol. Sci. 19 (2) (2018) 633.

[18] P. Agarwal, M. Singh, J. Singh, R. Singh, Microbial tyrosinases: a novel enzyme, structural features, and applications, Applied Microbiology and Bioengineering, Elsevier, 2019, pp. 3–19.

[19] M. Fekry, K.K. Dave, D. Badgujar, E. Hamnevik, O. Aurelius, D. Dobritzsch, et al., The crystal structure of tyrosinase from verrucomicrobium spinosum reveals it to be an atypical bacterial tyrosinase, Biomolecule 13 (9) (2023) 1360.

[20] C.W.G. van Gelder, W.H. Flurkey, H.J. Wichers, Sequence and structural features of plant and fungal tyrosinases, Phytochemistry 45 (7) (1997) 1309–1323.

[21] N.J. Kus, M.B. Dolinska, K.L. Young, E.K. Dimitriadis, P.T. Wingfield, Y.V. Sergeev, Membrane-associated human tyrosinase is an enzymatically active monomeric glycoprotein, PLoS One 13 (6) (2018) e0198247.

[22] S. Kumari, S.T.G. Thng, N.K. Verma, H.K. Gautam, Melanogenesis inhibitors, Acta Derm. Vener. 98 (10) (2018) 924–931.

[23] G.H. Chen, W.M. Chen, Y.C. Huang, S.T. Jiang, Expression of recombinant mature human tyrosinase from Escherichia coli and exhibition of its activity without phosphorylation or glycosylation, J. Agric. Food Chem. 60 (11) (2012) 2838–2843.

[24] C. Ariöz, P. Wittung-Stafshede, Folding of copper proteins: role of the metal? Q. Rev. Biophys. 51 (2018) e4.

[25] D. Seruggia, S. Josa, A. Fernandez, L. Montoliu, The structure and function of the mouse tyrosinase locus, Pigment. Cell Melanoma Res. 34 (2) (2021) 212–221.

[26] K.L. Young, C. Kassouf, M.B. Dolinska, D.E. Anderson, Y.V. Sergeev, Human tyrosinase: temperature-dependent kinetics of oxidase activity, Int. J. Mol. Sci. 21 (3) (2020) 895.

[27] A.A. Saboury, S. Zolghadri, K. Haghbeen, A.A. Moosavi-Movahedi, The inhibitory effect of benzenethiol on the cresolase and catecholase activities of mushroom tyrosinase, J. Enzyme Inhib. Med. Chem. 21 (6) (2006) 711–717.

[28] E. Amin, A.A. Saboury, H. Mansouri-Torshizi, S. Zolghadri, A.K. Bordbar, Evaluation of p-phenylene-bis and phenyl dithiocarbamate sodium salts as inhibitors of mushroom tyrosinase, Acta Biochim. Pol. 57 (3) (2010) 277–283.

[29] T. Schweikardt, C. Olivares, F. Solano, E. Jaenicke, J.C. García-Borrón, H. Decker, A three-dimensional model of mammalian tyrosinase active site accounting for loss of function mutations, Pigment. Cell Res. 20 (5) (2007) 394–401.

[30] J.C. García-Borrón, F. Solano, Molecular anatomy of tyrosinase and its related proteins: beyond the histidine-bound metal catalytic center, Pigment. Cell Res. 15 (3) (2002) 162–173.

[31] C. Zou, W. Huang, G. Zhao, X. Wan, X. Hu, Y. Jin, et al., Determination of the bridging ligand in the active site of tyrosinase, Molecules 22 (2017) 1836.

[32] R. Haudecoeur, A. Gouron, C. Dubois, H. Jamet, M. Lightbody, R. Hardré, et al., Investigation of binding-site homology between mushroom and bacterial Tyrosinases by using Aurones as effectors, ChemBioChem 15 (9) (2014) 1325–1333.

[33] Á. Sánchez-Ferrer, J.N. Rodríguez-López, F. García-Cánovas, F. García-Carmona, Tyrosinase: a comprehensive review of its mechanism, Biochim. Biophys. Acta Protein Struct. Mol. Enzymol. 1247 (1) (1995) 1–11.

[34] H. Decker, T. Schweikardt, F. Tuczek, The first crystal structure of tyrosinase: all questions answered? Angew. Chem. Int. Ed. 45 (28) (2006) 4546–4550.

[35] N. Fujieda, K. Umakoshi, Y. Ochi, Y. Nishikawa, S. Yanagisawa, M. Kubo, et al., Copper–oxygen dynamics in the tyrosinase mechanism, Angew. Chem. 132 (32) (2020) 13487–13492.

[36] H. Decker, E. Solem, F. Tuczek, Are glutamate and asparagine necessary for tyrosinase activity of type-3 copper proteins? Inorg. Chim. Acta 481 (2018) 32–37.

[37] S.M. Prexler, M. Frassek, B.M. Moerschbacher, M.E. Dirks-Hofmeister, Catechol oxidase versus tyrosinase classification revisited by site-directed mutagenesis studies, Angew. Chem. Int. Ed. 58 (26) (2019) 8757–8761.

[38] S. Molloy, J. Nikodinovic-Runic, L.B. Martin, H. Hartmann, F. Solano, H. Decker, et al., Engineering of a bacterial tyrosinase for improved catalytic efficiency towards D-tyrosine using random and site directed mutagenesis approaches, Biotechnol. Bioeng. 110 (7) (2013) 1849–1857.

[39] M.B. Dolinska, E. Kovaleva, P. Backlund, P.T. Wingfield, B.P. Brooks, Y.V. Sergeev, Albinism-causing mutations in recombinant human tyrosinase alter intrinsic enzymatic activity, PLoS One 9 (1) (2014) e84494.

[40] A. Shahrisa, M. Nikkhah, H. Shirzad, R. Behzadi, M. Sadeghizadeh, Enhancing catecholase activity of a recombinant human tyrosinase through multiple strategies, Iran. J. Biotechnol. 18 (2) (2020) e2310.

[41] I. Kampatsikas, M. Pretzler, A. Rompel, Identification of amino acid residues responsible for C−H activation in type-III copper enzymes by generating tyrosinase activity in a catechol oxidase, Angew. Chem. Int. Ed. 59 (47) (2020) 20940–20945.

[42] M.B. Dolinska, N.J. Kus, S.K. Farney, P.T. Wingfield, B.P. Brooks, Y.V. Sergeev, Oculocutaneous albinism type 1: link between mutations, tyrosinase conformational stability, and enzymatic activity, Pigment. Cell Melanoma Res. 30 (1) (2017) 41–52.

[43] M. Fairhead, L. Thöny-Meyer, Role of the C-terminal extension in a bacterial tyrosinase, FEBS J. 277 (9) (2010) 2083–2095.

[44] R.A. Spritz, L. Ho, M. Furumura, V.J. Hearing Jr, Mutational analysis of copper binding by human tyrosinase, J. Invest. Dermatol. 109 (2) (1997) 207–212.

[45] M.P. Jackman, A. Hajnal, K. Lerch, Albino mutants of Streptomyces glaucescens tyrosinase, Biochem. J. 274 (3) (1991) 707–713.

[46] C. Kaintz, S.G. Mauracher, A. Rompel, Type-3 copper proteins: recent advances on polyphenol oxidases, Adv. Protein Chem. Struct. Biol. 97 (2014) 1–35.

[47] H. Noh, S.J. Lee, H.-J. Jo, H.W. Choi, S. Hong, K.-H. Kong, Histidine residues at the copper-binding site in human tyrosinase are essential for its catalytic activities, J. Enzyme Inhib. Med. Chem. 35 (1) (2020) 726–732.

[48] I. Kampatsikas, A. Rompel, Similar but still different: which amino acid residues are responsible for varying activities in type-III copper enzymes? ChemBioChem 22 (7) (2021) 1161–1175.

[49] M. Goldfeder, M. Kanteev, N. Adir, A. Fishman, Influencing the monophenolase/diphenolase activity ratio in tyrosinase, Biochim. Biophy Acta Proteins Proteom. 1834 (3) (2013) 629–633.

[50] E. Beltran, M.R. Serafini, I.A. Alves, D.M. Aragón Novoa, Novel synthesized tyrosinase inhibitors: a systematic patent review (2012–present), Curr. Med. Chem. 31 (3) (2024) 308–335.

[51] J.-H. Xu, J. Lee, S.-J. Yin, W. Wang, Y.-D. Park, Inhibitory effect of acarbose on tyrosinase: application of molecular dynamics integrating inhibition kinetics, J. Biomol. Struct. Dyn. 42 (1) (2024) 314–325.

[52] C.A. Ramsden, P.A. Riley, Tyrosinase: the four oxidation states of the active site and their relevance to enzymatic activation, oxidation and inactivation, Bioorg Med. Chem. 22 (8) (2014) 2388–2395.

[53] A. Ayuhastuti, I.S.K. Syah, S. Megantara, A.Y. Chaerunisaa, Nanotechnology-enhanced cosmetic application of kojic acid dipalmitate, a kojic acid derivate with improved properties, Cosmetics 11 (1) (2024) 21.

[54] T. Li, N. Zhang, S. Yan, S. Jiang, H. Yin, A novel tyrosinase from Armillaria ostoyae with comparable monophenolase and diphenolase activities suffers substrate inhibition, Appl. Environ. Microbiol. 87 (12) (2021) e00275-21.

[55] J.N. Rodríguez-López, L.G. Fenoll, P.A. García-Ruiz, R. Varón, J. Tudela, R.N. Thorneley, et al., Stopped-flow and steady-state study of the diphenolase activity of mushroom tyrosinase, Biochemistry 39 (34) (2000) 10497–10506.

[56] S. Zolghadri, M. Beygi, T.F. Mohammad, M. Alijanianzadeh, T. Pillaiyar, P. Garcia-Molina, et al., Targeting tyrosinase in hyperpigmentation: current status, limitations and future promises, Biochem. Pharmacol. (2023) 115574.

[57] S. Zolghadri, A. Bahrami, M.T. Hassan Khan, J. Munoz-Munoz, F. Garcia-Molina, F. Garcia-Canovas, et al., A comprehensive review on tyrosinase inhibitors, J. Enzyme Inhib. Med. Chem. 34 (1) (2019) 279–309.

[58] A. Bijelic, M. Pretzler, C. Molitor, F. Zekiri, A. Rompel, The structure of a plant tyrosinase from walnut leaves reveals the importance of "substrate-guiding residues" for enzymatic specificity, Angew. Chem. Int. Ed. 54 (49) (2015) 14677–14680.

[59] M. Vaezi, Structure and inhibition mechanism of some synthetic compounds and phenolic derivatives as tyrosinase inhibitors: review and new insight, J. Biomol. Struct. Dyn. 41 (10) (2023) 4798–4810.

[60] P. García-Molina, J.L. Munoz-Munoz, J.A. Ortuño, J.N. Rodríguez-López, P.A. García-Ruiz, F. García-Cánovas, et al., Considerations about the continuous assay methods, spectrophotometric and spectrofluorometric, of the monophenolase activity of tyrosinase, Biomolecules 11 (9) (2021) 1269.

[61] I. Kampatsikas, A. Rompel, Similar but still different: which amino acid residues are responsible for varying activities in type-III copper enzymes? ChemBioChem 22 (7) (2021) 1161–1175.

[62] F. Panis, I. Kampatsikas, A. Bijelic, A. Rompel, Conversion of walnut tyrosinase into a catechol oxidase by site directed mutagenesis, Sci. Rep. 10 (1) (2020) 1659.

[63] M. Goldfeder, M. Kanteev, S. Isaschar-Ovdat, N. Adir, A. Fishman, Determination of tyrosinase substrate-binding modes reveals mechanistic differences between type-3 copper proteins, Nat. Commun. 5 (1) (2014) 4505.

[64] M. Rolff, J. Schottenheim, H. Decker, F. Tuczek, Copper–O 2 reactivity of tyrosinase models towards external monophenolic substrates: molecular mechanism and comparison with the enzyme, Chem. Soc. Rev. 40 (7) (2011) 4077–4098.

[65] H. Decker, F. Tuczek, Tyrosinase/catecholoxidase activity of hemocyanins: structural basis and molecular mechanism, Trends Biochem. Sci. 25 (8) (2000) 392–397.

[66] R.J. Deeth, C. Diedrich, Structural and mechanistic insights into the oxy form of tyrosinase from molecular dynamics simulations, J. Biol. Inorg. Chem. 15 (2010) 117–129.

[67] I. Kipouros, A. Stańczak, J.W. Ginsbach, P.C. Andrikopoulos, L. Rulíšek, E.I. Solomon, Elucidation of the tyrosinase/O2/monophenol ternary intermediate that dictates the monooxygenation mechanism in melanin biosynthesis, Proc. Natl. Acad. Sci. 119 (33) (2022) e2205619119.

[68] P. García-Molina, F. García-Molina, J.A. Teruel-Puche, J.N. Rodríguez-López, F. García-Cánovas, J.L. Muñoz-Muñoz, Considerations about the kinetic mechanism of tyrosinase in its action on monophenols: a review, Mol. Catal. 518 (2022) 112072.

[69] M. Pérez-Gilabert, A. Morte, M. Honrubia, F. García-Carmona, Monophenolase activity of latent Terfezia claveryi tyrosinase: characterization and histochemical localization, Physiol. Plant. 113 (2) (2001) 203–209.

[70] J.R. Ros, J.N. Rodríguez-López, R. Varón, F. García-Cánovas, Kinetics study of the oxidation of 4-tert-butylphenol by tyrosinase, Eur. J. Biochem. 222 (2) (1994) 449–452.

[71] L.G. Fenoll, J.N. Rodríguez-López, F. García-Sevilla, J. Tudela, P.A. García-Ruiz, R. Varón, et al., Oxidation by mushroom tyrosinase of monophenols generating slightly unstable o-quinones, Eur. J. Biochem. 267 (19) (2000) 5865–5878.

[72] N. Gheibi, N. Taherkhani, A. Ahmadi, K. Haghbeen, D. Ilghari, Characterization of inhibitory effects of the potential therapeutic inhibitors, benzoic acid and pyridine derivatives, on the monophenolase and diphenolase activities of tyrosinase, Iran. J. Basic Med. Sci. 18 (2) (2015) 122.

[73] L.L. Ingraham, Variation of the Michaelis constant in polyphenol oxidase catalyzed oxidations: substrate structure and concentration, J. Am. Chem. Soc. 79 (3) (1957) 666–669.

[74] H.W. Duckworth, J.E. Coleman, Physicochemical and kinetic properties of mushroom tyrosinase, J. Biol. Chem. 245 (7) (1970) 1613–1625.

[75] S. Gutteridge, D. Robb, The catecholase activity of Neurospora tyrosinase, Eur. J. Biochem. 54 (1) (1975) 107–116.

[76] J.N. Rodríguez-López, J.R. Ros, R. Varón, F. García-Cánovas, Oxygen Michaelis constants for tyrosinase, Biochem. J. 293 (3) (1993) 859–866.

[77] K. Bagherzadeh, F. Shirgahi Talari, A. Sharifi, M.R. Ganjali, A.A. Saboury, M. Amanlou, A new insight into mushroom tyrosinase inhibitors: docking, pharmacophore-based virtual screening, and molecular modeling studies, J. Biomol. Struct. Dyn. 33 (3) (2015) 487–501.

[78] K. Haghbeen, A.A. Saboury, F. Karbassi, Substrate share in the suicide inactivation of mushroom tyrosinase, Biochim. Biophys. Acta Gen. Subj. 1675 (1-3) (2004) 139–146.

[79] F. Karbassi, K. Haghbeen, A. Saboury, B. Ranjbar, A. Moosavi-Movahedi, Activity, structural and stability changes of mushroom tyrosinase by sodium dodecyl sulfate, Colloids Surf. B Biointerfaces 32 (2) (2003) 137–143.

[80] S.S. Borojerdi, K. Haghbeen, A.A. Karkhane, M. Fazli, A.A. Saboury, Successful resonance Raman study of cresolase activity of mushroom tyrosinase, Biochem. Biophys. Res. Commun. 314 (4) (2004) 925–930.

[81] N. Gheibi, A. Saboury, K. Haghbeen, A. Moosavi-Movahedi, Activity and structural changes of mushroom tyrosinase induced by n-alkyl sulfates, Colloids Surf. B Biointerfaces 45 (2) (2005) 104–107.

[82] A. Saboury, F. Karbassi, K. Haghbeen, B. Ranjbar, A. Moosavi-Movahedi, B. Farzami, Stability, structural and suicide inactivation changes of mushroom tyrosinase after acetylation by N-acetylimidazole, Int. J. Biol. Macromol. 34 (4) (2004) 257–262.

[83] P. Garcia-Molina, F. Garcia-Molina, J.A. Teruel-Puche, J.N. Rodriguez-Lopez, F. Garcia-Canovas, J.L. Muñoz-Muñoz, The relationship between the IC50 values and the apparent inhibition constant in the study of inhibitors of tyrosinase diphenolase activity helps confirm the mechanism of inhibition, Molecules 27 (10) (2022) 3141.

[84] L. Liu, J. Li, L. Zhang, S. Wei, Z. Qin, D. Liang, et al., Conformational changes of tyrosinase caused by pentagalloylglucose binding: Implications for inhibitory effect and underlying mechanism, Food Res. Int. 157 (2022) 111312.

[85] C.V. Ortiz-Ruiz, J. Berna, M. Del Mar Garcia-Molina, J. Tudela, V. Tomas, F. Garcia-Canovas, Identification of p-hydroxybenzyl alcohol, tyrosol, phloretin and its derivate phloridzin as tyrosinase substrates, Bioorg Med. Chem. 23 (13) (2015) 3738–3746.

[86] T.S. Chang, An updated review of tyrosinase inhibitors, Int. J. Mol. Sci. 10 (6) (2009) 2440–2475.

[87] M. N. Masum, K. Yamauchi, T. Mitsunaga, Tyrosinase inhibitors from natural and synthetic sources as skin-lightening agents, Rev. Agric. Sci. 7 (2019) 41–58.

[88] A. Saboury, Enzyme inhibition and activation: a general theory, J. Iran. Chem. Soc. 6 (2009) 219–229.

[89] L.G. Fenoll, J.N. Rodríguez-López, F. García-Molina, F. García-Cánovas, J. Tudela, Michaelis constants of mushroom tyrosinase with respect to oxygen in the presence of monophenols and diphenols, Int. J. Biochem. Cell Biol. 34 (4) (2002) 332–336.

[90] J.L. Muñoz-Muñoz, F. Garcia-Molina, P.A. García-Ruiz, M. Molina-Alarcon, J. Tudela, F. Garcia-Canovas, et al., Phenolic substrates and suicide inactivation of tyrosinase: kinetics and mechanism, Biochem. J. 416 (3) (2008) 431–440.

[91] J.L. Muñoz-Muñoz, F. Garcia-Molina, R. Varon, P.A. Garcia-Ruíz, J. Tudela, F. Garcia-Cánovas, et al., Suicide inactivation of the diphenolase and monophenolase activities of tyrosinase, IUBMB Life 62 (7) (2010) 539–547.

[92] J. Munoz-Munoz, J. Acosta-Motos, F. Garcia-Molina, R. Varon, P. Garcia-Ruíz, J. Tudela, et al., Tyrosinase inactivation in its action on dopa, Biochim. Biophys. Acta Proteins Proteom. 1804 (7) (2010) 1467–1475.

[93] P.J. Fernandez-Julia, J. Tudela-Serrano, F. Garcia-Molina, F. Garcia-Canovas, A. Garcia-Jimenez, J.L. Munoz-Munoz, Study of tyrosine and dopa enantiomers as tyrosinase substrates initiating l- and d-melanogenesis pathways, Biotechnol. Appl. Biochem. 68 (4) (2021) 823–831.

[94] J.L. Muñoz-Muñoz, Md.M. García-Molina, F. García-Molina, R. Varon, P.A. García-Ruiz, J.N. Rodríguez-López, et al., Indirect inactivation of tyrosinase in its action on 4-tert-butylphenol, J. Enzyme Inhib. Med. Chem. 29 (3) (2014) 344–352.

[95] J.L. Muñoz-Muñoz, F. Garcia-Molina, J.R. Acosta-Motos, E. Arribas, P.A. Garcia-Ruíz, J. Tudela, et al., Indirect inactivation of tyrosinase in its action on tyrosine, Acta Biochim. Pol. 58 (4) (2011).

[96] J.L. Muñoz-Muñoz, J. Berna, F. Garcia-Molina, P.A. Garcia-Ruiz, J. Tudela, J.N. Rodriguez-Lopez, et al., Unravelling the suicide inactivation of tyrosinase: a discrimination between mechanisms, J. Mol. Catal. B Enzym. 75 (2012) 11–19.

[97] J. Rodriguez-Lopez, J. Ros, R. Varon, F. Garcia-Canovas, Oxygen Michaelis constants for tyrosinase, Biochem. J. 293 (3) (1993) 859–866.

[98] Garcia-Molina MdM, J. Berna, J.L. Muñoz-Muñoz, P.A. García-Ruiz, M.G. Moreno, J.R. Martinez, et al., Action of tyrosinase on hydroquinone in the presence of catalytic amounts of o-diphenol. A kinetic study, React. Kinet. Mech. Catal. 112 (2) (2014) 305–320.

[99] M. Del Mar García-Molina, J.L.M. Muñoz, F. Martinez-Ortiz, J.R. Martinez, P.A. García-Ruiz, J.N. Rodriguez-López, et al., Tyrosinase-catalyzed hydroxylation of hydroquinone, a depigmenting agent, to hydroxyhydroquinone: a kinetic study, Bioorg Med. Chem. 22 (13) (2014) 3360–3369.

[100] C.V. Ortiz-Ruiz, Md.M. Garcia-Molina, J.T. Serrano, V. Tomas-Martinez, F. Garcia-Canovas, Discrimination between alternative substrates and inhibitors of tyrosinase, J. Agric. Food Chem. 63 (8) (2015) 2162–2171.

CHAPTER THREE

Structural characterization of tyrosinases and an update on human enzymes

Luigi Franklin Di Costanzo*
Department of Agriculture, Department of Excellence, University of Naples Federico II, Palace of Portici,
Piazza Carlo di Borbone, Portici NA, Italy
*Corresponding author. e-mail address: luigi.dicostanzo4@unina.it

Contents

Abstract

Tyrosinase, a pivotal enzyme in melanin biosynthesis, orchestrates the pigmentation process in humans, affecting skin, hair, and eye color. This chapter examines the three-dimensional structure and functional aspects of tyrosinases from various sources, highlighting their di-metal ion coordination crucial for catalytic activity. I explore the biochemical pathwayscheme catalyzed by tyrosinase, specifically the oxidation of L-tyrosine to L-dopaquinone, a precursor in melanin synthesis. Detailed structural analyses, including 3D structures obtained from X-ray crystallography and computational modeling, reveal key insights into the enzyme's active site, variations among tyrosinases, and substrate binding mechanisms. Furthermore, the chapter investigates the role of human tyrosinase variants, their inhibitors, essential for developing therapeutic and cosmetic applications targeting

The Enzymes, Volume 56
ISSN 1874-6047, https://doi.org/10.1016/bs.enz.2024.06.004

hyperpigmentation disorders. Structural characterizations of tyrosinase-inhibitor complexes provide a foundation for designing effective inhibitors, with compounds like kojic acid, L-mimosine, and (S)-3-amino-tyrosine demonstrating significant inhibitory potential. This comprehensive examination of the structure, function, and inhibition mechanisms of tyrosinase offers avenues for innovative treatments in biotechnology, health, and beyond.

1. Structure and function of the tyrosinases

1.1 Unveiling pigmentation: tyrosinase as the central character

Melanin takes central stage as the main character, allowing pigmentation like a painter upon our skin, hair, and eyes, shaping our identities with its rich palette [1]. A pivotal process in biochemical synthesis occurs within specialized cells known as melanocytes, where the enzyme **tyrosinase (Ty)** plays a central role. This enzyme catalyzes in a stepwise manner a two oxidation processes, transforming the amino acid L–tyrosine into L–dopaquinone [2], as shown in Fig. 1. L–tyrosine is first converted into L–dopa. Then, L–dopa quickly transforms into L–dopaquinone within the same active site. Finally, L–dopaquinone is even more rapidly and spontaneously converted into dopachrome [3]. This last compound is processed enzymatically and serves building block for the biochemical synthesis of melanin [4]. Classified as oxidoreductase (EC.1.14.18.1),

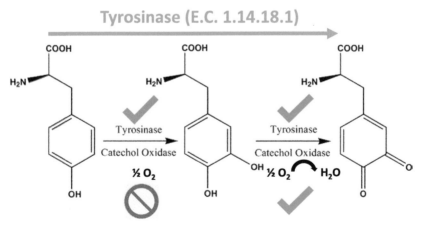

Fig. 1 Two-step reactions catalyzed by tyrosinase (TYR), a member of the protein family phenoloxidases (PPOs). Tyrosinase catalyzes both reactions: (1) the *o*-hydroxylation of *o*-monophenols (present in substrates like L-tyrosine) to *o*-diphenols, and (2) the oxidation of *o*-diphenols to the corresponding *o*-quinones. In contrast, a similar PPO, catechol oxidase, catalyzes only the conversion of *o*-diphenols to *o*-quinones.

Ty belong to the family of monophenol monooxygenases [5–7]. Human tyrosinase and its various isoforms are found in tissues, including skin, eyes, and hair follicles, and is particularly abundant in melanocytes, the cells responsible for melanin production [8]. Considering its function, mutations or variations in tyrosinase genes can lead to conditions such as albinism, characterized by a lack of melanin production [9]. Melanin also serves various functions, as protecting the skin from harmful UV radiation while absorbing this radiation.

Tyrosinase role beyond pigmentation is related to the oxidation of dopamine to quinone and its neuro-physiological consequences [10]. The widespread presence of tyrosinase in diverse kingdoms highlights its significance in various biological processes related to pigmentation and not. For example, in plants, tyrosinase contributes to pigment formation in flowers, seeds, and fruits [11,12]. Melanin has also been found in archeological samples including well preserved dinosaur fossils, and prehistoric bird feathers [13,14]. Microbial pigmentation, such as melanin synthesis, serves as a virulence factor [15]. For instance, *P. aeruginosa* exhibits pyoverdine, a major virulence factor, that utilizes a tyrosinase as a step for the molecule's biochemical synthesis [16]. The role of tyrosinases is not only limited to the biology and function of pigmentation within organisms, but can also have an impact on wetland ecosystems through the oxidative processes of phenolic compounds [17].

From a **bioinorganic chemistry** perspective, tyrosinases are members of the type-III copper enzyme family a bifunctional metalloenzymes that execute two sequential enzymatic reactions: first, they chemically hydroxylate monophenols to *o*-diphenols (termed monophenolase activity), and second, they oxidize the *o*-diphenols to the corresponding *o*-quinones (referred to as diphenolase or catecholase activity), as shown in Fig. 1 [18,19]. Upon binding of molecular oxygen, tyrosinase is activated to catalyze both monophenolase reaction cycle (reaction 1) and a diphenolase reaction cycle (reaction 2), see Fig. 1 [18].

1.2 The type-III di-copper cluster

Tyrosinases typically feature a **di-copper ions** cofactor cluster, crucial for catalytic activity [19–21]. These copper ions play a crucial role in the redox reactions facilitating oxygen transfer to the organic substrate. The interplay of tyrosinase's copper oxidation states between Cu (I) ($3d^9$ electronic configuration) and Cu (II) ($3d^{10}$), along with the element redox potential, substantiate their prevalence over other transition metal ions for oxygen activation. Moreover, copper-containing enzymes like tyrosinases adopt a strained or so called "entatic" state, intensifying their reactivity and enzyme catalytic efficiency [22]. This state involves a transiently distorted enzyme structure, in complex with substrate, during the catalytic mechanism, deviating from its equilibrium state in order to optimize its function [23,24]. More recent structural characterization has revealed the presence of **di-zinc ions cluster** (see Table 1), in the active site of tyrosynases including human tyrosinase [25].

1.3 Reaction mechanism in the context of enzyme family

Within the active site of tyrosinases, a distinctive configuration of the di-copper cluster offers a coordination geometry ideally suited for binding oxygen molecules and various small molecules as substrates or inhibitor ligands as those shown in Fig. 2, and discussed below [7]. This arrangement significantly contributes to the enzyme's functionality in hydroxylating and oxidizing substrates, such as L-tyrosine or simpler molecules like *o*-phenol.

In general, the catalytic reactions catalyzed by tyrosinases can be simplified as follows. In the monophenolase cycle, the enzyme transfers one of the bound oxygen atoms to a monophenol (e.g., L-tyrosine), creating a stable *o*-diphenol intermediate (see Fig. 1). This intermediate is then oxidized to an *o*-quinone, released along with a water molecule. The enzyme remains in an inactive deoxy state until a new oxygen molecule restores it to the active oxy state (see mechanism in the next section).

During the diphenolase cycle, the enzyme binds an external diphenol molecule (e.g., L-dopa) and oxidizes it to an *o*-quinone, releasing it along with a water molecule and entering the intermediate met state. The second reaction mirrors that catalyzed by the related enzyme catechol oxidase (EC 1.10.3.1). However, the latter cannot catalyze the hydroxylation or monooxygenation of monophenols.

Facilitating this intricate mechanism, the di-copper center undergoes coordinative geometrical rearrangements (refer to the following sections). Therefore, investigating the three-dimensional structure of tyrosinases has been vital for comprehending the enzyme's function and capturing atomic details in the deoxy, oxy, and phenolase states.

Despite their fundamental similarities and functions, tyrosinases exhibit notable variations in three-dimensional structures, activities, substrate specificities. Hence, exploring their 3D structures significantly contributes to refining our understanding of how their mechanisms are fine-tuned and regulated.

Finally, the 3D structure of tyrosinases has provided a foundation for the design of drugs or binding compounds that can modulate the enzyme activity. Insights into the structure-function relationship of tyrosinases contribute to advancements in fields like dermatology and cosmetics.

Table 1 The 3D structures of tyrosinases from various source organisms across all kingdoms of life were obtained from the RCSB PDB (RCSB.org).

Source organism	PDB ID	Ligand name	Ligand code	Release date	References
Agaricus bisporus	5M6B	Oxygen atom group	O, CU	5/17/2017	[26]
Agaricus bisporus	2Y9X	Tropolone	0TR	7/6/2011	[27]
Agaricus bisporus	4OUA	Copper (I), 6-tungstotellurate (VI)	CU1, TEW	6/25/2014	[28]
Aspergillus oryzae	3W6W	Copper (II)	CU	6/19/2013	[29]
Aspergillus oryzae	6JU4	L-DOPA	DAH	05/01/2020	[30]
Hahella	8B74			10/11/2023	
Homo sapiens (TYRP1)	9EY5	2-hydroxy-L-tyrosine	OTY	5/1/2024	[31]
Homo sapiens (TYRP1)	9EY6	Zinc (II)	Zn	5/1/2024	[31]
Homo sapiens (TYRP1)	9EY8	(S)-3-amino-tyrosine	TY2	5/1/2024	[31]
Homo sapiens (TYRP1)	5M8M	Kojic Acid	KOJ		
Homo sapiens (TYRP1)	5M8S	N-phenylthiourea	URS	1/31/2018	[32]
Homo sapiens (TYRP1)	5M8T	Tropolone	KOJ	7/12/2017	[25]
Homo sapiens (TYRP1)	5M8R	Mimosine	MMS	7/12/2017	[25]

(continued)

Table 1 The 3D structures of tyrosinases from various source organisms across all kingdoms of life were obtained from the RCSB PDB (RCSB.org). (cont'd)

Source organism	PDB ID	Ligand name	Ligand code	Release date	References
Homo sapiens (TYRP1)	5M8P	L–Tyrosine	TYR	7/12/2017	[25]
Homo sapiens (TYRP1)	5M8O	Tropolone	0TR	7/12/2017	[25]
Homo sapiens (TYRP2)	P40126-F1				AlphaFold
Homo sapiens (TYR)	P14679-F1				AlphaFold
Juglans regia (*walnut*)	5CE9	Oxygen atom group	O, CU	10/28/2015	[33]
Malus domestica (*apple*)	6ELS	Oxygen atom group	O, CU	3/20/2019	[34]
Priestia megaterium	6QXD	JKB inhibitor	JKB	6/19/2019	[35]
Priestia megaterium	5OAE	SVF inhibitor	SVF	4/25/2018	[36]
Priestia megaterium	5I3A	Benzene-1,4-diol (Hydroquinone)	HQE	10/12/2016	[37]
Priestia megaterium	5I3B	Benzene-1,4-diol (Hydroquinone)	HQE	10/12/2016	[37]
Priestia megaterium	5I38	Kojic Acid	KOJ	10/12/2016	[37]
Priestia megaterium	4P6T	*p*–Tyrosol	YRL	7/30/2014	[38]
Priestia megaterium	4P6S	L-DOPA	DAH	7/30/2014	[38]

Priestia megaterium	4P6R	L–Tyrosine	TYR	7/30/2014	[38]
Priestia megaterium	4J6V	Copper (II)	CU	12/25/2013	[39]
Priestia megaterium	3NTM	Copper (II)	CU	11/17/2010	[40]
Priestia megaterium	3NQ1	Kojic Acid	KOJ	11/17/2010	[40]
Priestia megaterium	8HPI	Zinc (II)	ZN	12/13/2023	
Ralstonia	7XIO	Phosphate ion	PO4	4/19/2023	
Solanum lycopersicum (Tomato)	6HQI	Oxygen atom polyphenol oxidase	O, CU	3/20/2019	[41]
Streptomyces avermitilis	6J2U	Zinc (II)	ZN	2/12/2020	
Streptomyces castaneoglobisporus	7CIY	Hydrogen peroxide	PEO	6/16/2021	[42]
Streptomyces castaneoglobisporus	3AX0	Copper (II)	CU	6/29/2011	[43]
Streptomyces castaneoglobisporus	2AHL	Copper (I)	CU1	1/31/2006	[44]
Streptomyces castaneoglobisporus	2AHK	Copper (II)	CU	1/31/2006	[44]
Streptomyces castaneoglobisporus	1WXC	Nitrate ion	NO3	1/31/2006	[44]
Streptomyces castaneoglobisporus	1WX2	Peroxide anion	PER	1/31/2006	[44]

(continued)

Table 1 The 3D structures of tyrosinases from various source organisms across all kingdoms of life were obtained from the RCSB PDB (RCSB.org). (cont'd)

Source organism	PDB ID	Ligand name	Ligand code	Release date	References
Thermothelomyces thermophilus (Thermophilic fungus)	6Z1S	Copper (II) *Polyphenol Oxidase*	CU	3/24/2021	[45]
Verrucomicrobium spinosum	8BBR	Copper (II)	CU	9/20/2023	[46]
Vitis vinifera (Grapevine)	2P3X	Oxygen atom	C2O	3/11/2008	[47]

Entries were selected based on the availability of complexes with key ligands of interest. Each entry's 3D structure and protein annotations can be explored by searching the provided PDB code in the second column and browsing the link below. Related entries can be retrieved from the structure summary page of each entry, as indicated in the 'PDB code ID' column. The 'ligand name' column specifically refers to the ligand of interest if bound to the structure, excluding any solvent molecules. Ligands can be explored using the link provided along with the appropriate 'ligand code'. The link is: https://www.rcsb.org, accessed in May 2024. Additionally, the structural models of human TYR and TYRP2 are available from the AlphaFold project using the Uniprot codes indicated [48].

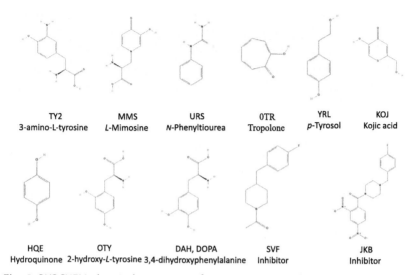

TY2	MMS	URS	0TR	YRL	KOJ
3-amino-L-tyrosine	L-Mimosine	N-Phenyltiourea	Tropolone	p-Tyrosol	Kojic acid

HQE	OTY	DAH, DOPA	SVF	JKB
Hydroquinone	2-hydroxy-L-tyrosine	3,4-dihydroxyphenylalanine	Inhibitor	Inhibitor

Fig. 2 PUBCHEM chemical structures of main tyrosinase inhibitors, for which structural studies of tyrosinase-complex are available. The 3-letter codes identify each ligand of interest, that can be explored from the RCSB PDB (link: www.rcsb.org). For clarity, chemical structures of substrates including L-DOPA and p-tyrosol, discussed in this chapter are also included.

1.4 Exploring tyrosinase: insights from available 3D structures

The 3D structure of tyrosinases is significant in understanding the function of the enzyme at the molecular level. The structure is essential for elucidating the catalytic mechanism and identifying key amino acid residues involved in substrate binding and catalysis. Most of tyrosinases 3D structure are based on experimentally determined structures by X-ray crystallography. A search for sequence similarity in the Protein Data Bank (accessed on May 20, 2024), the repository for macromolecular information, reveals 122 experimental crystal structures providing the atomic 3D details [49]. The four alphanumerical characters mentioned in this chapter serve as identifiers for the experimental structure retrieved from the Protein Data Bank (PDB) website at rcsb.org [50].

The available experimental tyrosinase structures are derived from several organisms and share less or ~30% sequence identity compared to the 273-amino acid sequence (Uniprot ID Q83WS2) of tyrosinase from *Streptomyces castaneoglobisporus*. This tyrosinase represents the first three-dimensional (3D) structure solved by Masanori Sugiyama and coworkers [44]. Table 1 includes a description of available tyrosinase three-dimensional models obtained from experimental and computational determination, sourced from the Protein

Data Bank [48] and AlphaFold [51]. This selection represent complex of tyrosinases with key ligands of interest and include representatives from all different domains of life: bacterial, plant, fungal and human.

Despite the apparent abundance of structural information, the crucial steps in the catalytic mechanism of tyrosinases often remain elusive. This challenge arises because trapping intermediates in catalytic reactions is not easy, limiting our ability to directly observe them. As a result, researchers still rely on inferences and computational studies to understand these key steps in the catalytic mechanism. Furthermore, tyrosinases, including human variants, often undergo glycosylation, posing challenges for structural analysis due to difficulties in producing these enzymes in a pure form. Tyrosinases are sensitive to oxygen, and exposure to air can result in the oxidation of the copper active sites, potentially affecting the enzyme's structure and activity.

As noted, the tyrosinase fold is a distinctive feature of T3Cu copper proteins, characterized by a binuclear copper site. This site serves both for the transport of molecular oxygen and the molecule activation. Alongside tyrosinases, other members of this protein class share a similar fold although serve different functions. Notable examples include the dioxygen transport protein hemocyanins (e.g., pdb entry 3QJO) [52,53], catechol oxidase (pdb entry 6GSG) [54,55], and aurone synthase (pdb entry 4Z11) [56]. Collectively, these proteins are referred as phenoloxidases (PPOs).

Structural insights have shed light on the role of second-shell residues at the entrance of the tyrosinase active site. These residues play a crucial role in the observed differences in activity, despite minimal variations in residues between plant PPOs [46,57].

2. Structure and mechanism of tyrosinase

The structure of a bacterial tyrosinase from S. *castaneoglobisporus* (pdb entry 2AHL) revealed an interesting partnership with a smaller protein, often referred to as the "caddie" protein, as shown in Fig. 3A [44]. This small companion protein, (Uniprot ID Q83WS1), is composed of 126 amino acids. Its specific binding to tyrosinase helps transport of copper ions to the active site of the tyrosinase enzyme, aiding in its catalytic function. This caddie protein located at the hydrophobic surface of the enzyme, modulates the active site by introducing a side chain of a tyrosine residue (Tyr98). The side chain of Tyr98 from the caddie protein partners with a hydrogen bond to the copper, which bridges the pair of ions (see Fig. 3A) [40]. This residue

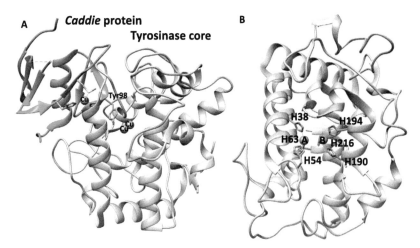

Fig. 3 Ribbon plot representation of tyrosinase. (A) Crystallographic structure of the copper-bound (blue spheres) *Streptomyces castaneoglobisporus* tyrosinase in complex with a "caddie" protein (pdb entry 2AHL) [44]. (B) Tyrosinase core domain indicating the two group of three histidine residues coordinating each copper ion.

has also been found in an oxidized quinone form (see below). This specific site is where the copper cluster resides and is involved in the potential modulation of substrate binding, particularly with the substrate tyrosine. However, among tyrosinase structures variation on the caddie theme are known. Several tyrosinase structures don't present a caddie protein, instead the tyrosinase structure from *Verrucomicrobium spinosum* presents a C-terminal extension domain (explore pdb entry 8BBR) [46]. The ancillary protein is also absent in the human tyrosinase variants (see section below).

Moving on to the overall structure of tyrosinase's core, it takes on a monomeric form, as seen in the PDB entry ID 2AHL. This core consists of a collection of alpha helices interconnected by loops of various lengths, resembling a canonical bundle (as shown in Fig. 3B). Despite this typical monomeric arrangement, the tyrosinase from *Aspergillus oryzae* (pdb entry ID 6JU9) does show a homodimeric arrangement and different occupancy at the copper sites [44].

The tyrosinase copper cluster, consisting of two ions, is situated at the protein's domain core, as shown in Fig. 3B. This arrangement involves two dimeric helices rotated in relation to each other, forming a configuration resembling a bundle of four helices. Within this bundle the coordinative di-copper ion cluster is hosted. The tyrosinase structure reveals two sets of three histidine residues coordinating each Cu^{2+} ion through their Nε atom:

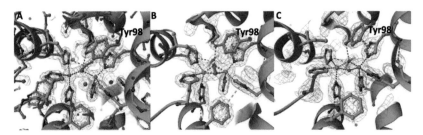

Fig. 4 View of the tyrosinase active site [44]. (A) Snapshot of the Cu(I) coordination as seen in pdb entry 2AHL and its 2Fo-Fc electron-density map (contour levels set at 1.2σ) of the dinuclear metal-binding site (**deoxy-Ty**). Notably, the Tyr98 (orange color) of the caddie protein donates a hydrogen bond to the bridging solvent molecule. (B) Snapshot of the Cu(II) coordination (**oxy-Ty**) as seen in pdb entry 1WX2 and its 2Fo-Fc electron-density map (contour levels set at 1.2σ) of the dinuclear metal-binding site. A copper peroxide side-on bridging configuration where the coordination type is μ-η2:η2. (C) Snapshot of the Cu(II) coordination (μ-OH-κ²O, **met-Ty**) as seen in pdb entry 2AHK where a μ-hydroxide ion, in addition to a solvent molecule, is observed.

one copper (CuA) coordinates with histidine residues at the N-terminal end of the sequence (His38, His54, and His63), while the other copper (CuB) coordinates the metal ion at the C-terminal end (His190, His194, and His216), (see Fig. 3B). All six histidine residues coordinating the copper or zinc ions are conserved among all enzymes with type-III copper proteins. Mutations in any of these residues can significantly or completely eliminate tyrosinase activity [57].

From a geometrical copper coordination perspective, the unliganded state, the tyrosinase structure displays two copper pairs separated by about ~3.2 Å, bridged by a water molecule. This water could also be interpreted as hydroxide anion. Each copper exhibits a pyramidal coordination geometry, illustrated in Fig. 3B. In general, in tyrosinase structures, the copper-copper distance—a typical measurement in the structural analysis of di-metal enzymes—exhibits a range of 3.2–4.0 Å in the various metal-complexed structures. Shorter distances are characteristic of the oxy form, where two oxygen atoms bridge the metal ions, while longer distances are associated with the deoxy form, where a water molecule and/or hydroxide ion bridges the two copper ions.

As revealed experimentally, the binding of the dioxygen molecule to the binuclear copper cluster can reveal in various structurally distinct modes, as shown in Fig. 4. This binding process also induces different copper oxidation states. In its unbound mode, where no oxygen molecule is present in the copper ions cluster, tyrosinase is designated as **deoxy-Ty**, as seen in the pdb

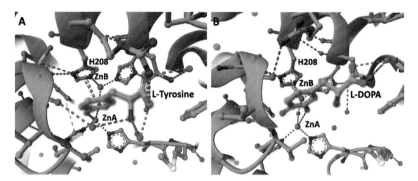

Fig. 5 *Bacillus megaterium* tyrosinase-zinc (II) reconstituted in complex with substrate L-tyrosine (pdb entry 4P6R) [38]. (A) Tyrosine side chain pack against the side chain of residue His208 through **π–π stacking interaction** (dashed line). H-bonding interactions are indicated. (B) Tyrosinase in complex with the L-DOPA (pdb entry 4P6S) reveals structural details guiding substrate orientation for catalysis. Notably, the side chain of arginine (Arg206) shows a different direction with respect the complex with L-tyrosine.

entry 2AHL, as seen in Fig. 4A. In this configuration, copper ions are in the lowest oxidation state, Cu(I) $(3d^9)$ (as shown in Fig. 4A). Upon binding of oxygen, a copper peroxide anion bridging form emerges, showcasing a side-on bridging configuration where the coordination type is μ-$\eta2$:$\eta2$ (PDB entry 1WX2 or 6JU4), and shown in Fig. 4B. In this state, the oxidized form of tyrosinase, known as **oxy-Ty**, presents both coppers in the Cu(II) $(3d^{10})$ oxidation state. Finally, upon substrate binding (see section below) and catalytic reaction, the product is released, and the copper cluster adopts a u-hydroxide bridging the two Cu(II) ions (pdb entry 2AHK) This specific configuration is designated as **met-Ty** where copper cluster is bridged as μ-OH-κ^2O (see Fig. 4C). Despite the atomic details and simplification offered by 3D structures, spectroscopic and theoretical studies are also shedding light on the catalytic cycle and the role of the elusive bis-μ-oxo di-copper(III), which is otherwise very rare in biology, in the catalytic pathway [23]. These studies have contributed to elucidating the steps of the tyrosinase catalytic mechanism, as detailed in Section 2.5.

2.1 Substrate binding mode

Crystal structures of tyrosinase in complex with substrates, reaction intermediates, and products have also helped to provide a glimpse into the productive orientation of these molecules with respect to the enzyme mechanism, as shown in Fig. 5. The tyrosinase in complex with the substrate L-tyrosine (pdb entry 4P6R) reveals structural details guiding substrate

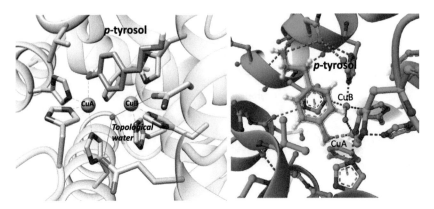

Fig. 6 Tyrosinase-copper in complex with substrate analogue *p*-tyrosol (pdb entry 4P6T) [38]. (A) The substrate aromatic group pack against the side chain of residue H208 through π–π stacking interaction (dashed line), H-bonding interactions are indicated. (B) Same complex exhibiting distinct hydrogen bonding and van der Waals interaction profiles.

orientation for catalysis, as seen in Fig. 5A [38]. In this structure of tyrosinase-L-tyrosine complex, zinc ions substitute copper ions, and the phenol ring of the substrate's sidechain packs closely with side chain histidine (His208) through **π–π stacking interaction**, (Fig. 5A). This noteworthy interaction is a common theme and is found in multiple compounds binding to the active site of Ty(s) from many source organisms. In addition, its side chain engages in ionic interactions with the sidechain the arginine residue (Arg206), few H-bonds interaction, and the –OH group does coordinate to one of the zinc ions (Fig. 5A). The side chain of the tyrosine is also sandwiched on the opposite side by the hydrophobic valine residue (Val215). Although the product of the reaction is difficult to capture due to the fast turnover of the enzyme, the tyrosinase in complex with the L-DOPA, the other stable intermediate from the reaction 1 (shown pdb entry 4P6S) reveals structural details guiding substrate orientation for catalysis (see Fig. 5B). This is evident from the crystal structure of tyrosinase-Zn (II) from *Bacillus megaterium* with L-DOPA in the active site.

2.2 Towards reaction product

Insightful suggestions are offered by analyzing the structure of the tyrosinase in complex with copper ions and the substrate analogue the *p*-tyrosol, shown in Fig. 6, another substrate analogue, bound to tyrosinase and in presence of copper ions (pdb entry 4P6T) [38]. A key understanding from the structure is represented by the orientation of the substrate side chain towards the metal cluster. In this structure, the metal ion CuA is found nearer to the −OH of the phenol group (see Fig. 6B). This orientation seems to suggest the optimal orientation to allow hydroxylation (from the metal ion) to occur since the substrate's aromatic ring π orbitals seems overlap with the σ^* orbital direction of the bridging peroxide ligand to the μ-$\eta2$:$\eta2$ copper centers [5,58]. Other interactions for substrate binding are observed from the structure of the tyrosinase-adduct complex, including the canonical interaction the coordinative side chain of His205.

2.3 A topological conserved water molecule

Among tyrosinase structures in complex with substrates and analogues, a water molecule is usually found near the phenolate, and is stabilized by hydrogen bond interactions with the side chains of a conserved glutamate, a residue bearing a negative charge, and a neutral asparagine residue (if present in the sequence). These residues are the so-called *waterkeeper*. This water molecule, shown in Fig. 6A, topologically conserved, when Asn is present in addition to Glu, could potentially serve as the accepting base for the phenolic proton, highlighting its significance in the catalytic mechanism and it is also characterizing the distinct activities between tyrosinase and catechol oxidase [42]. Interestingly, tyrosinase mutants of this glutamate residue with a neutral glutamine or aspartate, a negatively charged residue, have shown significant lower catalytic rate. On reverse, the mutation of this glutamate with a positively charged residue lysine completely abolish activity, illustrating the importance of this negatively charged residue [57]. The asparagine plays an important role since it properly orient the oxygen of a water molecule (Fig. 6A) [7]. This asparagine modulates the monophenolase activity of the enzyme. The basicity of the water molecule interacting with the two residues, discriminates between tyrosinase and catechol oxidase activity, and has been illustrated in the structure of *Streptomyces castaneoglobisporus* in complex with the caddie protein (pdb entry 7CIY) [42]. Notably, in this structure, the caddie protein Tyr98 (see section above) is oxidized to a quinone. Unlike tyrosinase, catechol oxidase lacks monophenolase activity due to the absence of specific amino acid side chains that increase the basicity of crucial Cu^{2+}-ligating histidine residues, which may also aid in substrate deprotonation [59]. This structural evidence has helped to formulate the tyrosinase mechanism for substrate deprotonation illustrated in Fig. 7 [7].

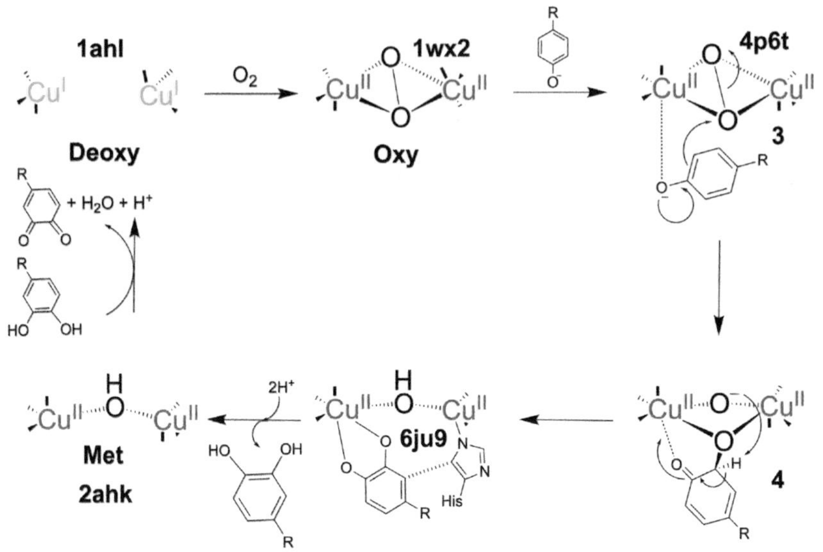

Fig. 7 Proposed tyrosinase mechanism. Representative PDB entry for each catalytic step has been indicated. Structural information elucidates key aspects of the mechanism, but the rapid electronic transitions in steps 3 and 4 remain challenging to capture experimentally. Notably, a histidine residue, highlighted in PDB entry 6ju9, packs against the side chain of the aromatic group, guiding the orientation of the substrate in the active site. *This figure has been adapted from N. Fujieda, K. Umakoshi, Y. Ochi, Y. Nishikawa, S. Yanagisawa, M. Kubo, et al., Copper–oxygen dynamics in the tyrosinase mechanism, Angew. Chem.—Int. Ed. 59 (32) (2020).*

2.4 Other relevant residues for tyrosinase activity

What other conserved residues have mutagenesis studies shown to play a role in the enzyme's catalytic mechanisms? Other relevant residues in the description of tyrosinase structures are represented by the so-called *gatekeeper* residues, are second shell residues, represented by sidechain of a usually conserved phenylalanine or a valine/leucine residue located within ~5–15 Å from the metal cluster, at the entrance of the active side. While the phenylalanine residue forms van der Waals interactions with the copper ion through charge-pi interaction, valine is usually found to be packed against the side chain of the aromatic substrate, or the enzyme inhibitor (see section below), and could therefore function as a guide for the substrate or design of novel inhibitors [21,25]. Several tyrosinase mutants of these residues have shown impaired catalytic activity showcasing the functional role of keeping the substrate (e.g. *p*-cresol, tyramine, or dopamine) in a precise orientation. Coordinating histidine residues also contribute to the fine-tuning of the oxidase reaction, beyond their basic role as coordination sphere residues. For instance, His63 or its equivalent residue in other tyrosinase structures, is often found as covalently bound with a cysteine residue. While this thioether link is commonly observed in eukaryotic tyrosinases, it is rarely encountered in enzymes of bacterial origin [46]. About the role of this linkage, it appears as to properly orient or stabilize the copper (CuA)-binding histidine, and therefore could play a role in the fine-tuning of the redox potential [29,60]. This linkage can be explored from the crystal structure of the holo-protyrosinase from *Asperugillus oryzae* (pdb entry 3W6W). Other variations are found by analyzing available structures and the more recent atomic detail determinations (see Table 1).

2.5 Catalytic mechanism

The analysis of the crystal structures solved in the past decades has revealed key insights to help to understand a tyrosinase mechanism. Briefly, tyrosinase processing of the substrate such as L-tyrosine or similar other molecules requires a monophenol group (like in *p*-cresol, tyramine, dopamine, among others). Upon substrate deprotonation, it gets coordinated to the metal cluster, and hydroxylated in its *o*-position, before leaving the active site (see Fig. 7). The structural studies provide insights into substrate interactions, particularly in pre catalytic step corresponding to the so called oxy-tyrosinase where the peroxide ligand is bridging the copper cluster in a μ-$\eta2$:$\eta2$ (side-on bridged) position. In other words, the structures while suggest a binding of the substrate reveals a specific phase in the catalytic cycle corresponding to the orientation of the substrate in the active site. Utilizing both structural evidence, computer simulations, molecular dynamics, including a 16 nanoseconds long-field molecular dynamics (LFMD) simulation and the copper-oxygen dynamics, researchers were able to delve into the movement and interactions of metal atoms within tyr- osinase [30,61]. Prof. Nobutaka Fujieda and colleagues by preparing several tyrosinase mutants were able to isolate complexes corresponding to a number of steps of a proposed tyrosinase mechanism as the one depicted in Fig. 7, providing valuable insights into the dynamics and functionality of tyrosinase within biological systems [30]. However, more recent structural elucidations (e.g., PDB entry 6ju9 or more recent ones) continue to provide new insights that may require refinements to the previously proposed mechanism (see Fig. 7).

Tyrosinase involves both monophenolase and diphenolase activities, each starting from the deoxy-form of the type-III copper center (Cu (I)-Cu (I)). The deoxy-form binds molecular oxygen, transitioning to the catalytically active oxy-form (Cu (II)-Cu (II)). For **monophenolase activity**, specific residues around the di-copper center enhance the basicity of conserved histidines, facilitating the deprotonation of monophenolic substrates. The substrate is therefore oriented and coordinated to the metal ion (Fig. 6B). The deprotonated monophenol interacts with the oxy-form copper center, undergoing *o*-hydroxylation and subsequent oxidation to produce an *o*-quinone and water, with the copper center reverting to its deoxy-form. In **diphenolase activity**, the di-copper center oxidizes diphenolic substrates to quinones, transitioning from the oxy- to the met-form. The met-form accepts more diphenolic substrates, converting them to quinones and reducing the copper center back to its deoxy-form, thus completing the catalytic cycle (Fig. 7).

3. Tyrosinase: structure and inhibitor complexes

Considering tyrosinase's role as a crucial enzyme involved in melanin synthesis and tissue pigmentation, it is a significant target for various health and biotechnological applications [62–64]. Structural complexes with inhibitors are particularly advantageous for improving inhibitor design. Over the past five decades, tyrosinase inhibitors have been studied extensively, including synthetic compounds, natural products, virtual screening, and molecular docking [65]. Among these studies, structural characterization of tyrosinase complexes offers insights into the inhibition mechanism provided by these compounds. Fig. 2 presents a summary of the main tyrosinase inhibitors, with available structural studies of macromolecules complexes. Refer to Table 1 for a list of macromolecular complexes.

Often, these compounds consist of small phenolic compounds such as resorcinol, hydroquinone, and guaiacol, or similar compounds adorned with a sugar group like arbutin or deoxyarbutin. The list also includes substrate, intermediate, and product analogues of L-tyrosine and L-3,4-dihydroxyphenylalanine (L-DOPA), the latter being a stable intermediate. Examples of these inhibitors/ligands of Ty include dopamine analogues such as L-mimosine, tyramine, α-methyl-*p*-tyrosine, and (*S*)-3-amino-tyrosine (see Fig. 2). In general, these compounds function by blocking the entrance of the substrate to the tyrosinase active site or by coordinating to one of the metal ions or bridging the metal cluster, thereby stopping further steps of the catalytic cycle. Among various chemical groups and other discussed below, kojic acid (containing a pyrone-ring) and its derivatives, is the most well-known tyrosinase inhibitors [65]. By reducing the activity of this enzyme, kojic acid can help lighten dark spots, age spots, and melasma, and it is commonly found in skin creams, lotions, and serums [66,67].

Despite the availability of these compounds often their efficacy is different when considering the human enzyme variants or tyrosinase from other source organisms. For instance, the resorcinol-thiazole derivative known as thiamidol exhibits potent inhibition of human tyrosinase, but shows only weak inhibition of mushroom tyrosinase [65,68]. Unfortunately, no structure is available of the tyrosinase-thiamidol complex as of this writing.

3.1 Human tyrosinases: TYR, TYRP1, and TYRP2

The structural biology of human tyrosinase involves three related tyrosinase-like enzymes: tyrosinase (TYR), and tyrosinase-related proteins 1 and 2 (TYRP1 and TYRP2) [69]. While TYRP1 is also known as "5,6-dihydroxyindole-2-carboxylic acid oxidase"; TYRP2 is identified as the "dopachrome tautomerase" (DCT/TYRP2) [70]. These enzymes, integral to melanocytes found in the skin, hair follicles, and eyes, play a crucial role in melanin production [9]. They share a sequence identity ranging from 40 to 70%. Human tyrosinases are glycoproteins and contain a single C-terminal transmembrane α-helix anchored to the melanosome membrane. These related enzymes are distinct proteins, each encoded by different genes with distinct functions. Like TYR and TYRP2, TYRP1 expression is primarily regulated by microphthalmia-associated transcription factor (MITF), but additional factors modulate their distinct expression patterns [69].

In exploring of their structures, the first crystal structure of human TYRP1 (Uniprot entry P17643) unveiled intriguing details (pdb entry 5M8L) [25]. The glycoprotein displayed a unique N-terminal Cys-rich domain (residues 25–127), exclusive to mammalian tyrosinases. Additionally, a tyrosinase-like domain housing a binuclear metal–ion-binding motif within a four-helical bundle structure was observed [32,71]. Unlike, most bacterial tyrosinases, a **di–zinc ion cluster** is present in place of copper ions, although their coordination geometry (with zinc ion a $3d^{10}$ element-configuration) is similar to that of copper (as shown in Fig. 7) [21]. Interestingly, since the chemistry of zinc ion involves plus-two cation only, it is unlike that human TYRP1 functions as a redox enzyme, prompting questions about its potential biochemical implications. Copper and zinc ions could have different affinities for the same human enzyme and may serve different functions, including modulating diverse glycosylation patterns upon binding [21].

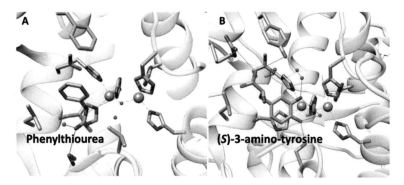

Fig. 8 Crystal structure of TYRP1-Zn (II) in complex with inhibitors. (A) Tyrosinase in complex with N-phenyltiourea (pdb entry 5M8S). (B) Tyrosinase in complex with (S)-3-amino-tyrosine (pdb entry 9EY8). These inhibitors show a different binding mode one located at the entrance of the active site, and the other more deeply anchored in the pocket.

Unfortunately, the structural characterization of TYR and TYRP2 is hindered by several glycosylation sites, impeding a comprehensive understanding of their functional biology. However, molecular models of TYR (Uniprot entry P14679) and TYRP2 (Uniprot entry P40126) are available from the Alphafold (see Table 1) [51]. Despite this scarcity of information, a significant revelation emerged from the studies of TYRP1 (see Table 1). The binding of sugars to five asparagine residues highlighted its glycosylation pattern, shedding light on its intricate structural features [25,32]. Notably, these structural insights remain a critical step toward unraveling the complete functional biology of these enzymes.

3.2 Structural studies of tyrosinase-inhibitor complexes

Among known inhibitors or compounds tested for the activity of tyrosinases, N-phenyltiourea, represented the most potent inhibitor of human tyrosinase [32]. This molecule contains an aromatic ring joined to a thiourea moiety ($NH–C(=S)–NH_2$) and it lacks a polar oxygen atom group that could allow metal coordination. The structure of the complex between an active mutant of human TYRP1 (pdb entry 5M8S) and N-phenylthiourea reveals the hydrophobic interaction of its phenyl group sandwiched between the side chains of an asparagine (Asn378), a phenylalanine residue (Phe362) and a leucine residue (Leu382), as shown in Fig. 8A. Notably, the inhibitor although displayed in the active site, is not anchored to the di-nuclear cluster. Therefore, it only functions as an access blocker for the substrate to the metal

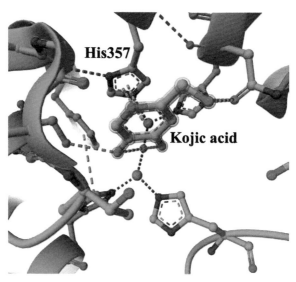

Fig. 9 Crystal structure of TYRP1-Zn (II) in complex with the kojic acid (PDB entry 5M8M).

cluster. Interestingly, *N*-phenylthiourea was found coordinated to the di-metal cluster in the structure of the complex between the plant catechol oxidase containing a di-copper center and showing a similarities with the structure of tyrosinase [32]. More recent crystal structure determinations are represented by complex between tyrosinase-Zn (II) and the substrate analogues (*S*)-3-amino-tyrosine (seen in Fig. 8B, pdb entry 9EY8) and 2-hydroxy-L-tyrosine (pdb entry 9EY5, Table 1) [31]. These inhibitors exhibit a canonical binding mode at the active site of Tyrosinase, with their aromatic ring snugly packed against the side chain of His357, while their polar atom groups engage in several hydrogen-bonding interactions, including interaction with the metal-bridging solvent.

Other compounds include kojic acid, a renowned ingredient in skin-care products for its skin-brightening properties, exhibited binding affinity to human TYRP1 [66]. Kojic acid, a chemical compound comprised of a pyrone ring with an –OH group at the 5-position and a –CH$_2$OH group at the 2-position, represents a classical inhibitor binding to TYRP1-Zn(II), illustrated in Fig. 9 (PDB entry 5M8M) [25]. This inhibitor demonstrates anchoring through hydrogen-bonding interactions involving its –OH group and C=O group to the di-Zinc(II) bridging solvent molecule, while the planar ring interacts with the side chain of His357 (see Fig. 9). This binding resembles that of mimosine in the structure of the TYRP1-mimosine complex (PDB entry 5M8R, Table 1) [25].

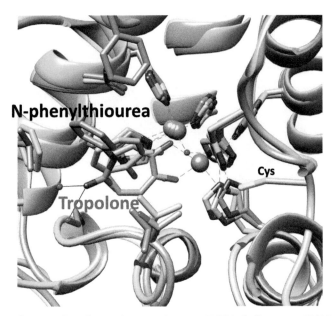

Fig. 10 Binding mode of tropolone in human TYRP1 (pdb entry 5M8O, purple) compared to N-phenylthiourea (PDB entry 5M8S, gray) and mushroom tyrosinase-tropolone complex (pdb entry 2Y9X, cyan). Tropolone shows different binding modes in tyrosinases from different sources.

Of particular interest is the comparison of the binding modes of various inhibitors to TYRP1 and mushroom tyrosinase. This includes the binding of the simple ring inhibitor tropolone to the structure of TYRP1, the complex of TYRP1-Zn (II) with *N*-phenylthiourea (pdb entry 5M8S), and the equivalent complex with mushroom tyrosinase-Cu (II)-tropolone complex (pdb entry 2Y9X), as shown in Fig. 10. Notably, while the tropolone ring binds deeper in the active site of TYRP1 (pdb entry 5M8O) and is coordinated to one metal ion, it binds differently in mushroom tyrosinase, as shown in Fig. 10. In the latter, the inhibitor is packed against a valine residue at the entrance of the active site, showcasing a different binding mode despite the similarities of the active site (see Fig. 10). Additionally, the aromatic ring position of L-mimosine to TYRP1 resembles that of tropolone binding in the complex with TYRP1 (pdb entry 5M8R). Therefore, this unexpected parallelism suggests underlying similarities in the molecular mechanisms of these enzymes, bridging the gap between different species and their respective tyrosinase-like proteins. In silico methods provide prediction for the binding modes

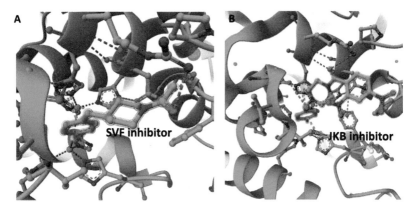

Fig. 11 Crystal structure of *Bacillus megaterium* Ty in complex with inhibitors containing a 4'-fluorobenzyl group. (A) Tyrosinase-Cu (II) in complex with SFV inhibitor (pdb entry 5OAE). (B) Tyrosinase-Cu (II) in complex with JKB inhibitor (pdb entry 6QXD). Notably, fluorine atom group of both inhibitors is coordinated to the copper Cu(A), while the aromatic ring engaging in π–π stacking interaction.

of more recent inhibitors such as thiamidol targeting both human and mushroom tyrosinase has no experimental structural studies are currently available [30,68].

Despite the atomic details available for more classical inhibitors such as kojic acid, these compounds are rather limited in terms of their relative human toxicity [72]. Recent advancements have therefore focused on selecting potent inhibitors with potentially lower toxicity, aiming to establish new reference compounds. Structural studies, aided by docking methods, have played a crucial role in identifying these leading compounds. Notably, the presence of a 4'-fluorobenzyl moiety embedded in the 4-position of the piperidine fragment has been observed in the most active inhibitors. Two selected inhibitors have demonstrated binding to *Bacillus megaterium* Ty, as illustrated in Fig. 11. Importantly, their terminal 4'-fluorobenzyl group exhibits a similar binding pattern, with the fluorine atom group coordinated to the copper Cu(A), while the aromatic ring is engaging in π–π stacking interaction with the side chain of His205 [36].

3.3 Comparison of TYR, TYRP1 and TYRP2 structural models

Despite experimental structures being available only for TYRP1, along with its complexes with various inhibitors/ligands (as listed in Table 1), the comparison of its active site with those of TYR and TYRP2 through computational modeling is noteworthy due to its relative accuracy. This approach

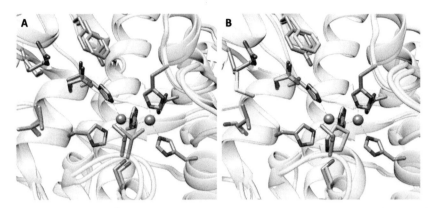

Fig. 12 Comparison of the active sites of human TYR, TYRP1, and TYRP2. (A) The crystal structure of TYRP1 (gray) is superimposed on TYR (purple, UniProt entry P14679). (B) The crystal structure of TYRP1 is superimposed on the computational model of TYRP2 (cyan, UniProt entry P40126). Various residues differentiate the three enzymes, potentially influencing the binding of inhibitors.

allows for a deeper understanding of the structural and functional similarities and differences among these proteins, which is crucial for developing specific inhibitors and understanding their roles in physiological and pathological processes. Despite the similarity of human tyrosinases, the enzymes exhibit different alignments of several residues within the active site, as depicted in Fig. 12. For example, TYRP2 contains a proline residue (Pro383), while at the equivalent position in TYRP1, there is a threonine residue (Thr391), and in TYR, a valine residue (Val377) (refer to Fig. 12). Similarly, at the entrance of the active site, TYRP1 features an arginine (Arg374), while TYR and TYRP2 have a serine (Ser360) and a methionine (Met366), respectively (see Fig. 10). Notably, the gatekeeper residue is a tyrosine (Tyr362) in TYRP1, whereas it is a phenylalanine (Phe347) in TYR and TYRP2 (Phe354). Other differences can also be observed by comparing these structural models. Therefore, these variations among human tyrosinases could have implications for the selectivity of different inhibitors.

4. Exploring tyrosinase: health and biotech

Tyrosinase, is a crucial enzyme involved in melanin synthesis and tissue pigmentation, and it is present across all life forms, making it a significant target for various health and biotechnological applications. In the food industry, effective and safe tyrosinase inhibitors can maintain the quality and extend the shelf life of fruits and vegetables [12,73]. However, challenges such as safety concerns, limited activity, and insufficient solubility hinder their integration into the industry, prompting ongoing research into novel inhibitors [68]. In addition, inhibitors that act as substrate analogues, such as hydroquinone (e.g., pdb entry 5I3A, Table 1), can function both as substrates and as inhibitors, revealing a complex inhibition pattern [37].

From a human health perspective, tyrosinase is extensively studied for its role in skin conditions, dermatoprotective treatments, free radicals, and pathologies linked to UV exposure. Recent inhibitors, such as thiamidol (inhibitor W630), a compound containing a thiazolyl and resorcinol groups, have shown significant efficacy in reducing hyperpigmentation conditions like melasma and post-inflammatory hyperpigmentation [74,75]. Thiamidol stands out for its high potency compared to other common inhibitors (e.g. arbutin, kojic acid, and hydroquinone), demonstrating improved skin evenness and melanin index scores with good tolerability and minimal side effects [76].

A search of the term 'tyrosinase' within the web-based resource ClinicalTrials.gov (accessed in May 2024) reveals a total of 76 "interventional studies" mainly targeting various stages of melanoma (especially affecting skin), skin aging, solar elastosis, melasma, and halitosis. Some of these studies specifically target the human enzyme TYRP1 [25].

In conclusion, the structures of tyrosinases reveal a blend of conserved features and unique variations, which are pivotal in understanding the enzyme's diverse functionalities [57]. These insights not only enhance our comprehension of tyrosinase enzymes but also open new pathways for skincare research and development. Despite advances in 3D modeling and ligand prediction, experimental structures of tyrosinase complexes with novel inhibitors remain essential for identifying efficient inhibitors. Continued exploration of tyrosinase structures promises to advance both biological and cosmetic applications, driven by the relationship between enzyme structure, function, and melanin production and it is vital for unlocking the full potential of tyrosinase in improving health and biotechnological outcomes.

References

[1] L. Weiner, W. Fu, W.J. Chirico, J.L. Brissette, Skin as a living coloring book: how epithelial cells create patterns of pigmentation, Pigment. Cell Melanoma Res. 27 (2014).

[2] S.H. Pomerantz, Separation, purification, and properties of two tyrosinases from hamster melanoma, J. Biol. Chem. 238 (1963).

[3] M. Mine, H. Mizuguchi, T. Takayanagi, Kinetic analyses of two-steps oxidation from L-tyrosine to L-dopaquinone with tyrosinase by capillary electrophoresis/dynamic frontal analysis, Anal. Biochem. 655 (2022).

[4] A.B. Lerner, T.B. Fitzpatrick, Biochemistry of melanin formation, Physiol. Rev. 30 (1) (1950).

[5] M. Rolff, J. Schottenheim, H. Decker, F. Tuczek, Copper-O2 reactivity of tyrosinase models towards external monophenolic substrates: molecular mechanism and comparison with the enzyme, Chem. Soc. Rev. 40 (7) (2011).

[6] E.I. Solomon, U.M. Sundaram, T.E. Machonkin, Multicopper oxidases and oxygenases, Chem. Rev. 96 (7) (1996).

[7] E. Solem, F. Tuczek, H. Decker, Tyrosinase versus catechol oxidase: one asparagine makes the difference, Angew. Chem.—Int. Ed. 55 (8) (2016).

[8] J.C. García-Borrón, F. Solano, Molecular anatomy of tyrosinase and its related proteins: beyond the histidine-bound metal catalytic center, Pigment. Cell Res. 15 (2002).

[9] M.B. Dolinska, E. Kovaleva, P. Backlund, P.T. Wingfield, B.P. Brooks, Y.V. Sergeev, Albinism-causing mutations in recombinant human tyrosinase alter intrinsic enzymatic activity, PLoS One 9 (1) (2014).

[10] Y. Xu, A.H. Stokes, R. Roskoski, K.E. Vrana, Dopamine, in the presence of tyrosinase, covalently modifies and inactivates tyrosine hydroxylase, J. Neurosci. Res. 54 (5) (1998).

[11] A.M. Mayer, E. Harel, Polyphenol oxidases in plants, Phytochemistry 18 (1979).

[12] K.M. Moon, E.Bin Kwon, B. Lee, C.Y. Kim, Recent trends in controlling the enzymatic browning of fruit and vegetable products, Molecules 25 (2020).

[13] H.E. Barden, R.A. Wogelius, D. Li, P.L. Manning, N.P. Edwards, B.E. Van Dongen, Morphological and geochemical evidence of eumelanin preservation in the feathers of the Early Cretaceous bird, Gansus yumenensis, PLoS One 6 (10) (2011).

[14] I.E. Pralea, R.C. Moldovan, A.M. Petrache, M. Ilieş, S.C. Hegheş, I. Ielciu, et al., From extraction to advanced analytical methods: the challenges of melanin analysis, Int. J. Mol. Sci. 20 (2019).

[15] G.Y. Liu, V. Nizet, Color me bad: microbial pigments as virulence factors, Trends Microbiol. 17 (2009).

[16] J.P. Wibowo, F.A. Batista, N. van Oosterwijk, M.R. Groves, F.J. Dekker, W.J. Quax, A novel mechanism of inhibition by phenylthiourea on PvdP, a tyrosinase synthesizing pyoverdine of Pseudomonas aeruginosa, Int. J. Biol. Macromol. 146 (2020).

[17] F. Panis, A. Rompel, Biochemical investigations of five recombinantly expressed tyrosinases reveal two novel mechanisms impacting carbon storage in wetland ecosystems, Environ. Sci. Technol. 57 (37) (2023).

[18] C.A. Ramsden, P.A. Riley, Tyrosinase: the four oxidation states of the active site and their relevance to enzymatic activation, oxidation and inactivation, Bioorg. Med. Chem. 22 (2014).

[19] M. Pretzler, A. Rompel, What causes the different functionality in type-III-copper enzymes? A state of the art perspective, Inorg. Chim. Acta 481 (2018).

[20] T. Tsang, C.I. Davis, D.C. Brady, Copper biology, Curr. Biol. 31 (9) (2021).

[21] F. Solano, On the metal cofactor in the tyrosinase family, Int. J. Mol. Sci. 19 (2018).

[22] B. Dicke, A. Hoffmann, J. Stanek, M.S. Rampp, B. Grimm-Lebsanft, F. Biebl, et al., Transferring the entatic-state principle to copper photochemistry, Nat. Chem. 10 (3) (2018).

[23] I. Kipouros, E.I. Solomon, New mechanistic insights into coupled binuclear copper monooxygenases from the recent elucidation of the ternary intermediate of tyrosinase, FEBS Lett. 597 (1) (2023).

[24] W. Keown, J.B. Gary, T.D.P. Stack, High-valent copper in biomimetic and biological oxidations, J. Biol. Inorg. Chem. 22 (2017).

[25] X. Lai, H.J. Wichers, M. Soler-Lopez, B.W. Dijkstra, Structure of human tyrosinase related protein 1 reveals a binuclear zinc active site important for melanogenesis, Angew. Chem.—Int. Ed. 56 (33) (2017).

[26] M. Pretzler, A. Bijelic, A. Rompel, Heterologous expression and characterization of functional mushroom tyrosinase (AbPPO4), Sci. Rep. 7 (1) (2017).

[27] W.T. Ismaya, H.J. Rozeboom, A. Weijn, J.J. Mes, F. Fusetti, H.J. Wichers, et al., Crystal structure of agaricus bisporus mushroom tyrosinase: identity of the tetramer subunits and interaction with tropolone, Biochemistry 50 (24) (2011).

[28] S.G. Mauracher, C. Molitor, R. Al-Oweini, U. Kortz, A. Rompel, Latent and active abPPO4 mushroom tyrosinase cocrystallized with hexatungstotellurate(VI) in a single crystal, Acta Crystallogr. Sect. D. Biol. Crystallogr. 70 (9) (2014).

[29] N. Fujieda, S. Yabuta, T. Ikeda, T. Oyama, N. Muraki, G. Kurisu, et al., Crystal structures of copper-depleted and copper-bound fungal pro-tyrosinase: insights into endogenous cysteine-dependent copper incorporation, J. Biol. Chem. 288 (30) (2013).

[30] N. Fujieda, K. Umakoshi, Y. Ochi, Y. Nishikawa, S. Yanagisawa, M. Kubo, et al., Copper–oxygen dynamics in the tyrosinase mechanism, Angew. Chem.—Int. Ed. 59 (32) (2020).

[31] C. Faure, Y. Min Ng, C. Belle, M. Soler-Lopez, L. Khettabi, M. Saïdi, N. Berthet, M. Maresca, C. Philouze, W. Rachidi, M. Réglier, A. du Moulinet d'Hardemare, H. Jamet Interactions of phenylalanine derivatives with human tyrosinase: lessons from experimental and theoretical studies. ChemBioChem (2024) e202400235.

[32] X. Lai, H.J. Wichers, M. Soler-Lopez, B.W. Dijkstra, Phenylthiourea binding to human tyrosinase-related protein 1, Int. J. Mol. Sci. 21 (3) (2020).

[33] A. Bijelic, M. Pretzler, C. Molitor, F. Zekiri, A. Rompel, The structure of a plant tyrosinase from Walnut leaves reveals the importance of "substrate-guiding residues" for enzymatic specificity, Angew. Chem.—Int. Ed. 54 (49) (2015).

[34] I. Kampatsikas, A. Bijelic, M. Pretzler, A. Rompel, A peptide-induced self-cleavage reaction initiates the activation of tyrosinase, Angew. Chem.—Int. Ed. 58 (22) (2019).

[35] L. Ielo, B. Deri, M.P. Germanò, S. Vittorio, S. Mirabile, R. Gitto, et al., Exploiting the 1-(4-fluorobenzyl)piperazine fragment for the development of novel tyrosinase inhibitors as anti-melanogenic agents: Design, synthesis, structural insights and biological profile, Eur. J. Med. Chem. 178 (2019).

[36] S. Ferro, B. Deri, M.P. Germanò, R. Gitto, L. Ielo, M.R. Buemi, et al., Targeting tyrosinase: development and structural insights of novel inhibitors bearing arylpiperidine and arylpiperazine fragments, J. Med. Chem. 61 (9) (2018).

[37] B. Deri, M. Kanteev, M. Goldfeder, D. Lecina, V. Guallar, N. Adir, et al., The unravelling of the complex pattern of tyrosinase inhibition, Sci. Rep. 6 (2016).

[38] M. Goldfeder, M. Kanteev, S. Isaschar-Ovdat, N. Adir, A. Fishman, Determination of tyrosinase substrate-binding modes reveals mechanistic differences between type-3 copper proteins, Nat. Commun. 5 (2014).

[39] M. Kanteev, M. Goldfeder, M. Chojnacki, N. Adir, A. Fishman, The mechanism of copper uptake by tyrosinase from Bacillus megaterium, J. Biol. Inorg. Chem. 18 (8) (2013).

[40] M. Sendovski, M. Kanteev, V.S. Ben-Yosef, N. Adir, A. Fishman, First structures of an active bacterial tyrosinase reveal copper plasticity, J. Mol. Biol. 405 (1) (2011).

[41] I. Kampatsikas, A. Bijelic, A. Rompel, Biochemical and structural characterization of tomato polyphenol oxidases provide novel insights into their substrate specificity, Sci. Rep. 9 (1) (2019).

[42] Y. Matoba, K. Oda, Y. Muraki, T. Masuda, The basicity of an active-site water molecule discriminates between tyrosinase and catechol oxidase activity, Int. J. Biol. Macromol. 183 (2021).

[43] Y. Matoba, N. Bando, K. Oda, M. Noda, F. Higashikawa, T. Kumagai, et al., A molecular mechanism for copper transportation to tyrosinase that is assisted by a metallochaperone, caddie protein, J. Biol. Chem. 286 (34) (2011).

[44] Y. Matoba, T. Kumagai, A. Yamamoto, H. Yoshitsu, M. Sugiyama, Crystallographic evidence that the dinuclear copper center of tyrosinase is flexible during catalysis, J. Biol. Chem. 281 (13) (2006).

[45] E. Nikolaivits, A. Valmas, G. Dedes, E. Topakas, M. Dimarogona, Considerations regarding activity determinants of fungal polyphenol oxidases based on mutational and structural studies, Appl. Environ. Microbiol. 87 (11) (2021).

[46] M. Fekry, K.K. Dave, D. Badgujar, E. Hamnevik, O. Aurelius, D. Dobritzsch, et al., The crystal structure of tyrosinase from *Verrucomicrobium spinosum* reveals it to be an atypical bacterial tyrosinase, Biomolecules 13 (9) (2023).

[47] V.M. Virador, J.P. Reyes Grajeda, A. Blanco-Labra, E. Mendiola-Olaya, G.M. Smith, A. Moreno, et al., Cloning, sequencing, purification, and crystal structure of grenache (*Vitis vinifera*) polyphenol oxidase, J. Agric. Food Chem. 58 (2) (2010).

[48] S.K. Burley, C. Bhikadiya, C. Bi, S. Bittrich, L. Chen, G.V. Crichlow, et al., RCSB protein data bank: powerful new tools for exploring 3D structures of biological macromolecules for basic and applied research and education in fundamental biology, biomedicine, biotechnology, bioengineering and energy sciences, Nucleic Acids Res. (2021).

[49] C. Markosian, L. Di Costanzo, M. Sekharan, C. Shao, S.K. Burley, C. Zardecki, Analysis: analysis of impact metrics for the protein data bank, Sci. Data 5 (2018).

[50] D.S. Goodsell, C. Zardecki, L. Di Costanzo, J.M. Duarte, B.P. Hudson, I. Persikova, et al., RCSB protein data bank: enabling biomedical research and drug discovery, Protein Sci. (2020).

[51] J. Jumper, R. Evans, A. Pritzel, T. Green, M. Figurnov, O. Ronneberger, et al., Highly accurate protein structure prediction with AlphaFold, Nature 596 (7873) (2021).

[52] E. Jaenicke, K. Büchler, H. Decker, J. Markl, G.F. Schröder, The refined structure of functional unit h of keyhole limpet hemocyanin (KLH1-h) reveals disulfide bridges, IUBMB Life (2011).

[53] S. Kato, T. Matsui, C. Gatsogiannis, Y. Tanaka, Molluscan hemocyanin: structure, evolution, and physiology, Biophys. Rev. 10 (2018).

[54] L. Penttinen, C. Rutanen, J. Jänis, J. Rouvinen, N. Hakulinen, Unraveling substrate specificity and catalytic promiscuity of *Aspergillus oryzae* catechol oxidase, ChemBioChem 19 (22) (2018).

[55] C. Eicken, B. Krebs, J.C. Sacchettini, Catechol oxidase—structure and activity, Curr. Opin. Struct. Biol. 9 (1999).

[56] C. Molitor, S.G. Mauracher, A. Rompel, Aurone synthase is a catechol oxidase with hydroxylase activity and provides insights into the mechanism of plant polyphenol oxidases, Proc. Natl. Acad. Sci. U. S. A. 113 (13) (2016).

[57] I. Kampatsikas, A. Rompel, Similar but still different: which amino acid residues are responsible for varying activities in type-III copper enzymes? ChemBioChem 22 (2021).

[58] O. Sander, A. Henß, C. Näther, C. Würtele, M.C. Holthausen, S. Schindler, et al., Aromatic hydroxylation in a copper bis(imine) complex mediated by a µ-η2:η2 peroxo dicopper core: a mechanistic scenario, Chem.—Eur. J. 14 (31) (2008).

[59] H. Decker, T. Schweikardt, F. Tuczek, The first crystal structure of tyrosinase: all questions answered? Angew. Chem.—Int. Ed. 45 (2006).

[60] T. Klabunde, C. Eicken, J.C. Sacchettini, B. Krebs, Crystal structure of a plant catechol oxidase containing a dicopper center, Nat. Struct. Biol. 5 (1998).

[61] R.J. Deeth, C. Diedrich, Structural and mechanistic insights into the oxy form of tyrosinase from molecular dynamics simulations, J. Biol. Inorg. Chem. 15 (2) (2010).

[62] M. He, J. Zhang, N. Li, L. Chen, Y. He, Z. Peng, et al., Synthesis, anti-browning effect and mechanism research of kojic acid-coumarin derivatives as anti-tyrosinase inhibitors, Food Chem. X 21 (2024).

[63] Z. Peng, G. Wang, Q.H. Zeng, Y. Li, H. Liu, J.J. Wang, et al., A systematic review of synthetic tyrosinase inhibitors and their structure-activity relationship, Crit. Rev. Food Sci. Nutr. Vol. 62 (2022).

[64] S. Carradori, F. Melfi, J. Rešetar, R. Şimşek, Tyrosinase enzyme and its inhibitors: an update of the literature, Metalloenzymes: From Bench to Bedside, Elsevier, 2023.

[65] S. Zolghadri, A. Bahrami, M.T. Hassan Khan, J. Munoz-Munoz, F. Garcia-Molina, F. Garcia-Canovas, et al., A comprehensive review on tyrosinase inhibitors, J. Enzyme Inhib. Med. Chem. 34 (2019).

[66] M. Saeedi, M. Eslamifar, K. Khezri, Kojic acid applications in cosmetic and pharmaceutical preparations, Biomed. Pharmacother. 110 (2019).

[67] A. Ayuhastuti, I.S.K. Syah, S. Megantara, A.Y. Chaerunisaa, Nanotechnology-enhanced cosmetic application of kojic acid dipalmitate, a kojic acid derivate with improved properties, Cosmetics 11 (2024).

[68] F. Ricci, K. Schira, L. Khettabi, L. Lombardo, S. Mirabile, R. Gitto, et al., Computational methods to analyze and predict the binding mode of inhibitors targeting both human and mushroom tyrosinase, Eur. J. Med. Chem. 260 (2023).

[69] A. Gautron, M. Migault, L. Bachelot, S. Corre, M.D. Galibert, D. Gilot, Human TYRP1: two functions for a single gene? Pigment. Cell Melanoma Res. 34 (2021).

[70] H. Sugimoto, M. Taniguchi, A. Nakagawa, I. Tanaka, M. Suzuki, J. Nishihira, Crystallization and preliminary X-ray analysis of human D-dopachrome tautomerase, J. Struct. Biol. 120 (1) (1997).

[71] H. Decker, F. Tuczek, The recent crystal structure of human tyrosinase related protein 1 (HsTYRP1) solves an old problem and poses a new one, Angew. Chem.—Int. Ed. 56 (46) (2017).

[72] A. Gunia-Krzyżak, J. Popiol, H. Marona, Melanogenesis inhibitors: strategies for searching for and evaluation of active compounds, Curr. Med. Chem. 23 (31) (2016).

[73] Z. Peng, G. Wang, Y. He, J.J. Wang, Y. Zhao, Tyrosinase inhibitory mechanism and anti-browning properties of novel kojic acid derivatives bearing aromatic aldehyde moiety, Curr. Res. Food Sci. 6 (2023).

[74] T. Mann, C. Scherner, K.H. Röhm, L. Kolbe, Structure-activity relationships of thiazolyl resorcinols, potent and selective inhibitors of human tyrosinase, Int. J. Mol. Sci. 19 (3) (2018).

[75] K. Shimizu, R. Kondo, K. Sakai, Inhibition of tyrosinase by flavonoids, stilbenes and related 4- substituted resorcinols: structure-activity investigations, Planta Med. 66 (1) (2000).

[76] C. Arrowitz, A.M. Schoelermann, T. Mann, L.I. Jiang, T. Weber, L. Kolbe, Effective tyrosinase inhibition by thiamidol results in significant improvement of mild to moderate melasma, J. Invest. Dermatol. 139 (8) (2019).

CHAPTER FOUR

Natural products as tyrosinase inhibitors

Aslınur Doğan and Suleyman Akocak*

Department of Pharmaceutical Chemistry, Faculty of Pharmacy, Adıyaman University, Adıyaman, Türkiye
*Corresponding author. e-mail address: sakocak@adiyaman.edu.tr

Contents

Abstract

Tyrosinase is a crucial copper-containing enzyme involved in the production of melanin. Melasma, age spots, and freckles are examples of hyperpigmentation diseases caused by excess production of melanin. Inhibiting tyrosinase activity is a crucial method for treating these disorders along with various applications such as cosmetics, food technology, and medicine. Natural products have proven a rich source of tyrosinase inhibitors, with several molecules from plant, marine, and microbial sources showing potential inhibitory action. This chapter provides a complete overview of natural compounds that have been found as tyrosinase inhibitors, with emphasis on their structures, modes of action, and prospective applications.

The Enzymes, Volume 56
ISSN 1874-6047, https://doi.org/10.1016/bs.enz.2024.06.002

1. Introduction

Skin pigmentation is a significant human phenotypic trait, the full regulation of which remains incompletely understood despite recent advancements [1,2]. Melanocytes utilize a complicated process called melanogenesis to create melanin, the main component of skin pigment, inside melanosomes. Melanocytes are involved in interactions with the immunological, inflammatory, endocrine, and central neurological systems. In addition, external elements like chemicals and UV radiation have an impact on their activity [3]. Melanosomes are transferred from melanocytes to neighboring keratinocytes, and variations in their number, size, distribution, and composition can lead to different skin pigmentation patterns. Melanin serves as the primary determinant of skin, hair, and eye color, and plays crucial roles in maintaining skin homeostasis by offering protection against harmful ultraviolet radiation and by detoxifying harmful drugs and chemicals [4]. However, the abnormal accumulation of melanin in specific areas of the skin, resulting in hyperpigmented spots, can lead to aesthetic concerns.

Tyrosinase (E.C. 1.14.18.1) is a crucial copper-containing oxidase enzyme with multifunctional significance in living organisms due to its pivotal role in melanin pigment biosynthesis. It catalyzes the o-hydroxylation of polyphenolic substrates, leading to the formation of diphenol (catechol) derivatives, which further oxidize to o-quinone products. As a rate-limiting enzyme, tyrosinase mediates the hydroxylation of L-tyrosine to L-DOPA and 3,4-dihydroxyphenylalanine to dopaquinones, resulting in the accumulation of melanin pigment in the outermost layer of the skin [5,6].

Tyrosinase inhibition is the ability of certain agents to downregulate the activity of tyrosinase in the process of converting tyrosine into melanin. This would impede overpigmentation and also offer preventive measures for normal skin pigmentation turning into tanned skin. Tyrosinase inhibitors have clinical and cosmetic functions, as many people have been looking for ways to promote fairer skin. In clinical settings, standard tyrosinase inhibitors (TIs) such as kojic acid, tropolone, hydroquinone, and mercury have proven effective for facial aesthetic treatments and dermatological disorders associated with melanin hyperpigmentation. Moreover, these inhibitors are utilized in cosmetics as skin-whitening agents and in agriculture as bioinsecticides. Inhibition of tyrosinase could also help cure hyperpigmentation traces that are leftover from wounds, inflammations, or UV radiations. Tyrosinase inhibitors would inhibit the repair of pigmentation and gradually fade the pigments into normal skin. Although tyrosinase has beneficial functions for the skin and body, its inhibitors have become a rising star in dermatological research around the world [7–9].

Tyrosinases are metallic enzymes mostly found in pigmented cells. Their function is to convert tyrosine into melanin through several phases of chemical reactions. Tyrosinase overactivity could cause an increase in the synthesis and accumulation of melanin pigments. This condition could lead to several dermatological problems such as freckles, age spots, melasmas, and actinic lentiginosis. Currently used depigmenting agents like hydroquinone, kojic acid, and azelaic acid have been shown to inhibit the activity of tyrosinase, although they do have fatal side effects. Recent searches for natural tyrosinase inhibitors are progressing due to the need for safer ways to inhibit tyrosinase [10].

Though there is no physical harm, hyperpigmentation often causes psychological impact in many individuals, which could affect job performance, work prospects, marital and emotional life. Presently, the primary depigmenting agents like hydroquinone, kojic acid, and glucocorticoids are inhibitors of melanin synthesis and function through the inhibition of tyrosinase activity. But given the side effects of these molecules or certain restrictions such as the ban on hydroquinone in the EU, there is an enormous interest and urgent need to look for new molecules, particularly those derived from natural sources, which can inhibit tyrosinase to eradicate hyperpigmentation with fewer adverse side effects. This can lead to curing patients of tyrosinase-related neurogenic disorders without any side effects. Therefore, tyrosinase inhibitors are very important in today's context and will be even more important in the future.

2. Natural sources of tyrosinase inhibitors

Natural organisms serve as a vital source of tyrosinase enzyme inhibitors, offering protection against the harmful effects of ultraviolet radiation and chemicals. Plants, bacteria, and fungi, in particular, exhibit tyrosinase inhibitory activity due to their diverse bioactive components, making them compelling subjects for research.

2.1 Plants

It is known that various phytochemical agents found in plants have anti-tyrosinase activity [11]. In studies on the subject, especially *Achyrocline satureioides, Artemisia verlotiorum, Cotoneaster glaucophylla, Dalea elegans, Flourensia campestris, Jodina rhombifolia, Kageneckia lanceolata, Lepechinia floribunda, Lepe-chinia meyenii, Lithrea molleoides, Porlieria microphylla, Pterocaulon alopecuroid es, Ruprechtia apetala, Sen. na aphylla, Sida rhombifolia, Solanum argentinum, Tagetes minuta and Thalictrum decipiens* have been shown to have anti-tyrosinase activity. In addition, it has been determined that plants belonging to the *Moraceae* family, including *Morus species, Artocarpus, Maclura (Cudrania), Broussonetia, Milicia (Chlorophora)*, and *Ficus genera*, have inhibitory activity against this enzyme [12]. At the same time, *Hypericum laricifolium Juss, Taraxacum officinale*, and *Muehlenbeckia vulcanica* have been shown to inhibit high levels of tyrosinase [13]. On the other hand, studies have shown that ethanol and methanol extracts of *Ardisia elliptica Thunb, Phyllanthus acidus (L.) Skeels, Rhinacanthus nasutus L. Kurz, Arbutus andrachne* L. plants have anti-tyrosinase activity [14,15].

2.2 Bacteria

In the literature, it is reported that some bacterial species and various metabolites of these bacteria mediate tyrosinase inhibition. In particular, bacterial strains such as *Streptomyces hiroshimensis TI-C3.* [16], *Streptomyces swartbergensis sp.* Nov. [17] and *Streptomyces roseolilacinus NBRC 12815* [18] have been proven anti-tyrosinase activities. Studies have found that some probiotics such as *Lactobacillus* species [19] show anti-tyrosinase activity. Researches has shown that the activities of fermented plant extracts are higher than those of unfermented ones and their cytotoxic activities are lower than those of unfermented extracts [20]. Additionally, it emphasizes that the lactic acid bacteria strain isolated from dairy cow feces might act as a tyrosinase inhibitor [21].

2.3 Fungi

Research suggests that various fungi species may exhibit tyrosinase enzyme inhibitory activity. Studies *Apergillus sp.* [22], *Trichoderma sp.* [23], and *Paecilomyces sp.* [24] indicate that various fungal species such as *Phellinus linteus* [25], *Daedalea dickinsii* [26], *Dictyophora indusiata* [27] and *Neolentinus lepideus* [28] can be sources of anti-tyrosinase through various bioactive compounds. Additionally, in the literature it is emphasized that various marine mushroom species such as *Myrothecium sp.* may also show tyrosinase inhibitor activity [29,30].

3. Natural tyrosinase inhibitors and derivatives

3.1 Flavanoids

Flavonoids are a specific kind of naturally occurring plant polyphenolic secondary metabolite that are distributed extensively throughout the plant kingdom and therefore common in the human diet. Flavonoids, a type of polyphenolic molecules, are among the most extensively investigated plant derivatives. They comprise a combination of phenolic and pyranic rings and can be found in a broad range of plant parts, including leaves, seeds, bark, and flowers [31]. Considering roughly 4000 identified members, flavonoids serve an important function in protecting against UV radiation, infections, and herbivores. Notably, flavonoids are the biggest class of natural tyrosinase inhibitors known to date. They are divided into several main categories including: flavanols, flavones, flavonols, flavanones, isoflavones, and anthocyanidins (Fig. 1). Additionally, minor flavonoid subclasses encompass dihydroflavones, flavan-3,4-diols, coumarins, chalcones, dihydrochalcones, and aurones. Furthermore, prenylated and vinylated flavonoids, including flavonoid glycosides, represent additional subclasses within the broader category of flavonoids [9,31].

Research indicates that flavonoids like kaempferol [32], quercetin [33], and morin [34] inhibit the tyrosinase enzyme (Fig. 2). Additionally, flavonoid structures such as catechin and rhamnetin function as substrates,

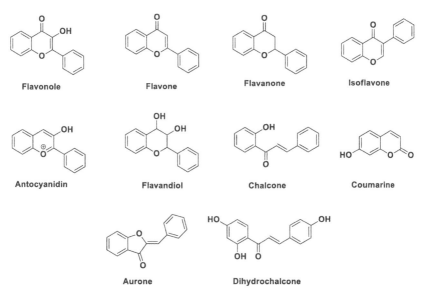

Fig. 1 General chemical structures of the main classes of flavonoids.

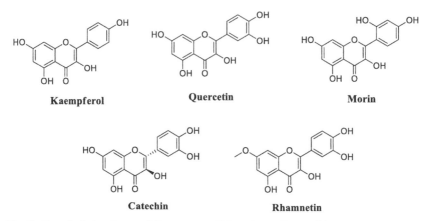

Fig. 2 Chemical structures of the some anti-tyrosinase flavonoids.

with catechin [35] acting as a cofactor and rhamnetin [36] serving as a free radical scavenger, thereby suppressing tyrosinase activity.

3.2 Flavones

The most common flavones include luteolin, apigenin, chrysin, baicalein, and their glycosides (such as apigetrin, baicalin, and vitexin) (Fig. 3) [37]. Research has demonstrated that many hydroxyflavone derivatives, including baicalein, 6-hydroxyapigenin, 6-hydroxygalangin, 6-hydroxy-kaempferol [38], and tricin (5,7,4'-trihydroxy-3',5' dimethoxyflavone) (Fig. 3), exhibit anti-tyrosinase activity [39]. One study specifically investigating the mechanistic effects of baicalein on tyrosinase found that this natural compound (IC_{50} = 0.11 mM) effectively inhibits the enzyme [40]. Similarly, morusone, a flavone derived from the branches of *Morus alba* L., showed significant tyrosinase inhibitory activity (IC_{50} = 290.00 ± 7.90 μM) [41].

3.3 Flavonoles

Myricetin, kaempferol, quercetin, morin, isorhamnetin, galangin and their glycosides (routine, quercitrin and astragalin) are the most commonly found predominant flavonoles (Fig. 4) [37]. So far, various flavonols such as kaempferol from *H. laricifolium Juss* [42] and *Crocus sativus* L. [43], quercetin from *Olea europaea* L. [44], quercetin-4'-O-beta-D-glucoside from *Potentilla bifurca* [45], quercetin-3-O-(6-O-malonyl)-b-D-glucopyranoside and kaempferol-3-O-(6-O-malonyl)-b-D-glucopyranoside from *mulberry leaves* [45], galangin [46], morin [47], 3-cis-dihydromorin (IC_{50} =31.1 μM) and 2,3-transdihydromorin (IC_{50} =21.1 μM) from *Cudrania cochinchinensis* [48], were defined as tyrosinase inhibitors.

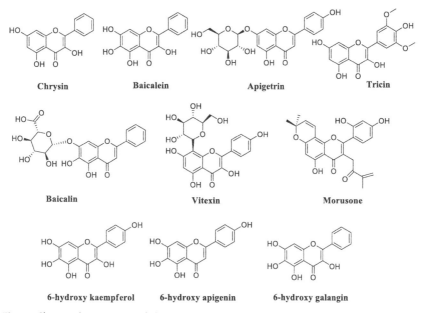

Fig. 3 Chemical structures of the some anti-tyrosinase flavones.

Research shows that morine reversibly inhibits tyrosinase [47]. Additionally, several natural molecules such as flavonoles derivatives galangin [46], and kaempferol [43] have been shown to inhibit tyrosinase through the copper chelation mechanism. It is thought that the chelation mechanism of flavonoles may be through the free 3-hydroxyl group [43]. At the same time, 8-prenylkaempferol, as a competitive tyrosinase inhibitor, was investigated together with Kushenol A (non-competitive) isolated from *Sophora flavescens*, at concentrations lower than IC_{50} values such as 10 μM [49]. Additionally, many flavonol derivatives competitively inhibit tyrosinase by chelating copper, usually via the 3-hydroxy-4-keto moiety in the active site [43]. This information shows that flavonol derivatives have anti-tyrosinase activity, and can be used as skin whitening or browning inhibitors in the food industry.

3.4 Isoflavones

Isoflavones such as daidzein, genistein, glycitin, formononetin, and their glycosides (genistin, and daidzin) are abundant in medicinal plants (Fig. 5) [37]. In a study, 7,8,4′-trihydroxyisoflavone (11.21 ± 0.8 μM) and 7,3′,4′-trihydroxyisoflavone (5, 23 ± 0.6 μM IC_{50}) molecules were

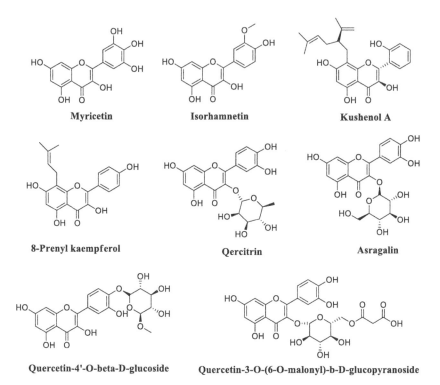

Fig. 4 Chemical structures of the some anti-tyrosinase flavonoles.

found to show anti-tyrosinase activity [50]. Similarly, in another study, 6,7,4′-trihydroxyisoflavone derivatives were determined to be 6 times more potent tyrosinase inhibitors than kojic acid [51]. However, other isoflavone analogues, glycitein, daidzein and genistein, have been shown to have very low anti-tyrosinase activity. These results indicate that the C-6 and C-7 hydroxyl groups in the isoflavone skeleton may be the main part responsible for the tyrosinase inhibitory activity [52]. On the other hand, research has shown that extract from *Glycyrrhiza glabra* root shows that glabridin is a highly effective inhibitor on tyrosinase. This powerful effect has led to the molecule being considered as a new approach in the treatment of hyperpigmentation [53]. Similarly, studies have shown that mircoin (IC_{50} = 5 μM) isolated from *Maackia fauriei* has very strong anti-tyrosinase activity [54]. Lupinalbin, an isoflavonoid analogue recently obtained from *Apios americana* (IC_{50} = 39.7 ± 1.5 μM) and 2-hydro-xygenistein-7-O-gentibiocidin (IC_{50} = 50.0 ± 3.7 μM) revealed that it competitively inhibits the tyrosinase enzyme [55].

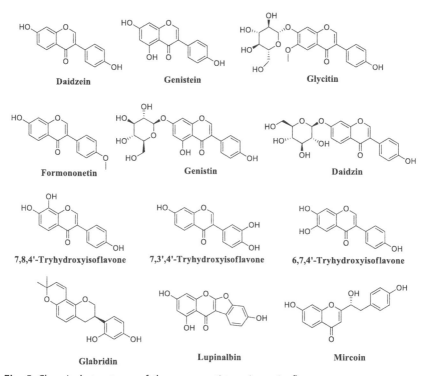

Fig. 5 Chemical structures of the some anti-tyrosinase isoflavones.

3.5 Anthocyanidins

Anthocyanidins, including cyanidin, malvidin, delphinidin, peonidin, pelargonidin and their glycosides, are found in many medicinal plants (Fig. 6) [56]. Literature studies show that the amount of anthocyanins found in natural sources is closely related to anti-tyrosinase activities [57].

3.6 Aurons

It is stated in the literature that Z-benzylidenebenzofuran-3(2H)-one (Aurone) structures inhibit tyrosinase. Researchers have shown that aurons are weak inhibitors, especially their derivatives with two or three hydroxyl groups at the 4,6 and 4′ positions, which inhibit tyrosinase to a high degree. 4,6,4′-trihydroxyaurone, a very potent auron derivative, causes 75% tyrosinase inhibition at 0.1 mM concentration [58]. In addition to synthetic compounds, various natural compounds such as, (2′R)-2′,3′-dihydro-2′-(1-hydroxy-1-methylethyl)-2,6′-bibenzofuran-6,4′-diol [59] and 2-arylbenzofurans isolated from *Morus notabilis* [60] and *Morus yunnanensis* [61], benzofuran flavonoids such as mulberrofuran G (MG) and

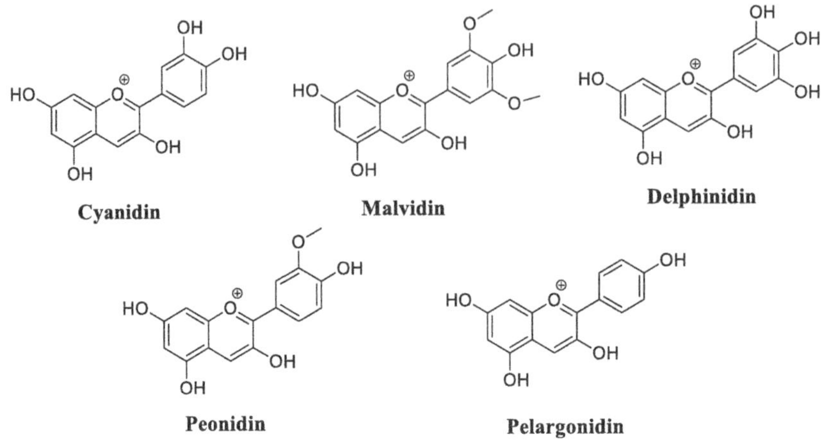

Fig. 6 Chemical structures of the some anti-tyrosinase anthocyanidines.

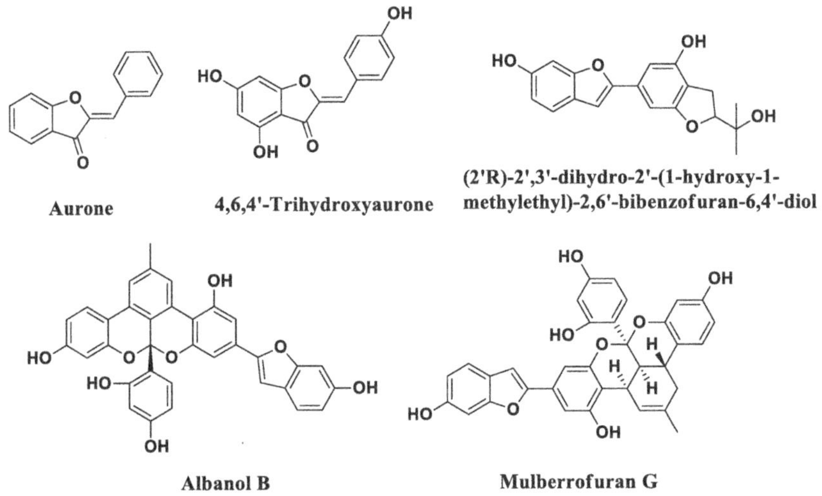

Fig. 7 Chemical structures of the some anti-tyrosinase aurones.

albanol B (AB) isolated from *Morus* sp. [62] and macrourins E isolated from *Morus macroura* (IC_{50} =0.39 μM) and are potent tyrosinase inhibitors among other aurones [63] (Fig. 7).

3.7 Hydroquinones

Hydroquinones derivatives, commonly found in wheat, tea and various fruits, are components used in the treatment of hyperpigmentation [64,65]. This component has long been considered the standard treatment for

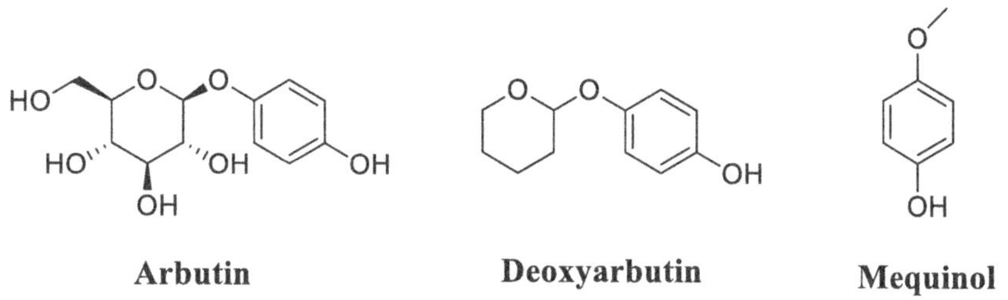

Arbutin   **Deoxyarbutin**   **Mequinol**

Fig. 8 Chemical structures of the some anti-tyrosinase hydroquinones.

hyperpigmentation due to its high effectiveness [66]. Hydroquinones bind covalently to the histone structure and competitively inhibit melanin synthesis by serving as a substrate for the tyrosinase enzyme [67]. However, radicals that do not contain semiquinone can cause permanent and serious damage to melanocytes synthesized through this enzymatic reaction [66]. On the other hand, it has been determined that over time, this substance is transported quite quickly through the epidermis layer and is detoxified by the liver [68]. These effects of the compound cause side effects such as permanent depigmentation and exogenous ochronosis in its long-term use, and therefore have led to most hydroquinone derivatives being banned from cosmetic use in various countries. Despite all these negative effects, hydroquinone derivatives such as arbutin, deoxyarbutin and mequinol are still used in the cosmetic industry as skin whitening agents (Fig. 8) [69]. It has been shown that β–arbutin, a β-D-glucopyranoside derivative of hydroquinone found especially in bearberry, cranberry, blueberry and pear leaves, inhibits the tyrosinase enzyme in a dose-dependent manner [70,71]. At the same time, studies have shown that arbutin is comparable to kojic acid in the treatment of hyperpigmentation [72]. Additionally, deoxyarbutin, a synthetic derivative of hydroquinone, has been found to inhibit tyrosinase in a dose-dependent manner and is also less cytotoxic than hydroquinone [73,74]. On the other hand, mequinol, a mono methyl ether derivative of hydroquinone, acts as a substrate for the tyrosinase enzyme and prevents melanin synthesis [75]. Current literature information shows that hydroquinone derivatives commonly found in natural sources inhibit the tyrosinase enzyme to a high extent.

3.8 Chalcones

Chalcones, considered precursor compounds to flavonoids and iso-flavonoids, are recognized for their diverse biological activities [76]. These compounds, characterized by their 1,3-diaryl-2-propen-1-one structure, exhibit specific biological functions due to their hydroxy, methoxy, and

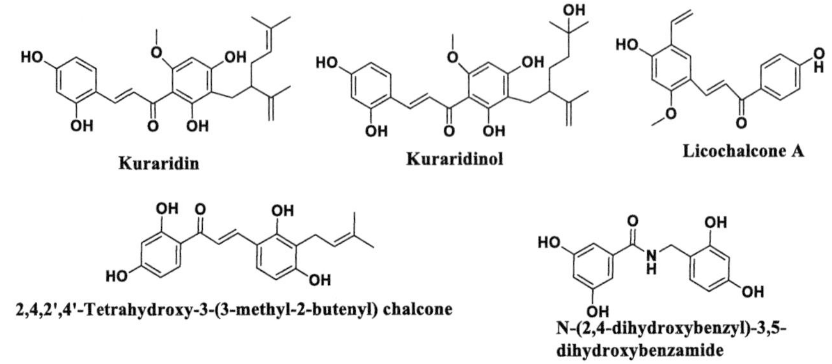

Kuraridin

Kuraridinol

Licochalcone A

2,4,2',4'-Tetrahydroxy-3-(3-methyl-2-butenyl) chalcone

N-(2,4-dihydroxybenzyl)-3,5-dihydroxybenzamide

Fig. 9 Chemical structures of the some anti-tyrosinase chalcones.

alkyl derivatives [77]. Many studies on the subject show that chalcone-derived compounds have anti-tyrosinase activities [78–81]. Kuraridin of the chalcone derivatives [78] (34-fold), kuraridinol [79] (18-fold), and 2,4,2',4'-tetrahydroxy-3-(3-methyl-2-butenyl) chalcone (TMBC) [80] has been reported that the inhibitory activity of their structures on the tyrosinase enzyme is higher than kojic acid (Fig. 9). Similarly, licochalcone derivative A, isolated from the roots of *Glycyrrhize* sp. has been shown to inhibit fungal tyrosinase activity at a level 5 times higher than kojic acid [81] (Fig. 9). Studies showing the high anti-tyrosinase activities of chalcones have led researchers to examine the structure-activity relationships of these compounds in detail. In a study, it was reported that chalcones containing resorcinol, 2,4-dihydrophenyl substituent in their structure, exhibited strong tyrosinase inhibitory activity and that the resorcinol structure in the B ring of the chalcone was the structure that contributed the most to the biological effect [82]. Similarly, in a study by Jun et al., the 2,4,2',4',6'-pentahydroxychalcone structure was found to be the derivative with the strongest anti-tyrosinase activity. In the study, it was determined that the activity of this derivative was 5 times higher than that of others [83]. At the same time, in a study conducted by Nguyen et al., it was determined that prenylated chalcone and flavone isolated from *Artocarpus heterophyllus* wood showed 3000 times higher activity (IC_{50} = 13 nM) than kojic acid (IC_{50} = 45 µM) [84]. Recently, in a study in which new phenylpropanoid amide derivatives with chalcone structure were developed, it was reported that the active compound, which is 10 times more active in tyrosinase inhibition than the positive control tert -butylhydroquinone, can be found in the garlic peel. The results of the study pointed out that garlic peels, which are industrial waste, can be used as a depigmentation agent [84].

This information has paved the way for the synthesis of many synthetic chalcone derivatives. One of the studies conducted for this purpose stated that the *N*-(2,4-dihydroxybenzyl)-3,5-dihydroxybenzamide structure showed 8 times stronger inhibitory activity than kojic acid used as a positive control in tyrosinase inhibition [85]. Similarly, in a study by Baek et al., it was reported that *N*-2,4-dihydroxybenzyl-2-hydroxy-4-methoxy-5-adamantyl benzamide derivative showed 40 times higher tyrosinase inhibitory activity than kojic acid [86].

3.9 Stilbenes

Stilbenes are compounds that have the structure of diarylethenes and are produced in the shikimate-acetate pathway [87]. In the literature, it has been reported that various stilbene derivatives such as polyoxygenated stilbenes [88], oxyresveratrol [89], chlorophorin [90] and andalasin A [91] obtained from natural sources such as resveratrol have an inhibitory effect on the tyrosinase enzyme (Fig. 10). Research indicates that oxyresveratrol, which has a 4-resorcinol moiety in the B ring and a 5-resorcinol moiety in the A ring, has a tyrosinase enzyme inhibitory effect that is 32 times higher than kojic acid and thus may be the most promising inhibitor possible [92]. Resveratrol, which lacks the 4-resorcinol moiety of this compound, has been shown to be a 50-fold less active tyrasinase inhibitor compared to oxyresveratrol. Additionally, the study also states that dihydroresveratrol is less active than resveratrol [93]. However, it has been shown that the compound exhibits an 8-fold stronger inhibitory effect than oxyresveratrol on the L-DOPA oxidase activity of tyrosinase, and this proves that the inhibitory activity on the enzyme, as well as the number of phenols, also plays a critical role in their arrangement in the space [94]. The high

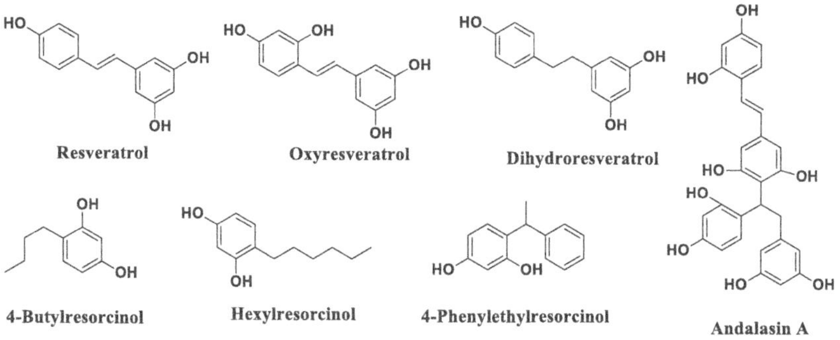

Fig. 10 Chemical structures of the some anti-tyrosinase stilbenes.

effectiveness of stilbene derivatives obtained from natural sources has mediated the synthesis of many synthetic derivatives [95–97]. For this purpose, in a study conducted by Bae et al., various azo-stilbene derivatives were synthesized. In the study, it was reported that the mono-tosylated derivative competitively inhibited tyrosinase and showed 3 times higher activity than kojic acid, which was the positive control [95]. Additionally, 4-methoxy- or 4-hydroxy-anilino-containing derivatives have been found to show stronger inhibition of the tyrosinase enzyme than 2-substituted analogues. It was determined that the strongest inhibitory compound, (*E*)-4-((4-hydroxyphenylimino)methyl)benzene-1,2-diol, inhibited the tyrosinase enzyme at a rate 3 times higher than kojic acid [96]. Additionally, another study found that 4-*n*-butylresorcinol is a tyrosinase inhibitor, and that alkylated resveratrol showed a stronger effect than kojic acid in inhibiting human tyrosinase (25-fold) and artificial skin model melanin (20-fold). Additionally, in the study, 4-butylresorcinol showed higher activity compared to 4-hexylresorcinol and 4-phenylethylresorcinol, which are derivatives with longer alkyl chains [97].

3.10 Kojic acid

Kojic acid is a metabolite obtained from fungi and is used in the cosmetic industry as an anti-browning and skin whitening agent [98]. This metabolite competitively inhibits tyrosinase enzyme activity by acting as a chelator for ions such as Cu^{+2} and Fe^{+3} [99]. Kojic acid is a positive control molecule used to compare the inhibitory activities of various compounds against the tyrosinase enzyme. However, the instability of this substance in the presence of aerobic environment and light limits its use. This has necessitated numerous studies to develop more stable synthetic derivatives of kojic acid [100] (Fig. 11). For this purpose, different derivatives of kojic acid such as polymer [101], 4-pyridone [102] and peptide-conjugate [103] were developed and the activities of these derivatives were examined. In a

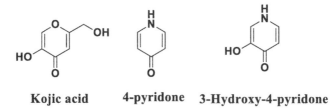

Kojic acid   4-pyridone   3-Hydroxy-4-pyridone

Fig. 11 Chemical structures of the some anti-tyrosinase kojic acids.

study conducted by Saghaie and colleagues on the subject, it was determined that 3-hydroxy-4-pyridone derivatives showed tyrosinase inhibitory activity, and at the same time, pyridone derivatives containing two free hydroxyl groups showed stronger inhibitory activity compared to derivatives containing one hydroxyl [104]. Similarly, hexadimeric derivatives of two kojic acids linked via an alkyl ester, an amide, or a sulfite linkage exhibited high anti-tyrosinase activity, and derivatives with the propane 1,3-dithioether structure showed tyrosinase inhibition activity 25 times higher than that of kojic acid [105]. Additionally, in a study conducted by Noh et al., mono- and dipeptide-conjugated kojic acid derivatives were synthesized and the anti-tyrosinase activities of these derivatives were examined. In the study, it was determined that the kojic acid-phenylalanine amide derivative had 6 times stronger anti-tyrosinase activity than kojic acid and that this derivative maintained its activity for three months [106].

3.11 Coumarins

Coumarins are heterocyclic compounds obtained from natural sources, consisting of an aromatic ring fused to a 6-membered lactone ring with the structure 2H-1-benzopyran-2-one (Fig. 1). Their broad biological activities have caused these heretocyclic compounds to attract great attention from researchers [107]. In one of the studies conducted for this purpose, it was shown that the 6,7-dihydroxylcoumarin derivative esculetin isolated from *Euphorbia lathyris* served as the substrate of the tyrosinase enzyme [108] (Fig. 12). In a comparative study, the coumarin derivatives 9-hydroxy-4-methoxypsoralen (6-fold increase) [109] from *Angelica dahurica* and 8′-epi-cleomyscosin A (13-fold increase) [110] from *Rhododendron collettianum* demonstrated significantly greater tyrosinase inhibitory activity than kojic acid (Fig. 12).

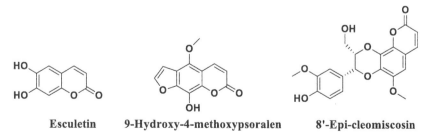

Esculetin   **9-Hydroxy-4-methoxypsoralen**   **8′-Epi-cleomiscosin**

Fig. 12 Chemical structures of the some anti-tyrosinase coumarines.

4. Other natural tyrosinase inhibitors

Aloesin, a hydroxychromone glucoside isolated from Aloe vera, is one of the natural tyrosinase inhibitors. It was determined that this herbal molecule showed greater inhibitory activity against mouse tyrosinase than fungal tyrosinase. Due to its effectiveness in skin care, aloesin is frequently preferred in topically applied cosmetics [111,112].

Anti-tyrosinase activities of metabolites such as naturally occurring benzaldehyde and benzoate derivatives [113], benzaldehyde [114], anisic acid [115], anisaldehyde [116], cinnamic acid [117], methoxycinnamic acid [118] and vanillic acid [119] are reported (Fig. 13). At the same time, it has been determined that gallic acid, which can be obtained from many plants, has tyrosinase inhibitory activity [120]. Various lipids and fatty acids such as trilinolein [121], soyacerebroside I [122] and trans-geranic acid [123] are molecules obtained from natural sources and exhibit tyrosinase inhibitory activity, albeit at a weak level. The proposed inhibitory mechanism for lipids due to lack of copper chelation could be free radical scavenging via unsaturated alkene or binding outside the catalytic site. Steroids are biologically important structural scaffolds that perform essential functions in human life. Many of them exhibited tyrosinase inhibitory activity, such as stigmast-5-ene-3β,26-diol [124], 3β,21,22,23-tetrahydroxycycloart-24 [26,31], 25-diene from *Trifolium balansae* [125], Amberboa ramose and arjunilic acid obtained from *R. collettianum* showed 12-, 13- and 17-fold greater tyrosinase inhibitory activity than kojic acid [126]. However, hydrophobic steroid or long-chain lipid molecules have the potential to develop as skin whitening agents due to their skin permeability. Yet they are less used cosmetically due to the lack of cellular or clinical assays to determine their depigmentation activity. Anthraquinones exhibit various pharmacological activities such as anti-inflammatory, wound healing, analgesic, antipyretic, antimicrobial, antitumor and anti-tyrosinase activities. Physcion, an anthraquinone, has shown similar tyrosinase inhibitory activity comparable to kojic acid [127]. Additionally, many lignans from *Vitex negundo* have shown higher tyrosinase inhibitory activity than kojic acid [128].

There are many new naturally occurring tyrosinase inhibitors that have been isolated and updated as natural organisms have evolved numerous methods to protect themselves from sunlight. In recent studies, Wang et al. [129] reported the isolation of (−)-N-formylonaine from the magnolia plants *Michelia alba* as a human tyrosinase inhibitor and antioxidant. In the fungal tyrosinase inhibition assay, (−)-N-formylanonaine showed comparable activity to kojic acid, but the compound exhibited higher inhibition against

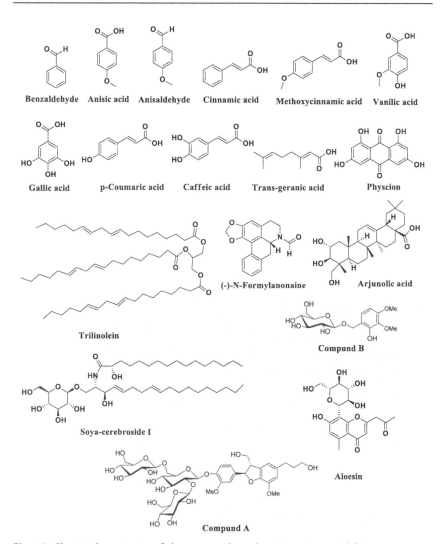

Fig. 13 Chemical structures of the some selected anti-tyrosinase inhibitors.

human tyrosinase. showed much less cytotoxicity than kojic acid. In homology modeling, the compound binds to the active site of tyrosinase by coordinating two Cu^{2+} ions. Additionally, Wu et al. [130] isolated three new lignin glycosides from *Castanea ehnryi* along with two known lignins. All of them were subjected to fungal tyrosinase inhibition assay and among them, 2,3-dihydro-2-[4-(β-glucopyranosyl (1 → 2)-[β -glucopyranosyl(1 → 6)]-β-glucopyranosyloxy)-3-methoxyphenyl]-3-(hydroxymethyl)-7-methoxy-5-benzofuranpropanol (Compound **A**) showed an o-diphenolase inhibitory activity comparable to kojic acid. Akihisa et al. [131] isolated an aromatic glycoside, 3-o-demethylnicoenoside (Compund **B**) (Fig. 13), and 11 known compounds from the root bark of *Acer buergerianum*. All of the isolated compounds were evaluated for their melanogenesis inhibitory activity in B16 melanoma cells stimulated with α-melanocyte-stimulating hormone (α-MSH), and 3-o-demethylnicoenoside exhibited comparable activity to kojic acid. Additionally, Mohd et al. isolated three xanthone derivatives from *Artocarpus obtusus FM Jarrett,* among which pyranocycloartobiloxanthone A inhibited fungal tyrosinase with activity comparable to kojic acid. The compound also exhibited a potent free radical scavenger against DPPH free radicals with an IC_{50} value of $2\,\mu g/mL$ and exhibited potent and moderate antimicrobial activity against MRSA and *Bacillus subtilis* (clinically isolated strain) with inhibition zones of 20 and 12 mm, respectively [132].

5. Conclusions

Tyrosinase, a copper-containing enzyme, is essential for melanin formation, and excessive production causes hyperpigmentation diseases including as melasma, age spots, and freckles. Inhibiting tyrosinase activity has emerged as an essential approach for treating these disorders, with significant implications for cosmetics, food technology, and medicine. Natural products have shown to be an excellent source of tyrosinase inhibitors, with substances produced from plants, marine organisms, and microbes showing promising inhibitory properties. This chapter provides a comprehensive overview of natural tyrosinase inhibitors, including their structures, mechanisms of action, and potential applications, emphasizing natural products' significant potential in the development of effective tyrosinase inhibitors for a wide range of applications.

Conflicts of interest

Hereby, we declare that we have no conflicts of interest.

References

[1] V.M. Dembitsky, A. Kilimnik, Anti-melanoma agents derived from fungal species, M. J. Pharma 1 (2016) 1–16.
[2] S. Maghsoudi, H. Adibi, M. Hamzeh, et al., Kinetic of mushroom tyrosinase inhibition by benzaldehyde derivatives, Journal of Reports in Pharmaceutical Sciences 2 (2013) 156–164.
[3] J.Y. Lin, D.E. Fisher, Melanocyte biology and skin pigmentation, Nature 445 (2007) 843–850.

[4] C.H. Lee, S.B. Wu, C.H. Hong, et al., Molecular mechanisms of UV-induced apoptosis and its effects on skin residential cells: The implication in UV-based phototherapy, International Journal of Molecular Sciences 14 (2013) 6414–6435.

[5] T.S. Chang, Natural melanogenesis inhibitors acting through the down-regulation of tyrosinase activity, Materials 5 (2012) 1661–1685.

[6] M.R. Loizzo, R. Tundis, F. Menichini, Natural and synthetic tyrosinase inhibitors as antibrowning agents: An update, Comprehensive Reviews in Food Science and Food Safety 11 (2012) 378–398.

[7] N.V. Thomas, S.K. Kim, Beneficial effects of marine algal compounds in cosmeceuticals, Marine Drugs 11 (2013) 146–164.

[8] T.S. Chang, An updated review of tyrosinase inhibitors, International Journal of Molecular Sciences 10 (2009) 2440–2475.

[9] Y.J. Kim, H. Uyama, Tyrosinase inhibitors from natural and synthetic sources: Structure, inhibition mechanism and perspective for the future, Cellular and Molecular Life Sciences 62 (2005) 1707–1723.

[10] C.A. Ramsden, P.A. Riley, Tyrosinase: The four oxidation states of the active site and their relevance to enzymatic activation, oxidation and inactivation, Bioorganic & Medicinal Chemistry 22 (2014) 2388–2395.

[11] S. Zolghadri, A. Bahrami, M.T. Hassan Khan, J. Munoz-Munoz, F. Garcia-Molina, F. Garcia-Canovas, et al., A comprehensive review on tyrosinase inhibitors, Journal of Enzyme Inhibition and Medicinal Chemistry 34 (1) (2019) 279–309.

[12] B. Burlando, M. Clericuzio, L. Cornara, Moraceae plants with tyrosinase inhibitory activity: A review, Mini-Reviews in Medicinal Chemistry 17 (2017) 108–121.

[13] Y.N. Quispe, S.H. Hwang, Z. Wang, S.S. Lim, Screening of peruvian medicinal plants for tyrosinase inhibitory properties: Identification of tyrosinase inhibitors in *Hypericum laricifolium* Juss, Molecules 22 (2017).

[14] R.A. Issa, F.U. Afifi, B.I. Amro, Studying the anti-tyrosinase effect of *Arbutus andrachne* L. extracts, International Journal of Cosmetic Science 30 (2008) 271–276.

[15] M. Chatatikun, A. Chiabchalard, Thai plants with high antioxidant levels, free radical scavenging activity, antityrosinase and anti-collagenase activity, BMC Complementary Medicine and Therapies 17 (2017) 487.

[16] T.S. Chang, M. Tseng, H.Y. Ding, S. Shou-Ku Tai, Isolation and characterization of *Streptomyces hiroshimensis* strain TI-C3 with anti-tyrosinase activity, Journal of Cosmetic Science 59 (2008) 33–40.

[17] M. le Roes-Hill, A. Prins, P.R. Meyers, *Streptomyces swartbergensis* sp. Nov., a novel tyrosinase and antibiotic producing actinobacterium, Antonie Van. Leeuwenhoek 111 (2018) 589–600.

[18] T. Nakashima, K. Anzai, N. Kuwahara, et al., Physicochemical characters of a tyrosinase inhibitor produced by *Streptomyces roseolilacinus* NBRC 12815, Biological and Pharmaceutical Bulletin 32 (2009) 832–836.

[19] K. Ji, Y.S. Cho, Y.T. Kim, Tyrosinase inhibitory and anti-oxidative effects of lactic acid bacteria isolated from dairy cow feces, Probiotics and Antimicrobial Proteins 10 (2018) 43–55.

[20] G.H. Wang, C.Y. Chen, T.H. Tsai, et al., Evaluation of tyrosinase inhibitory and antioxidant activities of *Angelica dahurica* root extracts for four different probiotic bacteria fermentations, Journal of Bioscience and Bioengineering 123 (2017) 679–684.

[21] A. Crozier, I.B. Jaganath, M.N. Clifford, Phenols, polyphenols and tannins: An overview, Plant. Secondary Metabolites (2007) 1–24.

[22] K.Y. Vasantha, C.S. Murugesh, A.P. Sattur, A tyrosinase inhibitor from *Aspergillus niger*, Journal of Food Science and Technology 51 (2014) 2877–2880.

[23] T. Tsuchiya, K. Yamada, K. Minoura, et al., Purification and determination of the chemical structure of the tyrosinase inhibitor produced by *Trichoderma viride* strain H1-7 from a marine environment, Biological and Pharmaceutical Bulletin 31 (2008) 1618–1620.

[24] R. Lu, X. Liu, S. Gao, et al., New tyrosinase inhibitors from *Paecilomyces gunnii*, Journal of Agricultural and Food Chemistry 62 (2014) 11917–11923.

[25] H.S. Kang, J.H. Choi, W.K. Cho, et al., A sphingolipid and tyrosinase inhibitors from the fruiting body of *Phellinus linteus*, Archives of Pharmacal Research 27 (2004) 742–750.

[26] K. Morimura, C. Yamazaki, Y. Hattori, et al., A tyrosinase inhibitor, daedalin a, from mycelial culture of *Daedalea dickinsii*, Bioscience, Biotechnology, and Biochemistry 71 (2007) 2837–2840.

[27] V.K. Sharma, J. Choi, N. Sharma, et al., In vitro anti-tyrosinase activity of 5-(hydroxymethyl)-2-furfural isolated from *Dictyophora indusiata*, Phytotherapy Research 18 (2004) 841–844.

[28] A. Ishihara, Y. Ide, T. Bito, et al., Novel tyrosinase inhibitors from liquid culture of *Neolentinus lepideus*, Bioscience, Biotechnology, and Biochemistry 82 (2018) 22–30.

[29] X. Li, M.K. Kim, U. Lee, et al., Myrothenones A and B, cyclopentenone derivatives with tyrosinase inhibitory activity from the marine-derived fungus Myrothecium sp, Chemical and Pharmaceutical Bulletin ((Tokyo)) 53 (2005) 453–455.

[30] B. Wu, X. Wu, M. Sun, M. Li, Two novel tyrosinase inhibitory sesquiterpenes induced by cucl2 from a marine-derived fungus Pestalotiopsis sp. Z233, Marine Drugs 11 (2013) 2713–2721.

[31] J.B. Harborne, C.A. Williams, Advances in flavonoid research since 1992, Phytochemistry 55 (2000) 481–504.

[32] I. Kubo, I. Kinst-Hori, S.K. Chaudhuri, et al., Flavonols from *Heterotheca inuloides*: Tyrosinase inhibitory activity and structural criteria, Bioorganic & Medicinal Chemistry 8 (2000) 1749–1755.

[33] I. Kubo, I. Kinst-Hori, K. Ishiguro, et al., Tyrosinase inhibitory flavonoids from *Heterotheca inuloides* and their structural functions, Bioorganic & Medicinal Chemistry Letter 4 (1994) 1443–1446.

[34] L.P. Xie, Q.X. Chen, H. Huang, et al., Inhibitory effects of some flavonoids on the activity of mushroom tyrosinase, Biochemistry 68 (2003) 487–491.

[35] J.K. No, D.Y. Soung, Y.J. Kim, et al., Inhibition of tyrosinase by green tea components, Life Sciences 65 (1999) PL241–PL246.

[36] Y.J. Kim, Rhamnetin attenuates melanogenesis by suppressing oxidative stress and pro-inflammatory mediators, Biological and Pharmaceutical Bulletin 36 (2013) 1341–1347.

[37] W.Y. Huang, Y.Z. Cai, Y. Zhang, Natural phenolic compounds from medicinal herbs and dietary plants: Potential use for cancer prevention, Nutrition and Cancer 62 (2010) 1–20.

[38] H. Gao, J. Nishida, S. Saito, J. Kawabata, Inhibitory effects of 5,6,7-trihydroxy-flavones on tyrosinase, Molecules 12 (2007) 86–97.

[39] Y. Mu, L. Li, S.Q. Hu, Molecular inhibitory mechanism of tricin on tyrosinase, Spectrochimica Acta Part A: Molecular and Biomolecular Spectroscopy 107 (2013) 235–240.

[40] N. Guo, C. Wang, C. Shang, et al., Integrated study of the mechanism of tyrosinase inhibition by baicalein using kinetic, multispectroscopic and computational simulation analyses, International Journal of Biological Macromolecules 118 (2018) 57–68.

[41] L. Zhang, G. Tao, J. Chen, Z.P. Zheng, Characterization of a new flavone and tyrosinase inhibition constituents from the Twigs of *Morus alba* L, Molecules 21 (9) (2016) 1130.

[42] Y.N. Quispe, S.H. Hwang, Z. Wang, S.S. Lim, Screening of peruvian medicinal plants for tyrosinase inhibitory properties:identification of tyrosinase inhibitors in *Hypericum laricifolium* Juss, Molecules 22 (2017).

[43] I. Kubo, I. Kinst-Hori, Flavonols from saffron flower: Tyrosinase inhibitory activity and inhibition mechanism, Journal of Agricultural and Food Chemistry 47 (1999) 4121–4125.

[44] S.H. Omar, C.J. Scott, A.S. Hamlin, H.K. Obied, Biophenols: Enzymes (b-secretase, cholinesterases, histone deacetylas and tyrosinase) inhibitors from olive (*Olea europaea* L.), Fitoterapia 128 (2018) 118–129.

[45] Z. Yang, Y. Zhang, L. Sun, et al., An ultrafiltration highperformance liquid chromatography coupled with diode array detector and mass spectrometry approach for screening and characterising tyrosinase inhibitors from mulberry leaves, Analytica Chimica Acta 719 (2012) 87–95.

[46] Y.H. Lu, T. Lin, Z.T. Wang, et al., Mechanism and inhibitory effect of galangin and its flavonoid mixture from *Alpinia officinarum* on mushroom tyrosinase and B16 murine melanoma cells, Journal of Enzyme Inhibition and Medicinal Chemistry 22 (2007) 433–438.

[47] Y. Wang, G. Zhang, J. Yan, D. Gong, Inhibitory effect of morin on tyrosinase: Insights from spectroscopic and molecular docking studies, Food Chemistry 163 (2014) 226–233.

[48] Z.P. Zheng, Q. Zhu, C.L. Fan, et al., Phenolic tyrosinase inhibitors from the stems of *Cudrania cochinchinensis*, Food & Function Journal 2 (2011) 259–264.

[49] J.H. Kim, I.S. Cho, Y.K. So, et al., Kushenol a and 8-prenylkaempferol, tyrosinase inhibitors, derived from *Sophora flavescens*, Journal of Enzyme Inhibition and Medicinal Chemistr 33 (2018) 1048–1054.

[50] J.S. Park, D.H. Kim, J.K. Lee, et al., Natural ortho-dihydroxyisoflavone derivatives from aged Korean fermented soybean paste as potent tyrosinase and melanin formation inhibitors, Bioorganic & Medicinal Chemistry Letter 20 (2010) 1162–1164.

[51] T.S. Chang, H.Y. Ding, H.C. Lin, Identifying 6,7,4′-trihydroxyisoflavone as a potent tyrosinase inhibitor, Bioscience, Biotechnology, and Biochemistry 69 (2005) 1999–2001.

[52] T.S. Chang, Two potent suicide substrates of mushroom tyrosinase: 7,8,4′-trihydroxyisoflavone and 5,7,8,4′-tetrahydroxyisoflavone, Journal of Agricultural and Food Chemistry 55 (5) (2007) 2010.

[53] K. Deshmukh, S.S. Poddar, Tyrosinase inhibitor-loaded microsponge drug delivery system: New approach for hyperpigmentation disorders, Journal of Microencapsulation 29 (2012) 559–568.

[54] J.M. Kim, R.K. Ko, D.S. Jung, et al., Tyrosinase inhibitory constituents from the stems of *Maackia fauriei*, Phytotherapy Research 24 (2010) 70–75.

[55] J.H. Kim, H.Y. Kim, S.Y. Kang, et al., Chemical constituents from *Apios americana* and their inhibitory activity on tyrosinase, Molecules 23 (2018) 232.

[56] Y. Jiang, Z. Du, G. Xue, et al., Synthesis and biological evaluation of unsymmetrical curcumin analogues as tyrosinase inhibitors, Molecules 18 (2013) 3948–3961.

[57] J.K. Jhan, Y.C. Chung, G.H. Chen, et al., Anthocyanin contents in the seed coat of black soya bean and their anti-human tyrosinase activity and antioxidative activity, International Journal of Cosmetic Science 38 (2016) 319–324.

[58] S. Okombi, D. Rival, S. Bonnet, et al., Discovery of benzylidenebenzofuran- 3(2H)-one (aurones) as inhibitors of tyrosinase derived from human melanocytes, Journal of Medicinal Chemistry 49 (2006) 329–333.

[59] J.J. Zhu, G.R. Yan, Z.J. Xu, et al., Inhibitory effects of (2′R)-2′,3′- dihydro-2′-(1-hydroxy-1-methylethyl)-2,6′-bibenzofuran-6,4′-diol on mushroom tyrosinase and melanogenesis in B16- F10 melanoma cells, Phytotherapy Research 29 (2015) 1040–1045.

[60] X. Hu, M. Wang, G.R. Yan, et al., 2-Arylbenzofuran and tyrosinase inhibitory constituents of *Morus notabilis*, Journal of Asian Natural Products Research 14 (2012) 1103–1108.

[61] X. Hu, J.W. Wu, M. Wang, et al., 2-Arylbenzofuran, flavonoid, and tyrosinase inhibitory constituents of *Morus yunnanensis*, Journal of Natural Products 75 (2012) 82–87.

[62] P. Koirala, S.H. Seong, Y. Zhou, et al., Structure-activity relationship of the tyrosinase inhibitors kuwanon G, mulberrofuran G, and albanol B from Morus species: A kinetics and molecular docking study, Molecules 23 (2018) 1413.

[63] Y. Wang, L. Xu, W. Gao, et al., Isoprenylated phenolic compounds from *Morus macroura* as potent tyrosinase inhibitors, Planta Medica 84 (2018) 336–343.

[64] C.R. Denton, A.B. Lerner, T.B. Fitzpatrick, Inhibition of melanin formation by chemical agents, Journal of Investigative Dermatology 18 (1952) 119–135.

[65] M. Amer, M. Metwalli, Topical hydroquinone in the treatment of some hyperpigmentary disorders, International Journal of Dermatology 37 (1998) 449–450.

[66] G.H. Findlay, Ochronosis following skin bleaching with hydroquinone, Journal of the American Academy of Dermatology 6 (1982) 1092–1093.

[67] Z.D. Draelos, Skin lightening preparations and the hydroquinone controversy, Dermatologic Therapy 20 (2007) 308–313 27.

[68] R.C. Wester, J. Melendres, X. Hui, et al., Human in vivo and in vitro hydroquinone topical bioavailability, metabolism, and disposition, Journal of Toxicology and Environmental Health, Part A 54 (1998) 301–317.

[69] R. Sarkar, P. Arora, K.V. Garg, Cosmeceuticals for hyperpigmentation: What is available? Journal of Cutaneous and Aesthetic Surgery 6 (2013) 4–11.

[70] I. Parejo, F. Viladomat, J. Bastida, C. Codina, A single extraction step in the quantitative analysis of arbutin in bearberry (*Arctostaphylos uva-ursi*) leaves by high-performance liquid chromatography, Phytochemical Analysis 12 (2001) 336–339.

[71] D.H. Seo, J.H. Jung, J.E. Lee, et al., Biotechnological production of arbutins (a- and b-arbutins), skin-lightening agents, and their derivatives, Applied Microbiology and Biotechnology 95 (2012) 1417–1425.

[72] A. Garcia, J.E. Fulton, The combination of glycolic acid and hydroquinone or kojic acid for the treatment of melasma and related conditions, Dermatologic Surgery 22 (1996) 443–447.

[73] R.E. Boissy, M. Visscher, M.A. DeLong, Deoxyarbutin: A Novel reversible tyrosinase inhibitor with effective in vivo skin lightening potency, Experimental Dermatology 14 (8) (2005) 601.

[74] S. Chawla, M.A. deLong, M.O. Visscher, et al., Mechanism of tyrosinase inhibition by deoxyArbutin and its second-generation derivatives, British Journal of Dermatology 159 (2008) 1267–1274.

[75] Z.D. Draelos, Cosmetic therapy, in: S.E. Wolverton (Ed.), Comprehensive dermatologic drug therapy, 2nd edn., Saunders, Philadelphia (PA), 2007, pp. P761–P774.

[76] N.K. Sahu, S.S. Balbhadra, J. Choudhary, D.V. Kohli, Exploring pharmacological significance of chalcone scaffold: A review, Current Medicinal Chemistry 19 (2014) 209–225.

[77] D.I. Batovska, I.T. Todorova, Trends in utilization of the pharmacological potential of chalcones, Current Clinical Pharmacology 5 (2010) 1–29.

[78] S.J. Kim, K.H. Son, H.W. Chang, et al., Tyrosinase inhibitory prenylated flavonoids from *Sophora flavescens*, Biological and Pharmaceutical Bulletin 26 (2003) 1348–1350.

[79] S.K. Hyun, W.H. Lee, M. Jeong, et al., Inhibitory effects of kurarinol, kuraridinol, and trifolirhizin from *Sophora flavescens* on tyrosinase and melanin synthesis, Biological and Pharmaceutical Bulletin 31 (2008) 154–158.

[80] X. Zhang, X. Hu, A. Hou, H. Wang, Inhibitory effect of 2,4,20,40- tetrahydroxy-3-(3-methyl-2-butenyl)-chalcone on tyrosinase activity and melanin biosynthesis, Biological and Pharmaceutical Bulletin 32 (2009) 86–90.

[81] B. Fu, H. Li, X. Wang, et al., Isolation and identification of flavonoids in licorice and a study of their inhibitory effects on tyrosinase, Journal of Agricultural and Food Chemistry 53 (2005) 7408–7414.

[82] S. Khatib, O. Nerya, R. Musa, et al., Chalcones as potent tyrosinase inhibitors: The importance of a 2,4-substituted resorcinol moiety, Bioorganic & Medicinal Chemistry 13 (2005) 433–441.

[83] N. Jun, G. Hong, K. Jun, Synthesis and evaluation of 20,40,60- trihydroxychalcones as a new class of tyrosinase inhibitors, Bioorganic & Medicinal Chemistry 15 (2007) 2396–2402.

[84] Z. Wu, L. Zheng, Y. Li, et al., Synthesis and structure-activity relationships and effects of phenylpropanoid amides of octopamine and dopamine on tyrosinase inhibition and antioxidation, Food Chemistry 134 (2012) 1128–1131.

[85] S.J. Cho, J.S. Roh, W.S. Sun, et al., N-Benzylbenzamides: A new class of potent tyrosinase inhibitors, Bioorganic & Medicinal Chemistry Letter 16 (2006) 2682–2684.

[86] H.S. Baek, Y.D. Hong, C.S. Lee, et al., Adamantyl N-benzylbenzamide: New series of depigmentation agents with tyrosinase inhibitory activity, Bioorganic & Medicinal Chemistry Letter 22 (2012) 2110–2113.

[87] K. Likhitwitayawuid, Stilbenes with tyrosinase inhibitory activity, Current Science India 94 (2008) 44–53.

[88] B.B. Aggarwal, A. Bhardwaj, R.S. Aggarwal, et al., Role of resveratrol in prevention and therapy of cancer: Preclinical and clinical studies, Anticancer Research 24 (2004) 2783–2840.

[89] N.H. Shin, S.Y. Ryu, E.J. Choi, et al., Oxyresveratrol as the potent inhibitor on dopa oxidase activity of mushroom tyrosinase, Biochemical and Biophysical Research Communications 243 (3) (1998) 801.

[90] K. Shimizu, R. Kondo, K. Sakai, et al., The inhibitory components from *Artocarpus incisus* on melanin biosynthesis, Planta Medica 64 (1998) 408–412.

[91] K. Likhitwitayawuid, B. Sritularak, A new dimeric stilbene with tyrosinase inhibitory activity from *Artocarpus gomezianus*, Journal of Natural Products 64 (2001) 1457–1459.

[92] M.R. Stratford, C.A. Ramsden, P.A. Riley, Mechanistic studies of the inactivation of tyrosinase by resorcinol, Bioorganic & Medicinal Chemistry 21 (2013) 1166–1173.

[93] K. Ohguchi, T. Tanaka, T. Ito, et al., Inhibitory effects of resveratrol derivatives from Dipterocarpaceae plants on tyrosinase activity, Bioscience, Biotechnology, and Biochemistry 67 (2003) 1587–1589.

[94] K. Likhitwitayawuid, A. Sornsute, B. Sritularak, P. Ploypradith, Chemical transformations of oxyresveratrol (trans-2,4,30,50-tetrahydroxystilbene) into a potent tyrosinase inhibitor and a strong cytotoxic agent, Bioorganic & Medicinal Chemistry Letter 16 (2006) 5650–5653.

[95] S.J. Bae, Y.M. Ha, J.A. Kim, et al., A novel synthesized tyrosinase inhibitor: (*E*)-2-((2,4-dihydroxyphenyl)diazenyl)phenyl 4-methylbenzenesulfonate as an azo-resveratrol analog, Bioscience, Biotechnology, and Biochemistry 77 (2013) 65–72.

[96] S.J. Bae, Y.M. Ha, Y.J. Park, et al., Design, synthesis, and evaluation of (*E*)-N-substituted benzylidene-aniline derivatives as tyrosinase inhibitors, European Journal of Medicinal Chemistry 57 (2012) 383–390.

[97] L. Kolbe, T. Mann, W. Gerwat, et al., Stab F, 4-n-butylresorcinol, a highly effective tyrosinase inhibitor of the topical treatment of hyperpigmentation, JEADV 27 (2013) 19–23.

[98] R. Bentley, From miso, sake and shoyu to cosmetics: A century of science for kojic acid. Natural Product Reports 23 (2006) 1046–1062.

[99] V. Kahn, N. Ben-Shalom, V. Zakin, Effect of kojic acid on the oxidation of N-acetyldopamine by mushroom tyrosinase, Journal of Agricultural and Food Chemistry 45 (1997) 4460–4465.

[100] A.F. Lajis, M. Hamid, A.B. Ariff, Depigmenting effect of kojic acid esters in hyperpigmented B16F1 melanoma cells, Journal of Biomedicine and Biotechnology 2012 (2012) 952452.

[101] I. Tomita, K. Mitsuhashi, T. Endo, Synthesis and radical polymerization of styrene derivative bearing Kojic acid moieties, Journal of Polymer Science. A1 34 (1996) 271–276.

[102] J.J. Molenda, M.A. Basinger, T.P. Hanusa, M.M. Jones, Synthesis and iron (III) binding properties of 3-hydroxypyrid-4-ones derived from Kojic acid, Journal of Inorganic Biochemistry 55 (1994) 131–146.

[103] H. Kim, J. Choi, J.K. Cho, et al., Solid-phase synthesis of kojic acidtripeptides and their tyrosinase inhibitory activity, storage stability, and toxicity, Bioorganic & Medicinal Chemistry Letter 14 (2004) 2843–2846.

[104] L. Saghaie, M. Pourfarzam, A. Fassihi, B. Sartippour, Synthesis and tyrosinase inhibitory properties of some novel derivatives of kojic acid, Research in Pharmaceutical Sciences 8 (2013) 233–242.

[105] H.S. Rho, H.S. Baek, S.M. Ahn, et al., Synthesis of new antimelanogenic compounds containing two molecules of kojic acid, Bulletin of the Korean Chemical Society 29 (2008) 1569–1571.

[106] J.M. Noh, S.Y. Kwak, H.S. Seo, et al., Kojic acid-amino acid coujugates as tyrosinase inhibitors. Bioorganic & Medicinal Chemistry Letter 19 (2009) 5586–5589.

[107] G.B. Bubols, D.R. Vianna, A. Medina-Remon, et al., The antioxidant activity of coumarins and flavonoids, Mini-Reviews in Medicinal Chemistry 13 (2013) 318–334.

[108] Y. Masamoto, H. Ando, Y. Murata, et al., Mushroom tyrosinase inhibitory activity of esculetin isolated from seeds of *Euphorbia lathyris* L, Bioscience, Biotechnology, and Biochemistry 67 (2003) 631–634.

[109] X.L. Piao, S.H. Baek, M.K. Park, J.H. Park, Tyrosinase-inhibitory furanocoumarin from *Angelica dahurica*, Biological and Pharmaceutical Bulletin 27 (2004) 1144–1146.

[110] V.U. Ahmad, F. Ullah, J. Hussain, et al., Tyrosinase inhibitors from *Rhododendron collettianum* and their structure-activity relationship (SAR) studies, Chemical and Pharmaceutical Bulletin 52 (2004) 1458–1461.

[111] K. Jones, J. Hughes, M. Hong, et al., Modulation of melanogenesis by aloesin: A competitive inhibitor of tyrosinase, Pigment Cell & Melanoma Research 15 (2002) 335–340.

[112] S. Choi, S.K. Lee, J.E. Kim, et al., Aloesin inhibits hyperpigmentation induced by UV radiation, Clinical and Experimental Dermatology 27 (2002) 513–515.

[113] H.W. Duckworth, J.E. Coleman, Physicochemical and kinetic properties of mushroom tyrosinase, Journal of Biological Chemistry 245 (1970) 1613–1625.

[114] S. Maghsoudi, H. Adibi, M. Hanzeh, et al., Kinetic of mushroom tyrosinase inhibition by benzaldehyde derivarives, Journal of Reports in Pharmaceutical Sciences 2 (2013) 156–164.

[115] I. Kubo, Q.X. Chen, K. Nihei, et al., Tyrosinase inhibition kinetics of anisic acid, Zeitschrift für Naturforschung C 58 (2003) 713–718.

[116] T.J. Ha, S. Tamura, I. Kubo, Effects of mushroom tyrosinase on anisaldehyde, Journal of Agricultural and Food Chemistry 53 (2005) 7024–7028.

[117] Y. Shi, Q.X. Chen, Q. Wang, et al., Inhibitory effects of cinnamic acid and its derivatives on the diphenolase activity of mushroom (*Agaricus bisporus*) tyrosinase, Food Chemistry 92 (2005) 707–712.

[118] H.S. Lee, Tyrosinase inhibitors of *Pulsatilla cernua* root-derived materials, Journal of Agricultural and Food Chemistry 50 (2002) 1400–1403.

[119] M. Miyazawa, T. Oshima, K. Koshino, et al., Tyrosinase inhibitor from black rice bran, Journal of Agricultural and Food Chemistry 51 (2003) 6953–6956.

[120] I. Kubo, I. Kinst-Hori, Y. Kubo, et al., Molecular design of antibrowning agents, Journal of Agricultural and Food Chemistry 48 (2000) 1393–1399.

[121] H.J. Jeon, M. Noda, M. Maruyama, et al., Identification and kinetic study of tyrosinase inhibitors found in sake lees, Journal of Agricultural and Food Chemistry 54 (2006) 9827–9833.

[122] A.A. Magid, L. Voutguenne-Nazabadioko, G. Bontemps, et al., Tyrosinase inhibitors and sesquiterpene diglycosides from *Guioa villosa*, Planta Medica 74 (2008) 55–60.

[123] T. Masuda, Y. Odaka, N. Ogawa, et al., Identification of geranic acid, a tyrosinase inhibitor in lemongrass (*Cymbopogon citratus*), Journal of Agricultural and Food Chemistry 56 (2008) 597–601.

[124] T. Sabudak, H.T.M. Khan, M.I. Choudhary, S. Oksuz, Potent tyrosinase inhibitors from *Trifolium balansae*, Natural Product Research 20 (2006) 665–670.

[125] M.T. Khan, S.B. Khan, A. Ather, Tyrosinase inhibitory cycloartane type triterpenoids from the methanol extract of the whole plant of *Amberboa ramosa* Jafri and their structure-activity relationship, Bioorganic & Medicinal Chemistry 14 (2006) 938–943.

[126] F. Ullah, H. Hussain, J. Hussain, et al., Tyrosinase inhibitory pentacyclic triterpenes and analgesic and spasmolytic activities of methanol extracts of *Rhododendron collettianum*, Phytotherapy Research 21 (2007) 1076–1081.

[127] Y.L. Leu, T.L. Hwang, J.W. Hu, J.Y. Fang, Anthraquinones from *Polygonum cuspidatum* as tyrosinase inhibitors for dermal use, Phytotherapy Research 22 (2008) 552–556.

[128] Azhar-Ul-Haq, A. Malik, M.T. Khan, et al., Tyrosinase inhibitory lignans from the methanol extract of the roots of *Vitex negundo* Linn, and their structure-activity relationship, Phytomedicine 13 (2006) 255–260.

[129] H.M. Wang, C.Y. Chen, C.Y. Chen, et al., (-)-N-Formylanonaine from *Michelia alba* as a human tyrosinase inhibitor and antioxidant, Bioorganic & Medicinal Chemistry 18 (2010) 5241–5247.

[130] B. Wu, X. Zhang, X. Wu, New lignan glucosides with tyrosinase inhibitory activities from exocarp of *Castanea henryi*, Carbohydrate Research 355 (2012) 45–49.

[131] T. Akihisa, M. Orido, H. Akazawa, et al., Melanogenesis-inhibitory activity of aromatic glycosides from the stem bark of *Acer buergerianum*, Chemistry and Biodiversity 10 (2013) 167–175.

[132] N.M. Hashim, M. Rahmani, G.C. Ee, et al., Antioxidant, antimicrobial and tyrosinase inhibitory activities of xanthones isolated from *Artocarpus obtusus* F.M. Jarrett, Molecules 17 (2012) 6071–6082.

CHAPTER FIVE

Drug design of tyrosinase inhibitors

Francesco Melfi[a], Simone Carradori[a,*], Arianna Granese[b], Amar Osmanović[c], and Cristina Campestre[a]

[a]Department of Pharmacy, G. d'Annunzio University of Chieti-Pescara, Chieti, Italy
[b]Department of Drug Chemistry and Technology, "Sapienza" University of Rome, Rome, Italy
[c]Faculty of Pharmacy, University of Sarajevo, Sarajevo, Bosnia and Herzegovina
*Corresponding author. e-mail address: simone.carradori@unich.it

Contents

Abstract

This copper-containing enzyme catalyzes the rate-limiting step for the melanin skin pigment bioproduction. Tyrosinase inhibitors can be exploited as skin whitening agents and food preservatives, opening new scenarios in food, cosmetics, agriculture and medicine. Despite the availability of natural inhibitors (hydroquinone, α-arbutin, kojic acid, retinoids, azelaic acid, resveratrol, caftaric acid, valonea tannin, chrysosplenetin and phenylethyl resorcinol), several synthetic compounds were proposed to overcome side effects and to improve the efficacy of natural agents. This chapter will gather the recent advances about synthetic tyrosinase inhibitors from the MedChem perspective, providing new suggestions for the scaffold-based design of innovative compounds.

1. Introduction

Tyrosinase (EC 1.14.18.1) is a binuclear copper enzyme that is multifunctional and widely distributed across nature. It is crucial for melanin biosynthesis, the primary pigment found in bacteria, fungi, plants, and animals [1,2]. Tyrosinases from various species are type-3 metalloenzymes

ISSN 1874-6047, https://doi.org/10.1016/bs.enz.2024.06.001

with a conserved catalytic domain containing six histidine residues and two copper ions (Cu-A and Cu-B), as revealed by X-ray structures from *Bacillus megaterium* (*Bm*Tyr) [3], *Streptomyces castaneoglobisporus* (*Sc*Tyr), *Aspergillus oryzae*, *Agaricus bisporus* [4], and *Juglans regia* [5]. *Bm*Tyr shares approximately 42% identity with mushroom tyrosinase (*m*Tyr). Despite differences in second shell residues among tyrosinases, their overall structure, binding site, and catalytic mechanism are similar [6,7]. The binuclear copper site resides at the bottom of a spacious cavity, central to two pairs of antiparallel α-helices. Cu-A is coordinated by His61, His85, and His94, while Cu-B is coordinated by His259, His263, and His296. The distance between Cu-A and Cu-B is 4.5 $\pm$ 0.2 Å, with a bridging water molecule or hydroxyl ion at 2.65 $\pm$ 0.2 Å to both coppers [4].

Natural substrates are believed to bind to the active site as phenolate, implying an enzyme-mediated deprotonation mechanism [8]. Numerous natural compounds act as tyrosinase inhibitors, including tropolone, kojic acid (KA), hydroquinone, arbutin, and ascorbic acid [9]. Human tyrosinase (*h*Tyr), a glycosylated membrane-bound protein, differs significantly from the tetrameric *m*Tyr in inhibitory activities. Many inhibitors effective against *m*Tyr are less potent or inactive against *h*Tyr. For instance, KA binds strongly to *m*Tyr (K_i = 4.3 μM) but weakly to *h*Tyr (K_i = 350 μM), and aesculetin, effective against *m*Tyr, shows no detectable activity against *h*Tyr. Other inhibitors, such as phenylthiourea, L-mimosine, cinnamic and benzoic acids, also exhibit significant differences in inhibitory efficacy between *m*Tyr and *h*Tyr [10].

Nowadays, it is assumed that human tyrosinase represents the ideal target to refer for the design and the development of chemical agents able to modulate melanin synthesis and treat browning skin phenomena that can grievously influence human being psychology, principally in some ethnic groups from Asia, Africa, and Middle East, where skin whitening practice is hugely common because of the social and cultural aspects that recognize lighter skin as a beauty and health condition [11,12].

In truth, human tyrosinase is costly and its high scale production is tough, due to the process of mammal-specific post-translational maturation. Indeed, a variety of both synthetic and natural molecules were tested on the cheapest and the most available mushroom tyrosinase from *A. bisporus* (*ab*Tyr), as well as the majority of the tyrosinase inhibitors frequently used as positive reference compounds (such as kojic acid, hydroquinone, arbutin). Unfortunately, these compounds showed lower efficacy against human tyrosinase, because of the structural differences in the interaction motifs of the active site [13].

The aim to produce new tyrosinase inhibitors is not only linked to dermo-cosmetic field, in which the available depigmentation agents are used in high concentrations, coming up against risks of side effects [14], but it is also recognized the central role of melanin in the resistance of melanoma to classical anticancer therapies. Thus, human tyrosinase inhibitors could also restore the sensitivity to radio- and/or chemotherapy [15,16].

The aim of this chapter is to gather the advances in the past two years about synthetic tyrosinase inhibitors from the MedChem perspective as well as insights from the crystallographic structures reported in the Protein Data Bank, providing new suggestions for the scaffold-based design of innovative compounds, despite the availability of well-recognized natural compounds (e.g., hydroquinone, α-arbutin, kojic acid, retinoids, azelaic acid, resveratrol, caftaric acid, valonea tannin, chrysosplenetin and phenylethyl resorcinol) [17].

2. Recent compounds tested on human and mushroom tyrosinases

2.1 Heterocyclic derivatives

The (E)-2-benzylidene-2,3-dihydro-1H-inden-1-one scaffold seems to be one of the most promising cores in novel human tyrosinase inhibitors (Fig. 1). The design of these derivatives, characterized by the heterocyclic nucleus directly connected to resorcinol was based on the structure of two well-known human tyrosinase inhibitors: rucinol (IC$_{50}$ = 21 μM) and thiamidol (K_i = 0.25 μM). With the aim to obtain more potent molecules, indanone was connected to resorcinol, through a vinyl linker, to mimic, structurally, the thiamidol shape. The first synthesized compound of this series, that resulted active in the inhibition of human tyrosinase, was the one bearing a hydroxy function as R, by showing a K_i = 0.25 μM, being as potent as thiamidol. Unfortunately, at the same time, it was less active in a cellular context, leading researchers to improve its structure [18,19]. Given that, several derivatives were synthesized, leading to a deeper comprehension of the biological activity of the indanone core. Indeed, electron-donating groups increased the inhibitory activity of the molecule, thoroughly the best results were given when R was a substituted amide, as NHAc (IC$_{50}$ = 0.14 ± 0.01 μM), NHCOEt (IC$_{50}$ = 0.15 ± 0.02 μM), NHCOPr (IC$_{50}$ = 0.18 ± 0.04 μM), NHCOiPr (IC$_{50}$ = 0.35 ± 0.06 μM), and NH-L-Ala (IC$_{50}$ = 0.85 ± 0.01 μM). Moreover, the primary amine (IC$_{50}$ = 0.72 ± 0.08 μM) and the N-methylamine (IC$_{50}$ = 0.89 ± 0.05 μM)

(E)-2-benzylidene-2,3-dihydro-1H-inden-1-one derivatives

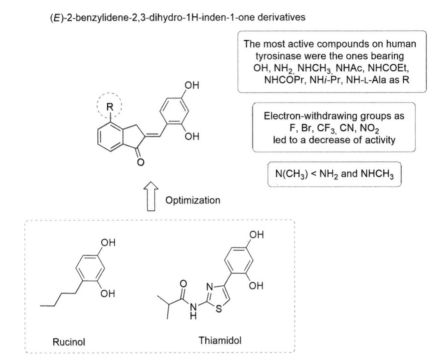

Fig. 1 Indanone derivatives.

derivatives showed good results as well, but it is interesting to note that the presence of N,N-dimethylamine as R led to a huge decrease of activity, with IC$_{50}$ value up to 24 μM, suggesting the importance of at least one H able to make a hydrogen bond with the carboxylate function of Glu203, as predicted in the molecular modeling. Moreover, predicted binding modes also suggested that the small aliphatic portion of the amide group can fit in a hydrophobic subpocket, nearby to Glu203. Instead, electron-withdrawing groups as F, Br, CF$_3$, CN, and NO$_2$ decreased the potency of indanone, resulting mostly inactive. All these compounds were tested on mushroom tyrosinase as well, with IC$_{50}$ values less than 0.6 μM, thus highlighting the structural and biological diversity between human and mushroom tyrosinase enzymes. Indanone derivatives were also tested as suppressor of melanogenesis in MNT-1 human melanoma cells for 4 and 14 days, recording results similar or even better, especially for amide derivatives, with respect to kojic acid and thiamidol. Derivatives with NH$_2$ (IC$_{50}$ = 3.8 μM for 4-days experiment and 0.53 μM for 14-days experiment) and NHCH$_3$ (IC$_{50}$ = 3.2 μM for 4-days experiment and 0.77 μM for 14-days experiment) effectively inhibited melanin production in melanoma cells [20].

In a previous study of synthesized inhibitors, among the 1H-indol-3-yl-2-(4-benzylpiperidin-1-yl)propan-1-one derivatives, the most potent compound was 1-(5,6-dimethoxy-1H-indol-3-yl)-2-(4-(4-fluorobenzyl) piperidin-1-yl)propan-1-one with the monophenolase activity of IC_{50} = 5.11 ± 0.36 µM and diphenolase activity of IC_{50} = 7.56 ± 1.90 µM, which is higher than reference compound KA. The (4-fluorobenzyl) piperidine moiety of this compound is positioned within a cavity, where it forms π-π interactions with the crucial His263 residue, cation-π interactions with His244, and van der Waals interactions with Val283. The fluorine atom is located between the copper ions, while the indole nucleus of the compound occupies the binding area near the entrance of the catalytic pocket cavity. The binding pose of the indole core is characterized by several specific contacts: the two oxygen atoms of the methoxy groups form hydrogen bonds with Asn81, and the indole NH group establishes an additional hydrogen bond with the carbonyl oxygen of the His85 backbone. The compound's piperidine ring is positioned in a lipophilic sandwich between Val248 and Val283. Docking studies suggest that the 4-fluorobenzyl portion of the compound overlaps with the inhibitor tropolone in the catalytic site of mushroom tyrosinase, sharing key interactions. Preliminary results of this structure in complex with bacterial tyrosinase support the predicted positioning of the 4-fluorobenzyl moiety within mTyr, where His263 corresponds to His208 in BmTyr [21].

Another scaffold that showed anti-melanogenic activity was (Z)-2-(benzylamino)-5-benzylidenethiazol-4(5H)-one, based on the structure of the molecule MHY2081 (Fig. 2), that showed inhibitory activity against mushroom tyrosinase (IC_{50} = 1.80 µM) [22]. Lee et al. inserted several substituents on ring A, like hydroxy, methoxy, ethoxy, and fluorine in the different positions (R^1-R^4), finding out that the most potent compounds in the inhibition of mushroom tyrosinase activity were **1**, characterized by two hydroxy functions as R^1 and R^3, IC_{50} = 0.27 ± 0.03 µM, and **2**, characterized by a hydroxy group as R^2 and a methoxy as R^3, IC_{50} = 10.0 ± 0.90 µM; whereas the other combinations of substituents gave poorer results up to 100 µM. Compound **1** exhibited no significant cytotoxicity in B16F10 cells (mouse melanoma) at concentrations of 20 µM, thus its ability to reduce melanin production of B16F10 cells was investigated as well. Kojic acid was taken as positive control, and 20 µM of it reduced extracellular melanin levels by 1.5-fold, compared to those in the untreated control, whereas **1** decreased extracellular melanin levels enhanced by stimulators like α-melanocyte-stimulating hormone (α-MSH)

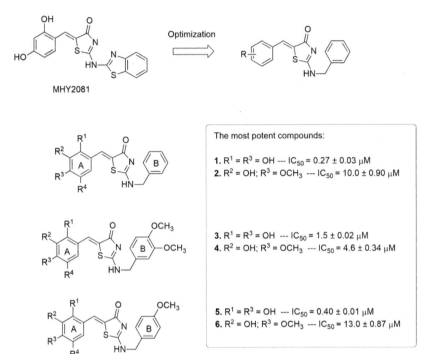

Fig. 2 (Z)-2-(benzylamino)-5-benzylidenethiazol-4(5H)-one derivatives.

and 3-isobutyl-1-methylxanthine (IBMX) to 1.22-, 1.03, and 1.00-fold at 5, 10, and 20 μM, respectively, confirming the ability to reduce melanin biosynthesis in a dose-dependent manner. Moreover, at concentration of 10 μM, compound **1** restored melanin levels that were increased by α-MSH and IBMX, showing a strong anti-melanogenic effect, stronger than kojic acid.

Other (Z)-2-(benzylamino)-5-benzylidenethiazol-4(5H)-one derivatives were synthesized as tyrosinase inhibitors, with methoxy substituents on ring B. It is interesting to note that in the case of 3,4-dimethoxybenzene (ring B) [23], the tolerated substitutions were the same of compounds **1** and **2**, while it was different when the molecule had just a methoxy in *para* position (Fig. 2) [24]. When tested on *ab*Tyr, compound **3**, with two hydroxy groups as R^1 and R^3 exhibited an IC_{50} of 1.5 ± 0.02 μM, while compound **4**, with a hydroxy group as R^2 and a methoxy function as R^3 resulted less potent, with an $IC_{50} = 4.6 ± 0.34$ μM (Fig. 2). Additionally, compounds **3** and **4** at 20 μM reduced the extracellular melanin contents by

1.27- and 1.19-fold, respectively, inhibiting melanin production more potently than kojic acid at 20 μM. The same results were given when there was only one *para* methoxy moiety as a substituent in ring B; in fact compound **5**, with two hydroxy groups, showed an IC_{50} value equal to 0.4 ± 0.01 μM, while compound **6** with a hydroxy group as R^1 and a methoxy as R^3 showed an IC_{50} equal to 13.0 ± 0.87 μM (Fig. 2). It is interesting to note that **5**, at 20 μM exhibited cellular tyrosinase activity like 20 μM of kojic acid, while **6**, despite showed a bigger IC_{50}, inhibited cellular tyrosinase activity to 44%, more strongly than kojic acid.

In compounds **7-9** the benzene ring is directly connected to the thiazole (Fig. 3) [25]. The most potent compound among them is **7**, characterized by two hydroxy groups in positions 1 and 3, with an IC_{50} equal to 0.1 ± 0.01 μM on mushroom tyrosinase. The presence of a single hydroxy function in position 3 led to a decrease of activity (**8** IC_{50} = 6.4 ± 0.52 μM) as well as the introduction of two hydroxy in positions 2 and 3 (**9** IC_{50} = 5.2 ± 0.32 μM). If on one hand compounds **7** and **8** did not exhibit any perceptible cytotoxicity at 20 μM, on the other hand derivative **9** showed the most potent inhibitory activity, but at the same time, it exhibited cytotoxicity in B16F10 cells at concentrations of 2 μM. Moreover, **7** and **8** were tested on B16F10 cells to evaluate the impact on

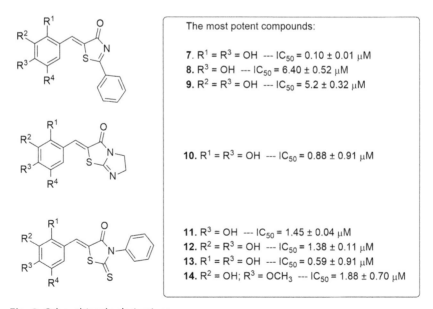

The most potent compounds:

7. $R^1 = R^3 = OH$ --- $IC_{50} = 0.10 ± 0.01$ μM
8. $R^3 = OH$ --- $IC_{50} = 6.40 ± 0.52$ μM
9. $R^2 = R^3 = OH$ --- $IC_{50} = 5.2 ± 0.32$ μM

10. $R^1 = R^3 = OH$ --- $IC_{50} = 0.88 ± 0.91$ μM

11. $R^3 = OH$ --- $IC_{50} = 1.45 ± 0.04$ μM
12. $R^2 = R^3 = OH$ --- $IC_{50} = 1.38 ± 0.11$ μM
13. $R^1 = R^3 = OH$ --- $IC_{50} = 0.59 ± 0.91$ μM
14. $R^2 = OH; R^3 = OCH_3$ --- $IC_{50} = 1.88 ± 0.70$ μM

Fig. 3 Other thiazole derivatives.

cellular tyrosinase activity. B16F10 were treated with the stimulators IBMX and α-MSH, increasing the activity of cellular tyrosinase by 3.3-fold, compared to the untreated ones; 20 μM of kojic acid decreased the enhanced tyrosinase activity by 2.4-fold, while compound **7** reached the same results of kojic acid at concentrations of 5 μM and derivative **8** displayed a similar profile of kojic acid.

It was also investigated the 5,6-dihydroimidazo[2,1-*b*]thiazol-3(2H)-one scaffold as mushroom tyrosinase inhibitor (Fig. 3) [26]. As the previous compounds, it is not surprising that the best compound was the one bearing two hydroxy groups in positions R^1 and R^3 (**10** IC$_{50}$ = 0.88 ± 0.91 μM), while the other combinations, or the introduction of methoxy groups and/or halogens led to a bigger increase of IC$_{50}$ values, suggesting that the adoption of this bicycle is not the first choice for tyrosinase inhibitors.

The substitution of the thiazole with a thioxothiazolidinone (**11-14**; Fig. 3) did not change a lot the inhibitory profile on mushroom tyrosinase [27]. In fact, the most potent compound was **13**, bearing two hydroxy functions in positions 1 and 3, with IC$_{50}$ = 0.59 ± 0.91 μM, while the other derivatives presented IC$_{50}$ values between 1.38 and 1.88 μM. *In vitro* tests demonstrated that **11-13** inhibited both cellular tyrosinase and melanin production on co-stimulated B16F10 cells with IBMX and α-MSH, more than kojic acid, with no perceptible cytotoxicity. Surprisingly, **12** inhibited the activity of B16F10 more than derivative **13**, and 5 μM of **12** were sufficient to duplicate the inhibition of melanin production of 20 μM of kojic acid.

Indeed, the scaffold thiazolidine-2-imine was taken into account to synthesize two mushroom tyrosinase inhibitors (Fig. 4A) [28]. Derivatives **15** and **16** presented a *para*-substituted phenyl ring linked to the imine group. The electronic features of the substituents influenced the activity of the two derivatives; in fact, **15**, with the electron-donating methyl group, showed an IC$_{50}$ = 1.15 μM, while **16**, with the electron-withdrawing bromine, exhibited a slight increase of IC$_{50}$ value (2.08 μM), suggesting that, for this scaffold, it was preferable to have electron-donating substitutes. Moreover, the structure of these derivatives is branched out, because all the components of the scaffold, when possible, presents a substitution like a vinyl benzene or aliphatic groups such as *tert*-butyl and propyl chains that surely contribute to make hydrophobic interactions.

Thiazolidine derivatives were tested on mushroom tyrosinase as well [29]. Compounds **17** and **18** are characterized by a thiazolidine directly

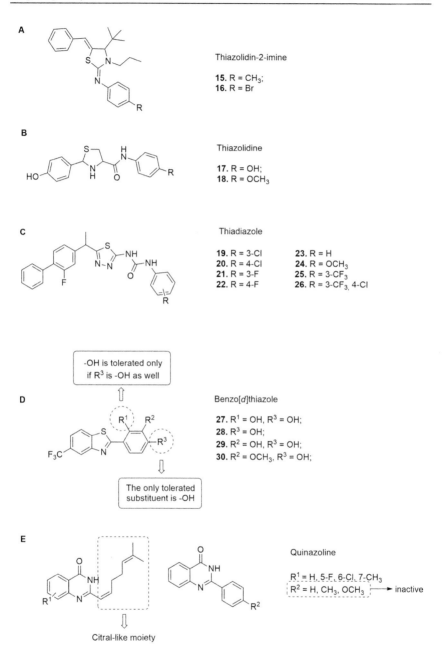

Fig. 4 (A) Thiazolidin-2-imine derivatives; (B) thiazolidine derivatives; (C) thiadiazole derivatives; (D) benzo[d]thiazole derivatives; (E) quinazoline derivatives.

connected to a phenol and a substituted benzene ring via an amide bridge (Fig. 4B). When tested on mushroom tyrosinase, they showed mid results in the inhibition of the enzyme, with IC_{50} values of 16.5 ± 0.37 and 44.8 ± 6.20 µM, respectively.

Compounds **19-26** present a central thiadiazole linked to a phenyl-fluorobenzene and a substituted benzene ring, connected to the thiadiazole through a urea linker (Fig. 4C) [30]. These derivatives did not show great results in the inhibition of mushroom tyrosinase, indeed the most potent derivative is **19**, with IC_{50} = 68 µM, while compounds **20** and **24-26** displayed values between 114 and 143 µM. Conversely, **21** and **22**, with a fluorine, and **23**, with no substitutions, displayed IC_{50} values up to 500 µM, resulting completely inactive. Moreover, these derivatives exhibited a huge cytotoxicity behavior, when tested on cells.

Benzothiazole has been considered in the past years for the synthesis of tyrosinase inhibitors, and recently, Hwang et al. designed and tested on *ab*Tyr benzothiazole-based molecules (Fig. 4D) [31]. Compounds **27-30** gave different results; interestingly if **27** was the most potent derivative, with IC_{50} value of 0.2 ± 0.01 µM, the other derivatives displayed weaker activity. Compound **28**, with a single hydroxy group as R^3 showed 54.2 µM as IC_{50} value, while the results for **29** and **30** were up to 300 µM. Thus, the hydroxy group was tolerated as R^1 only if R^3 is a hydroxy as well, and the only allowed substituent for R^3 is a hydroxy function too. They also tried to introduce bromine and methyl as substituents, but the results were not appreciable. Moreover, inhibitory effect on cellular tyrosinase activity of **27** and **28** was tested, by resulting similar. In detail, 10 µM of these two derivatives, individually tested, showed the same results of 20 µM of kojic acid on B16F10 cells, co-treated with IBMX and α-MSH to increase cellular tyrosinase activities.

Quinazolinone derivatives have been reported in the past as tyrosinase inhibitors and the analogue quinazoline scaffold has shown inhibitory activity on different enzymes, like glucuronidase, thymidine phosphorylase, xanthine oxidase, phosphodiesterase I, carbonic anhydrase II, and so on [32–34]. Huang et al. synthesized some quinazoline derivatives, substituted in position 2 with a citral moiety or a benzene ring, and tested them on mushroom tyrosinase (Fig. 4E) [35]. Among these derivatives, there are quinazoline-citral hybrids that displayed IC_{50} values between 105 and 255 µM, while the 2-phenyl quinazoline did not show any inhibitory activity.

Among the nitrogen-containing heterocyclic compounds, triazole is widely used in medicinal chemistry for antimicrobial, antiviral, anti-tubercular, anticancer activities and so on [36]. Hassan et al. synthesized a

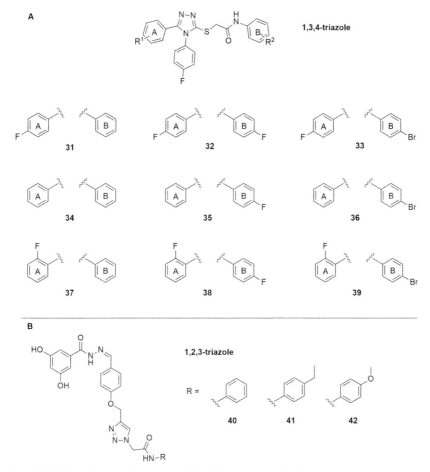

Fig. 5 (A) 1,3,4-triazole derivatives; (B) 1,2,3-triazole derivatives.

series of 1,3,4-triazole derivatives as mushroom tyrosinase inhibitors (Fig. 5A) [37]. The modifications on these molecules regarded the two opposite rings, that we designed as A and B. The results exhibited by these molecules were interesting. For compounds with a fluorine in *para* in ring A (**31-33**), only **31** exhibited good results in the inhibition of mushroom tyrosinase, with IC$_{50}$ value of $0.124 \pm 0.077\,\mu M$, whereas when ring B presented a fluorine (**32**) or bromine (**33**) in *para*, the derivatives did not exhibit any activity. Compounds **34-36** were characterized by no substitution on ring A and they exhibited similar activity (IC$_{50}$ = 0.142–$0.379\,\mu M$). The most captivating results were given by compounds **37-39**, characterized by an *ortho*-fluorobenzene as ring A. In fact, **37**, with no substituents in ring B showed excellent results in the inhibition of *ab*Tyr (IC_{50} = 0.111 ± 0.021 μM), and **38**, with a fluorine in *para* in ring B, resulted to be the most potent derivative of the class, with IC_{50} equal to 98 ± 9 nM. Compound **39**, presenting a bromine in *para* on ring B, did not exhibit any appreciable inhibitory data as the fluorine-containing **38**.

Despite the 1,2,3-triazole derivatives (Fig. 5B), synthesized by Bagheri et al., presented a resorcinol structure, well-recognized as tyrosinase inhibitor, compounds **40-42** did not show great results in the inhibition of mushroom tyrosinase. In detail, **40** and **41** showed IC_{50} values of 90.5 and 87.5 μM. Conversely **42**, despite being structurally similar to **41**, exhibited better results, with IC_{50} = 55.4 ± 4.9 μM [38].

Very recently, novel 2-aminothiazole-oxadiazole bearing N-arylated butanamides (**43-50**; Fig. 6) were tested on mushroom tyrosinase [39]. In general, these compounds exhibited good results on *ab*Tyr. The most active compounds were **48** and **50** with IC_{50} values of 31.1 ± 6.1 nM and 34.9 ± 9.8 nM, respectively, suggesting that the aryl portion was most tolerated in a single *ortho* position, while two methyl groups in *ortho* position (**46** IC_{50} = 1.61 ± 0.87 μM) and one methyl in *meta* position

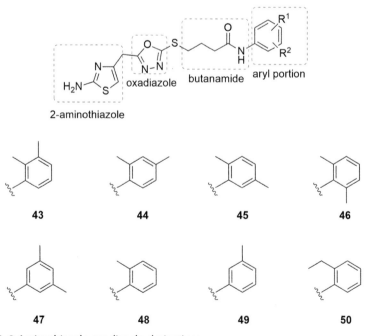

Fig. 6 2-Aminothiazole-oxadiazole derivatives.

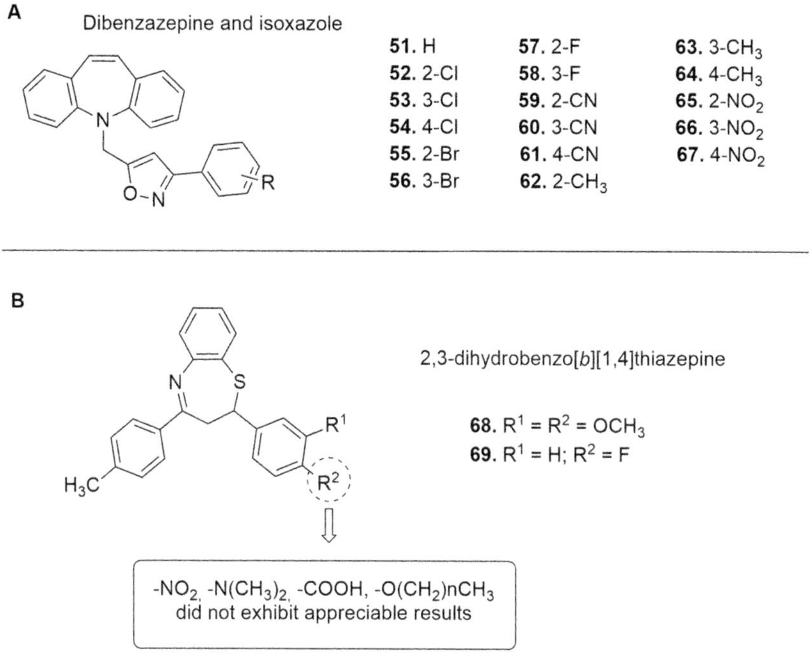

A

Dibenzazepine and isoxazole

51. H	**57.** 2-F	**63.** 3-CH$_3$
52. 2-Cl	**58.** 3-F	**64.** 4-CH$_3$
53. 3-Cl	**59.** 2-CN	**65.** 2-NO$_2$
54. 4-Cl	**60.** 3-CN	**66.** 3-NO$_2$
55. 2-Br	**61.** 4-CN	**67.** 4-NO$_2$
56. 3-Br	**62.** 2-CH$_3$	

B

2,3-dihydrobenzo[*b*][1,4]thiazepine

68. R^1 = R^2 = OCH$_3$
69. R^1 = H; R^2 = F

-NO$_2$, -N(CH$_3$)$_2$, -COOH, -O(CH$_2$)nCH$_3$
did not exhibit appreciable results

Fig. 7 (A) Dibenzazepine derivatives; (B) 2,3-dihydrobenzo[*b*][1,4]thiazepine.

(**49** IC$_{50}$ = 1.01 ± 0.61 μM) led to a slight decrease of the activity. Compounds **43–45** and **47** displayed a good inhibitory profile, with IC$_{50}$ values between 0.20 and 0.75 μM.

Several isoxazole analogues of dibenzazepine were synthesized and tested on mushroom tyrosinase (**51–67**; Fig. 7A) [40]. These compounds were characterized by a phenyl ring, directed connected to an isoxazole, differently substituted, except for **51**, with halogens (**52–58**) and other electron–withdrawing groups like cyano (**59–61**), and nitro (**65–67**), or with the electron–donating methyl function (**62–63**). The unsubstituted compound **51** did not exhibit inhibitory activity on mushroom tyrosinase, while, among the chlorine derivatives, **52** showed an IC$_{50}$ of 7.35 ± 0.89 μM and **54** 4.32 ± 0.31 μM, thus resulting the most potent compounds. On the contrary, when chlorine was in position *meta*, like in **53**, it did not show any appreciable result. For bromine derivatives, *ortho* bromine **55** was almost 1-fold most potent on mushroom tyrosinase than its *meta* regioisomer **56** (IC$_{50}$ = 6.38 ± 1.06 μM and 12.36 ± 0.61 μM, respectively). 2-Fluorine derivative, **57**, did not exhibit considerable result, while 3-fluorine, **58**, exhibited an IC$_{50}$ value of 29.30 ± 0.48 μM. As well as for cyano and methyl derivatives, *meta* position (**60** and **63**) was not convenient, while *ortho* (**59** IC_{50} = 10.35 ± 0.61 µM, **62** IC_{50} = 9.18 ± 1.36 µM) and *para* positions (**61** IC_{50} = 8.62 ± 0.67 µM, **64** IC_{50} = 41.08 ± 0.61 µM) provided some appreciable results. Lastly, independently of the position, nitro (**65-67**) derivatives did not display any considerable data.

From the studies conducted by Al-Rooqi et al., 1,5-benzothiazepine derivatives gave promising results in the inhibition of *ab*Tyr activity (Fig. 7B) [41]. Indeed, compounds **68** and **69** showed IC_{50} of 1.21 and 1.34 µM, respectively. Other R^2 substitutents were tested, like nitro, dimethylamine, carboxylic acid, and alkyl–ethers, but these latter derivatives were not capable of efficiently inhibiting mushroom tyrosinase.

Some 3-hydroxypyridin-4-one derivatives were synthesized to inhibit mushroom tyrosine (**70-79**; Fig. 8). [42]. Structurally, this scaffold mimicked kojic acid, that is a well-known tyrosinase inhibitor, but the researchers introduced a benzene ring, an arylidenhydrazide as linker and a terminal substituted benzene or thiophene. In general, these derivatives did not display strong inhibitory activity on mushroom tyrosinase with IC_{50} values between 25.29 and > 100 µM. The most potent compounds were **70**, **73**, and **78**, with IC_{50} of 28.57, 26.36, and 25.29 µM, respectively. It is interesting to note that the introduction in **70** of a methyl group in *ortho* to the nitro function, as in **75**, led to a dramatic increase of IC_{50} up to 100 µM, as well as the substitution of bromine in **78**, with a nitro function, as in **77**.

3-Phenyl-2-cyanomethylpyrrole derivatives (**80-90**; Fig. 9) displayed, in general, good activity when tested on mushroom tyrosinase [43]. The unsubstituted **80** was one of the most potent compounds, with IC_{50} of 7.42 µM, whereas the introduction of halogens gave different results; in fact, the presence of fluorine in *meta* position (**81**) led to a 3-fold increase of IC_{50}, while bromine was in general more tolerated. Compounds **82** and **84**

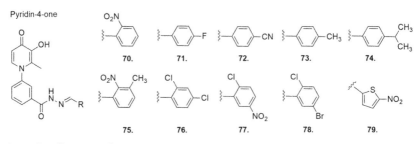

Fig. 8 Pyridin-4-one derivatives.

Pyrrole

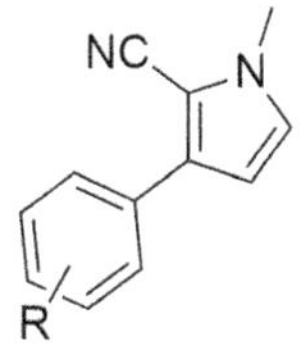

80. H	**84.** 4-Br	**88.** 2,3-diCl
81. 3-F	**85.** 2-OCH$_3$	**89.** 2-Br-6-OCH$_3$
82. 2-Br	**86.** 4-OCH$_3$	**90.** 2-vinyl
83. 3-Br	**87.** 4-CF$_3$	

Fig. 9 Pyrrole derivatives.

(*ortho* and *para* bromine derivatives) exhibited IC$_{50}$ values of 8.47 and 8.17 µM, while *meta* bromo derivative, **83**, showed a decreased activity by 2-fold to its regioisomers. Moreover, the presence of two chlorines, in *ortho* and *meta* positions like **88**, led to an increase of activity, with IC$_{50}$ equal to 4.83 µM. The presence of methoxy was deleterious, especially in *para* position (**85** IC$_{50}$ = 23.57 µM, **86** IC$_{50}$ = 89.15 µM). The introduction of a trifluoromethyl group in *para* (**87**) led to a slight increase of activity with respect to **80**, with IC$_{50}$ of 12.44 µM. Surprisingly the presence of a bromine in position 2 and a methoxy in 6, like **89**, and especially a vinyl function in 2, like **90**, markedly ameliorated the activity of 3-phenyl-2-cyanomethylpyrrole, with IC$_{50}$ of 2.11 and 0.97 µM, respectively. By being the most active molecule, the inhibitory tyrosinase activity of **90** in B16F10 melanoma cells was also evaluated, and at concentrations of 100 µM, it showed effective inhibition of 33.48%, very similar to kojic acid activity (39.81% at 100 µM).

The thiosemicarbazone **91** (Fig. 10), characterized by a terminal pyridine showed IC$_{50}$ = 5.82 µM on mushroom tyrosinase. To examinate the anti-browning potential on fruits and vegetables of this compound, it was tested on fresh-cut potato, in distilled water, over 10 days. From this test, **91** was proven to be a more potent anti-browning agent than kojic acid, with a browning index of 0.292 (browning index of kojic acid = 0.332) [44], whereas 1-(4-(4,4,6-trimethyl-2-thioxo-3,4-dihydropyrimidin-1(2H)-yl)phenyl)ethanone **92** (Fig. 10) was tested on mushroom tyrosinase, by giving IC$_{50}$ of 1.98 µM [45].

2.2 Aromatic derivatives

Mirabile et al. synthesized novel mushroom tyrosinase inhibitors with a central piperazine linked to a phenol and to another substituted benzene though an amide bridge (**93-126**) [46]. The unsubstituted **93** exhibited an IC$_{50}$ = 73.2 ± 2.98 µM, while the more hindered **94**, with a phenyl ring in *para* to the amide function, worsened the activity (IC$_{50}$ = 128.3 ± 4.1 µM). They functionalized the benzene ring with halogens (**95-106**), methyl (**107-110**),

4-Acetylpyridinethiosemicarbazone   Dihydropyrimidine-2(1H)-thione

91   **92**

Fig. 10 Structures of compounds 91 and 92.

trifluoromethyl (**111-114**), methoxy (**115-118**), nitro (**119-122**), or amine (**123-126**) groups in *ortho*, *meta*, *para*, and *ortho* and *para* positions. In general, independently of the moiety, the *ortho* position resulted to be always the best choice to obtain potent compounds. For halogen and nitro derivatives, substitution in both *ortho* and *para* positions led to potent activity over *meta* or *para* positions. Moreover, chlorine derivatives (**99-102**) were always better than fluorine (**95-98**) and bromine (**103-106**) ones. Compound **102** (2,4-difluorine) is the only disubstituted derivative that is more potent than the *ortho* derivative, i.e. **99**. Among the electron-donating derivatives, methoxy compounds were more potent than methyl analogs, and amino derivatives were almost inactive, if compared to the other molecules. Collectively, trifluoromethyl and nitro derivatives exhibited good results (Table 1).

Compound **102** was tested in melanogenesis inhibition in α-MSH stimulated B16F10 cells and it could inhibit synthesis of melanin in a dose-dependent manner: 44% of inhibition at concentration of 1 μM, 73% at 10 μM and 88% at 25 μM. Moreover, until 25 μM it showed no cytotoxic effects in cell viability assay.

In another previous study from the same research group, among the synthesized [4-(4-fluorobenzyl)piperazin-1-yl]methanone derivatives, the most potent compound was (2,4-dinitrophenyl)-[4-[(4-fluorophenyl) methyl]piperazin-1-yl]methanone, with diphenolase activity $IC_{50} = 0.96 \pm 0.21$ μM. The 4-fluorobenzyl moiety of this compound is stabilized by π-π interactions with His208 and is oriented toward Cu-A at a distance of 1.9 Å, similar to tyrosinase substrates and the *m*Tyr model. Additionally, a polar interaction is observed between Arg209 and the oxygen atom of the compound's carbonyl group, akin to the interaction in the compound [4-(4-fluorobenzyl)piperazin-1-yl](2-methylphenyl)methanone reported previously [47]. The new synthesized derivatives, particularly the compound with 2,4-dinitro substitutions, enabled new interactions in the *Bm*Tyr active site. This 2,4-dinitro-substituted compound is further stabilized by

Table 1 Phenol-piperazine derivatives as mushroom tyrosinase inhibitors with IC_{50} values.

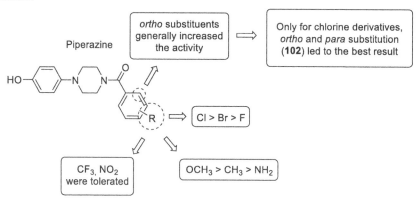

Compound	R	IC_{50} (μM)
93	H	73.2 ± 2.98
94	C_6H_5	128.3 ± 4.1
95	2-F	15.2 ± 0.9
96	3-F	22.6 ± 1.3
97	4-F	21.7 ± 1.2
98	2,4–diF	17.5 ± 1.3
99	2-Cl	2.6 ± 0.3
100	3-Cl	9.0 ± 1.0
101	4-Cl	8.9 ± 0.9
102	2,4–diCl	1.5 ± 0.1
103	2-Br	4.5 ± 0.6
104	3-Br	16.4 ± 1.9
105	4-Br	39.6 ± 0.9
106	2,4–diBr	9.5 ± 0.3
107	2-CH_3	29.9 ± 2.0
108	3-CH_3	31.7 ± 0.5

(continued)

Table 1 Phenol-piperazine derivatives as mushroom tyrosinase inhibitors with IC_{50} values. (cont'd)

109	4-CH$_3$	34.4 ± 0.5
110	2,4-diCH$_3$	33.7 ± 0.7
111	2-CF$_3$	4.6 ± 0.3
112	3-CF$_3$	5.5 ± 0.2
113	4-CF$_3$	35.1 ± 0.5
114	2,4-diCF$_3$	18.2 ± 2.4
115	2-OCH$_3$	3.5 ± 0.2
116	3-OCH$_3$	9.8 ± 0.4
117	4-OCH$_3$	14.9 ± 0.7
118	2,4-diOCH$_3$	16.7 ± 2.0
119	2-NO$_2$	7.4 ± 0.7
120	3-NO$_2$	21.8 ± 1.2
121	4-NO$_2$	23.2 ± 1.5
122	2,4-diNO$_2$	7.6 ± 0.05
123	2-NH$_2$	66.4 ± 2.1
124	3-NH$_2$	35.7 ± 1.05
125	4-NH$_2$	43.9 ± 3.4
126	2,4-diNH$_2$	82.4 ± 1.1

additional π-π interactions between the aroyl ring and Phe197. Moreover, the aroyl ring interacts with the amide group of Asn205. A significant movement of Arg209 is observed, forming a hydrogen bond with the carbonyl group, thus stabilizing the bulky compound in the active site [48].

Finally, Nazir et al. synthesized some acetophenone derivatives as mushroom tyrosinase inhibitors (**127-135**; Fig. 11) [49]. In general, these derivatives, independently of the substituent on the terminal benzene ring, did not exhibit good results in the inhibition of mushroom tyrosinase,

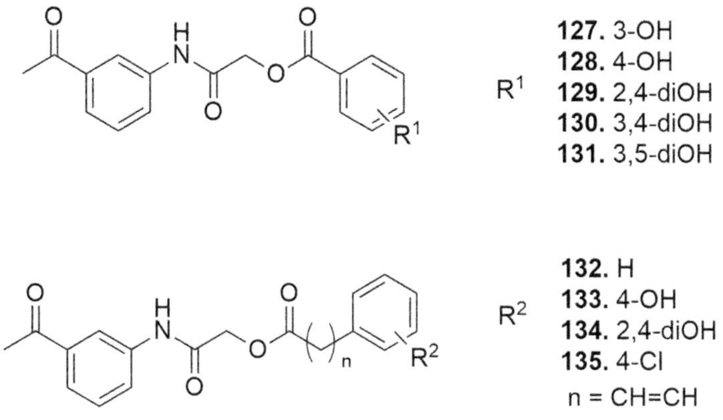
R1
127. 3-OH
128. 4-OH
129. 2,4-diOH
130. 3,4-diOH
131. 3,5-diOH
R2
132. H
133. 4-OH
134. 2,4-diOH
135. 4-Cl
n = CH=CH

Fig. 11 Acetophenone derivatives.

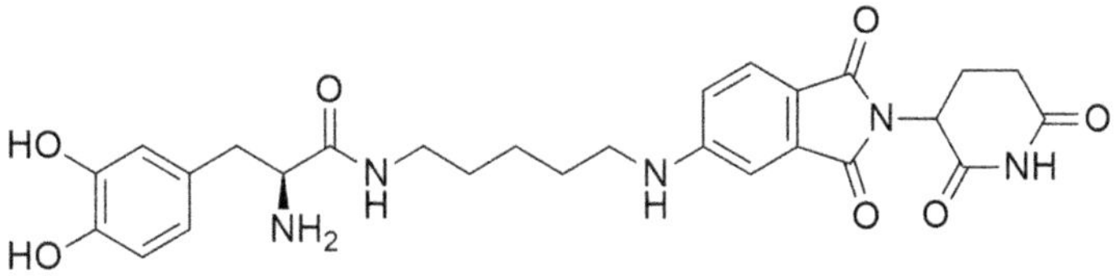
HO
HO
NH2
O
N
H
O
N
O
O
O
NH

136

Fig. 12 Structure of the first PROTAC human tyrosinase inhibitor.

except for **134**, that showed IC_{50} value of 0.002 ± 0.0002 μM. Its analog, **129**, displayed weaker activity (IC_{50} = 27.3 ± 3.6 μM). The other molecules showed IC_{50} values between 60 and 300 μM.

2.3 Tyrosinase-PROTAC hybrids

Fu et al. synthesized the first tyrosinase-PROTAC (PROteolysis TArgeting Chimeras) **136**, by using *L*-DOPA (tyrosinase ligand) and thalidomide as E3 ligase recruiter (Fig. 12) [50,51]. PROTACs are heterobifunctional molecules that bind a specific target and recruit an E3 ligase in order to mark the target for degradation by the proteasome. **136** was tested on human tyrosinase, exhibiting an IC_{50} of 113 μM.

2.4 Amino acid-based derivatives

The peptide scaffold was also used as a strategy to synthesize few novel tyrosinase inhibitors. After the screening of different amino acids able to inhibit the *ortho*-dihydroxyphenolase activity of mushroom tyrosinase, it was reported that *L*-cysteine could completely suppress dopachrome

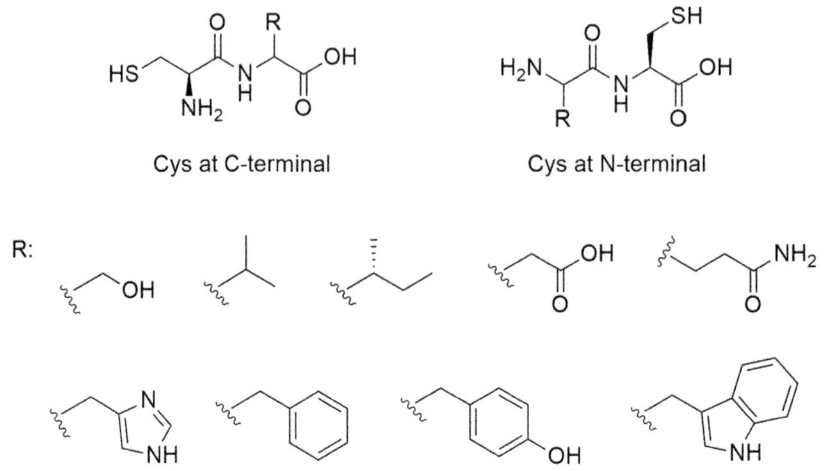

Fig. 13 Cysteine-dipeptide derivatives.

formation at 0.3 mM. In addition, among sulfur-containing di- and tri-peptides, those endowed with N-terminal L-cysteine exhibited more potent inhibitory effects against dopachrome formation, thus corroborating the crucial role of this nucleophilic amino acid in establishing a conjugate with dopaquinone [52]. More recently, the tetrapeptide Cys-Arg-Asp-Leu was developed by Joompang et al. and exhibited an IC_{50} value of 200 μM when tested on mushroom tyrosinase [53]. Consequently, Li et al. developed a series of L-cysteine-containing dipeptides as mushroom tyrosinase inhibitors (Fig. 13). These derivatives can be divided into two groups: the one bearing L-cysteine at C-terminal and the other one with L-cysteine at N-terminal. The peptide derivatives gave similar data, with pIC_{50} between 4.206 and 5.509. It is interesting to note that N-terminal derivatives always gave higher pIC_{50} values than the C-terminal analogs [54].

3. Conclusions and future perspectives

Despite the discovery of natural products as tyrosinase inhibitors, the interest for synthetic (hetero)aromatic and peptidic compounds is more and more increasing in the pharmaceutical field [55,56]. The tyrosinase enzyme has a crucial role in melanogenesis in all kingdoms of life. Given that, expression and purification protocols have improved the availability of the mushroom and human enzymes, furnishing high-resolution crystal structures for the rational optimization of the inhibitors. Unfortunately, despite inhibitors of tyrosinase have a prominent space in the cosmetic industry, their use has been often limited due to conflicting efficacy and potential toxicity. The aim of this chapter was to update the readers on the current knowledge about the design and discovery of recent chemotypes and strategies toward the clinical setting.

Acknowledgments

This work was partially supported by a grant from the Italian Ministry of University and Research (S.C. and C.C.).

References

[1] E.I. Solomon, U.M. Sundaram, T.E. Machonkin, Multicopper oxidases and oxygenases, Chemical Reviews 96 (7) (1996) 2563–2606.

[2] Á. Sánchez-Ferrer, J.N. Rodríguez-López, F. García-Cánovas, F. García-Carmona, Tyrosinase: A comprehensive review of its mechanism, Biochimica et Biophysica Acta (BBA) - Protein Structure and Molecular Enzymology 1247 (1) (1995) 1–11.

[3] B. Deri, M. Kanteev, M. Goldfeder, D. Lecina, V. Guallar, N. Adir, et al., The unravelling of the complex pattern of tyrosinase inhibition, Scientific Reports 6 (1) (2016) 1–10.

[4] W.T. Ismaya, H.J. Rozeboom, A. Weijn, J.J. Mes, F. Fusetti, H.J. Wichers, et al., Crystal structure of *Agaricus bisporus* mushroom tyrosinase: Identity of the tetramer subunits and interaction with tropolone, Biochemistry 50 (24) (2011) 5477–5486.

[5] A. Bijelic, M. Pretzler, C. Molitor, F. Zekiri, A. Rompel, The structure of a plant tyrosinase from walnut leaves reveals the importance of "Substrate-Guiding Residues" for enzymatic specificity, Angewandte Chemie International Edition 54 (49) (2015) 14677–14680.

[6] M. Kanteev, M. Goldfeder, A. Fishman, Structure–function correlations in tyrosinases, Protein Science 24 (9) (2015) 1360–1369.

[7] C.A. Ramsden, P.A. Riley, Tyrosinase: The four oxidation states of the active site and their relevance to enzymatic activation, oxidation and inactivation, Bioorganic & Medicinal Chemistry 22 (8) (2014) 2388–2395.

[8] M. Rolff, J. Schottenheim, H. Decker, F. Tuczek, Copper–O_2 reactivity of tyrosinase models towards external monophenolic substrates: molecular mechanism and comparison with the enzyme, Chemical Society Reviews 40 (7) (2011) 4077–4098.

[9] S. Zolghadri, A. Bahrami, M.T. Hassan Khan, J. Munoz-Munoz, F. Garcia-Molina, F. Garcia-Canovas, et al., A comprehensive review on tyrosinase inhibitors, Journal of Enzyme Inhibition and Medicinal Chemistry 34 (1) (2019) 279–309.

[10] T. Pillaiyar, V. Namasivayam, M. Manickam, S.H. Jung, Inhibitors of melanogenesis: An updated review, Journal of Medicinal Chemistry 61 (17) (2018) 7395–7418.

[11] B. Roulier, B. Pérès, R. Haudecoeur, Advances in the design of genuine human tyrosinase inhibitors for targeting melanogenesis and related pigmentations, Journal of Medicinal Chemistry 63 (22) (2020) 13428–13443.

[12] S. Carradori, F. Melfi, J. Rešetar, R. Şimşek, Tyrosinase enzyme and its inhibitors: An update of the literature, in: C.T. Supuran, W.A. Donald (Eds.), Metalloenzymes from bench to bedside, 1st edn.,, Elsevier, London, 2023, pp. 533–546.

[13] T. Mann, W. Gerwat, J. Batzer, K. Eggers, C. Scherner, H. Wenck, et al., Inhibition of human tyrosinase requires molecular motifs distinctively different from mushroom tyrosinase, Journal of Investigative Dermatology 138 (7) (2018) 1601–1608.

[14] B. Desmedt, P. Courselle, J.O. De Beer, V. Rogiers, M. Grosber, E. Deconinck, et al., Overview of skin whitening agents with an insight into the illegal cosmetic market in Europe, Journal of the European Academy of Dermatology and Venereology 30 (6) (2016) 943–950.

[15] A.A. Brozyna, L. VanMiddlesworth, A.T. Slominski, Inhibition of melanogenesis as a radiation sensitizer for melanoma therapy, International Journal of Cancer 123 (6) (2008) 1448–1456.

[16] E. Buitrago, R. Hardré, R. Haudecoeur, H. Jamet, C. Belle, A. Boumendjel, et al., Are human tyrosinase and related proteins suitable targets for melanoma therapy? Current Topics in Medicinal Chemistry 16 (27) (2016) 3033–3047.

[17] R. Logesh, S.R. Prasad, S. Chipurupalli, N. Robinson, S.K. Mohankumar, Natural tyrosinase enzyme inhibitors: A path from melanin to melanoma and its reported pharmacological activities, Biochimica et Biophysica Acta (BBA) - Reviews on Cancer 1878 (6) (2023) 188968.

[18] C. Dubois, R. Haudecoeur, M. Orio, C. Belle, C. Bochot, A. Boumendjel, et al., Versatile effects of aurone structure on mushroom tyrosinase activity, ChemBioChem 13 (4) (2012) 559–565.

[19] B. Roulier, I. Rush, L.M. Lazinski, B. Pérès, H. Olleik, G. Royal, et al., Resorcinol-based hemiindigoid derivatives as human tyrosinase inhibitors and melanogenesis suppressors in human melanoma cells, European Journal of Medicinal Chemistry 246 (2023) 114972.

[20] L.M. Lazinski, M. Beaumet, B. Roulier, R. Gay, G. Royal, M. Maresca, et al., Design and synthesis of 4-amino-2',4'-dihydroxyindanone derivatives as potent inhibitors of tyrosinase and melanin biosynthesis in human melanoma cells, European Journal of Medicinal Chemistry. 266 (2024) 116165.

[21] S. Ferro, L. De Luca, M.P. Germanò, M.R. Buemi, L. Ielo, G. Certo, et al., Chemical exploration of 4-(4-fluorobenzyl) piperidine fragment for the development of new tyrosinase inhibitors, European Journal of Medicinal Chemistry 125 (2017) 992–1001.

[22] J. Lee, Y.J. Park, H.J. Jung, S. Ullah, D. Yoon, Y. Jeong, et al., Design and synthesis of (Z)-2-(benzylamino)-5-benzylidenethiazol-4(5H)-one derivatives as tyrosinase inhibitors and their anti-melanogenic and antioxidant effects, Molecules 28 (2) (2023) 848.

[23] M.K. Kang, D. Yoon, H.J. Jung, S. Ullah, J. Lee, H.S. Park, et al., Identification and molecular mechanism of novel 5-alkenyl-2-benzylaminothiazol-4(5H)-one analogs as anti-melanogenic and antioxidant agents, Bioorganic chemistry 140 (2023) 106763.

[24] Y. Jung Park, H. Jin Jung, H. Jin Kim, H. Soo Park, J. Lee, D. Yoon, et al., Thiazol-4(5H)-one analogs as potent tyrosinase inhibitors: Synthesis, tyrosinase inhibition, antimelanogenic effect, antioxidant activity, and in silico docking simulation, Bioorganic & Medicinal Chemistry 98 (2024) 117578.

[25] D. Yoon, M.K. Kang, H.J. Jung, S. Ullah, J. Lee, Y. Jeong, et al., Design, synthesis, in vitro, and in silico insights of 5-(substituted benzylidene)-2-phenylthiazol-4(5H)-one derivatives: A novel class of anti-melanogenic compounds, Molecules 28 (8) (2023) 3293.

[26] H. Choi, I. Young Ryu, I. Choi, S. Ullah, H. Jin Jung, Y. Park, et al., Identification of (Z)-2-benzylidene-dihydroimidazothiazolone derivatives as tyrosinase inhibitors: Anti-melanogenic effects and in silico studies, Computational and Structural Biotechnology Journal 20 (2022) 899–912.

[27] Y. Jeong, S. Hong, H.J. Jung, S. Ullah, Y. Hwang, H. Choi, et al., Identification of a novel class of anti-melanogenic compounds, (Z)-5-(substituted benzylidene)-3-phenyl-2-thioxothiazolidin-4-one derivatives, and their reactive oxygen species scavenging activities, Antioxidants 11 (5) (2022) 948.

[28] S.A. Shehzadi, A. Saeed, F. Perveen, P.A. Channar, I. Arshad, Q. Abbas, et al., Identification of two novel thiazolidin-2-imines as tyrosinase inhibitors: Synthesis, crystal structure, molecular docking and DFT studies, Heliyon 8 (8) (2022) e10098.

[29] M.K. Zargaham, M. Ahmed, N. Akhtar, Z. Ashraf, M.A. Abdel-Maksoud, M. Aufy, et al., Synthesis, in silico studies, and antioxidant and tyrosinase inhibitory potential of 2-(substituted phenyl) thiazolidine-4-carboxamide derivatives, Pharmaceuticals 16 (6) (2023) 835.

[30] B. Zengin Kurt, Ö. Altundağ, M.N. Tokgöz, D. Öztürk Civelek, F.O. Tuncay, U. Cakmak, et al., Synthesis of flurbiprofen thiadiazole urea derivatives and assessment of biological activities and molecular docking studies, Chemical Biology and Drug Design 102 (6) (2023) 1458–1468.

[31] Y. Hwang, J. Lee, H.J. Jung, S. Ullah, J. Ko, Y. Jeong, et al., A novel class of potent anti-tyrosinase compounds with antioxidant activity, 2-(substituted phenyl)-5-(trifluoromethyl)benzo[d]thiazoles: In vitro and in silico insights, Antioxidants 11 (7) (2022) 1375.

[32] H. Zafar, S.M. Saad, S. Perveen, Arshia, R. Malik, A. Khan, et al., 2-Arylquinazolin-4(3H)-ones: Inhibitory activities against xanthine oxidase, Journal of Medicinal Chemistry 12 (1) (2016) 54–62.

[33] G. Benito, I. D'Agostino, S. Carradori, M. Fantacuzzi, M. Agamennone, V. Puca, et al., Erlotinib-containing benzenesulfonamides as anti-Helicobacter pylori agents through carbonic anhydrase inhibition, Future Medicinal Chemistry 15 (20) (2023) 1865–1883.

[34] S.M. Saad, M. Saleem, S. Perveen, M.T. Alam, K.M. Khan, M.I. Choudhary, Synthesis and biological potential assessment of 2-substituted quinazolin-4(3H)-ones as inhibitors of phosphodiesterase-I and carbonic anhydrase-II, Journal of Medicinal Chemistry 11 (4) (2015) 336–341.

[35] Y. Huang, J. Yang, Y. Chi, C. Gong, H. Yang, F. Zeng, et al., Newly designed quinazolinone derivatives as novel tyrosinase inhibitor: Synthesis, inhibitory activity, and mechanism, Molecules 27 (17) (2022) 5558.

[36] M.M. Matin, P. Matin, M.R. Rahman, T. Ben Hadda, F.A. Almalki, S. Mahmud, et al., Triazoles and their derivatives: Chemistry, synthesis, and therapeutic applications, Frontiers in Molecular Biosciences 9 (2022) 864286.

[37] M. Hassan, B.D. Vanjare, K.Y. Sim, H. Raza, K.H. Lee, S. Shahzadi, et al., Biological and cheminformatics studies of newly designed triazole based derivatives as potent inhibitors against mushroom tyrosinase, Molecules 27 (5) (2022) 1731.

[38] A. Bagheri, S. Moradi, A. Iraji, M. Mahdavi, Structure-based development of 3,5-dihydroxybenzoyl-hydrazineylidene as tyrosinase inhibitor; in vitro and in silico study, Scientific Reports 14 (1) (2024) 1540.

[39] H. Raza, A. Rehman Sadiq Butt, M. Athar Abbasi, S. Aziz-Ur-Rehman Zahra Siddiqui, M. Hassan, S. Adnan Ali Shah, et al., 2-aminothiazole-oxadiazole bearing N-arylated butanamides: Convergent synthesis, tyrosinase inhibition, kinetics, structure-activity relationship, and binding conformations, Chemistry and Biodiversity 20 (2) (2023) e202201019.

[40] M. Khan, S.A. Halim, L. Shah, A. Khan, I.E. Ahmed, A.N. Abdalla, et al., Isoxazole analogues of dibenzazepine as possible leads against ulcers and skin disease: In vitro and in silico exploration, Saudi Pharmaceutical Journal 31 (12) (2023) 101877.

[41] M.M. Al-Rooqi, A. Sadiq, R.J. Obaid, Z. Ashraf, Y. Nazir, R.S. Jassas, et al., Evaluation of 2,3-dihydro-1,5-benzothiazepine derivatives as potential tyrosinase inhibitors: In vitro and in silico studies, ACS Omega 8 (19) (2023) 17195–17208.

[42] B. Hassani, F. Zare, L. Emami, M. Khoshneviszadeh, R. Fazel, N. Kave, et al., Synthesis of 3-hydroxypyridin-4-one derivatives bearing benzyl hydrazide substitutions towards anti-tyrosinase and free radical scavenging activities, RSC Advances 13 (46) (2023) 32433–32443.

[43] Y.G. Hu, Z.P. Gao, Y.Y. Zheng, C.M. Hu, J. Lin, X.Z. Wu, et al., Synthesis and biological activity evaluation of 2-cyanopyrrole derivatives as potential tyrosinase inhibitors, Frontiers in Chemistry 10 (2022) 914944.

[44] N.A. Hassanuddin, E. Normaya, H. Ismail, A. Iqbal, M.B.M. Piah, S. Abd Hamid, et al., Methyl 4-pyridyl ketone thiosemicarbazone (4-PT) as an effective and safe inhibitor of mushroom tyrosinase and antibrowning agent, International Journal of Biological Macromolecules 255 (2024) 128229.

[45] A. Saeed, S.A. Ejaz, A. Khalid, P.A. Channar, M. Aziz, Q. Abbas, et al., Acetophenone-based 3,4-dihydropyrimidine-2(1H)-thione as potential inhibitor of tyrosinase and ribonucleotide reductase: Facile synthesis, crystal structure, in-vitro and in-silico investigations, International Journal of Molecular Sciences 23 (21) (2022) 13164.

[46] S. Mirabile, M.P. Germanò, A. Fais, L. Lombardo, F. Ricci, S. Floris, et al., Design, synthesis, and in vitro evaluation of 4-(4-hydroxyphenyl)piperazine-based compounds targeting tyrosinase, ChemMedChem 17 (21) (2022) e202200305.

[47] S. Ferro, B. Deri, M.P. Germanò, R. Gitto, L. Ielo, M.R. Buemi, et al., Targeting tyrosinase: Development and structural insights of novel inhibitors bearing arylpiperidine and arylpiperazine fragments, Journal of Medicinal Chemistry 61 (9) (2018) 3908–3917.

[48] L. Ielo, B. Deri, M.P. Germano, S. Vittorio, S. Mirabile, R. Gitto, et al., Exploiting the 1-(4-fluorobenzyl) piperazine fragment for the development of novel tyrosinase inhibitors as anti-melanogenic agents: Design, synthesis, structural insights and biological profile, European Journal of Medicinal Chemistry 178 (2019) 380–389.

[49] Y. Nazir, H. Rafique, S. Roshan, S. Shamas, Z. Ashraf, M. Rafiq, et al., Molecular docking, synthesis, and tyrosinase inhibition activity of acetophenone amide: Potential inhibitor of melanogenesis, BioMed Research International 2022 (2022) 1040693.

[50] D. Fu, D. Yuan, F. Qin, Y. Xu, X. Cui, G. Li, et al., Design, synthesis and biological evaluation of tyrosinase-targeting PROTACs, European Journal of Medicinal Chemistry 226 (2021) 113850.

[51] M.A. Baber, C.M. Crist, N.L. Devolve, J.D. Patrone, Tyrosinase inhibitors: A perspective, Molecules 28 (15) (2023) 5762.

[52] H.K. Lee, J.W. Ha, Y.J. Hwang, Y.C. Boo, Identification of L-cysteinamide as a potent inhibitor of tyrosinase-mediated dopachrome formation and eumelanin synthesis, Antioxidants 10 (8) (2021) 1202.

[53] A. Joompang, P. Anwised, S. Klaynongsruang, L. Taemaitree, A. Wanthong, K. Choowongkomon, et al., Rational design of an N-terminal cysteine-containing tetrapeptide that inhibits tyrosinase and evaluation of its mechanism of action, Current Research in Food Science 7 (2023) 100598.

[54] X. Li, F. Pan, Z. Yang, F. Gao, J. Li, F. Zhang, et al., Construction of QSAR model based on cysteine-containing dipeptides and screening of natural tyrosinase inhibitors, Journal of Food Biochemistry 46 (10) (2022) e14338.

[55] J. Li, C. Li, X. Peng, S. Li, B. Liu, C. Chu, Recent discovery of tyrosinase inhibitors in traditional Chinese medicines and screening methods, Journal of Ethnopharmacology 303 (2023) 115951.

[56] S. Ullah, S. Son, H.Y. Yun, D.H. Kim, P. Chun, H.R. Moon, Tyrosinase inhibitors: A patent review (2011-2015), Expert Opinion on Therapeutic Patents 26 (3) (2016) 347–362.

Peptide and peptidomimetic tyrosinase inhibitors ☆

Fosca Errante[a,b], **Lucrezia Sforzi**[a,b], **Claudiu T. Supuran**[b], **Anna Maria Papini**[a,c], **and Paolo Rovero**[a,b,*]

[a]Interdepartmental Laboratory of Peptide and Protein Chemistry and Biology, University of Florence, Sesto Fiorentino, Florence, Italy
[b]Department of Neurosciences, Psychology, Drug Research and Child Health, University of Florence, Sesto Fiorentino, Florence, Italy
[c]Department of Chemistry "Ugo Schiff", University of Florence, Sesto Fiorentino, Florence, Italy
*Corresponding author. e-mail address: paolo.rovero@unifi.it

Contents

Abstract

Melanin, which is produced by melanocytes and spread over keratinocytes, is responsible for human skin browning. There are several processes involved in melanogenesis, mostly prompted by enzymatic activities. Tyrosinase (TYR), a copper containing metalloenzyme, is considered the main actor in melanin production, as it catalyzes two crucial steps that modify tyrosine residues in dopaquinone. For this reason, TYR inhibition has been exploited as a possible mechanism of modulation of hyper melanogenesis. There are various types of molecules used to block TYR activity, principally used as skin whitening agents in cosmetic products, e.g., tretinoin, hydroquinone, azelaic acid, kojic acid, arbutin

☆THE ENZYMES Vol. 56 (2024)—TYROSINASES, Claudiu T. Supuran, Editor. Elsevier 2024.

The Enzymes, Volume 56
ISSN 1874-6047, https://doi.org/10.1016/bs.enz.2024.06.005

and peptides. Peptides are highly valued for their versatile nature, making them promising candidates for various functions. Their specificity often leads to excellent safety, tolerability, and efficacy in humans, which can be considered their primary advantage over traditional small molecules. There are several examples of tyrosinase inhibitor peptides (TIPs) operating as possible hypo-pigmenting agents, which can be classified according to their origin: natural, hybrid or synthetically produced. Moreover, the possibility of variating their backbones, introducing non-canonical amino acids or modifying one or more peptide bond(s), to obtain peptidomimetic molecules, is an added value to avoid or delay proteolytic activity, while the possibility of conjugation with other bioactive peptides or organic moieties can bring other specific activity leading to dual-functional peptides.

Abbreviations

α-MSH	alpha-melanocyte-stimulating hormone
αS-CN	αS-casein
Ala; A	alanine
Arg; R	arginine
Asn; N	asparagine
Asp; D	aspartic acid
BLG	β-lactoglobulin
CA	caffeic acid
Cys	C: cysteine
DWMP	defatted walnut meal protein
DWMPHs	defatted walnut meal protein hydrolysates
GA	gallic acid
Gln	Q: glutamine
Glu	E: glutamic Acid
Gly	G: glycine
GSH	glutathione
GSSG	oxidized glutathione
GT	gentisic acid
His	H: histidine
HPLC	high performance/pressure liquid chromatography
IC$_{50}$	half maximal inhibitory concentration
Ile	I: isoleucine
L-DOPA	L-3,4-dihydroxyphenylalanine
L-ERT	L-ergothioneine
Leu	L: leucine
LMW-GH	low molecular-weight gelatin hydrolysates
Lys	K: lysine
KA	kojic acid
KAE	kojic acid equivalents
MITF	melanocyte inducing transcription factor/microphthalmia-associated transcription factor
MC1R	melanocortin-1 receptor
Met	M: methionine

MSCP	milkfish scale peptide
MTT	3-(4,5-dimethylthiazol-2-yl)-2,5-diphenyltetrazolium bromide
PA	protocatechuic acid
Phe	F: phenylalanine
PPO	polyphenol oxidase
Pro	P: proline
RA	α-resocylic acid
RBAlb	rice bran albumin
RBP	rice bran protein
ROS	reactive oxygen species
SAP	stabilized ascorbyl pentapeptide
Ser	S: serine
SPPS	solid phase peptide synthesis
Thr	T: threonine
TILI-1	tyrosinase inhibitor Leucrocin I-1
TILI-2	tyrosinase inhibitor Leucrocin I-2
TIPs	tyrosinase inhibitor peptides
TRP-1	tyrosinase related protein-1
TRP-2	tyrosinase related protein-2
TSC	thiosemicarbazone
Tyr	Y: tyrosine
TYR	tyrosinase
Trp	W: tryptophan
UV-A	ultraviolet radiation A
UV-B	ultraviolet radiation B
Val	V: valine

1. Introduction

Tyrosinase (TYR), alternatively known as polyphenol oxidase (PPO), is a copper-containing metalloenzyme found across a diverse range of organisms, including mammals. TYR is crucial to the production of melanin, the pigment responsible for coloring skin, hair, and eyes in mammals. In fact, it catalyzes the oxidation of tyrosine to dihydroxyphenylalanine (DOPA) and subsequently to dopaquinone, which are key steps in the melanin synthesis pathway [1–3].

Due to their versatile nature, peptides are often regarded as promising compounds with ideal functions, especially as regulating and signaling molecules in stress, immunity and defense. Their specificity often results in excellent safety, tolerability, and efficacy in humans, which may be their main advantage over traditional small molecules, serving as a promising foundation for developing new therapeutics. Therefore, peptides occupy a unique position between small molecules and biopharmaceuticals in several respects [4,5]. Moreover, peptides present reduced ability to trigger an immune response and relatively low-cost production, making them interesting as both therapeutic agents and cosmeceutical compounds. In this respect, in fact, they can be used as antioxidant, anti-aging, moisturizing ingredients as well as promoters in collagen production or wound healing processes [6].

Accordingly, there are several examples of tyrosinase inhibitor peptides (TIPs) acting as possible hypo-pigmenting agents. Their mechanism of action can be classified in three groups:

1 Anti-oxidation.

Ultraviolet radiation is responsible of low-grade oxidative stress in melanocytes, causing the production of rapidly reactive free radicals that can cause abnormal tyrosinase activation resulting in increased melanogenesis. There are three ways in which TIPs can act on the enzyme: (i) directly scavenging of free radicals, (ii) activation of anti-oxidative enzymes, and (iii) regulation of gene pathways that decrease the amount of TYR.

2 Occupation of the bioactive site TYR.

TYR is a tetrameric enzyme with two copper ions (Cu^{2+}) in the active site, characterized by a hydrophilic exterior and a hydrophobic cavity near the active site. TIPs can interact with the protein in different ways such as direct chelation of the copper ions, reactions with amino acids residues in the cavity or near the catalytic center with formations of different types of interactions such as covalent bonds, hydrogen bonds, hydrophobic interactions, Van der Waals forces or π-π stacking. Chang and coworkers identified two types of the so called "true inhibitors" that can be found in this category: (i) suicide substrate: peptides that can form a covalent bond with the enzyme thus causing its complete inactivation; (ii) reversible inhibitors: compounds that can reversibly bind to TYR causing a decrease in its catalytic capability [7].

"True inhibitors" can also be classified according to their inhibition type: (i) competitive inhibitors; (ii) uncompetitive inhibitors; (iii) mixed type, competitive/uncompetitive inhibitors and (iv) noncompetitive inhibitors.

3 Regulation of related gene expression.

There are several signaling pathways connected to the expression of tyrosinase and thus to melanogenesis production. TIPs can interact with many enzymes, mediators and other chemical entities.

Considering that peptides are composed of amide bonds that are labile in physiological condition, as they can be cleaved by specific enzymes, several strategies can be employed to modify the peptide backbone, in order to overcome the disadvantages due to possible low bioavailability. These strategies led to the so-called peptidomimetics. In this context, chemical synthesis plays a particularly important role, as it offers access to a much wider chemical diversity than peptide derivatives produced by recombinant technologies. Peptidomimetic compounds are very similar to parent peptides, and they are intended to mimic their biological effects. Very often they contain nonpeptide structural elements, such as peptide bond surrogates that resist cleavage from peptidases, a feature also found in some natural peptidase inhibitors. In addition to permanent alterations of the structure, it is also possible to introduce temporary changes using a pro-drug approach. A peptide prodrug can be obtained by combining a biologically active peptide with additional elements, which give the entire molecule greater resistance against enzymatic hydrolysis and/or bioavailability [8].

This chapter is dedicated to a deep literature investigation on peptides and peptidomimetics as TYR inhibitors deriving from natural sources, hybrid, or totally synthetically approach.

2. Amino acids with anti-tyrosinase properties

In 2018, Park et al. investigated the effect of D-tyrosine on melanin synthesis in melanocyte-derived cells such as human melanoma cells, human melanocytes and 3D human skin model [9]. Each amino acids, except for glycine, exist in two enantiomeric forms. However, our organism uses only L-amino acids to accomplish its functions, thus, D-amino acids could potentially act as inhibitor for L-form dependent enzymes. Indeed, the results of the experiments done by these authors evidenced that D-tyrosine, which did not affect cell proliferation, is instead able to reduce in a dose-dependent manner the amount of melanin produced by the cells whereas L-tyrosine, as previously reported, is able to induce melanogenesis [10]. The performed trials highlighted the ability of D-tyrosine to directly interfere the enzymatic activity of TYR in a dose-dependent manner, acting as competitive inhibitors. Interestingly, D-tyrosine can inhibit not only the melanogenic function of L-tyrosine but also melanogenesis induced by alpha-melanocyte-stimulating hormone (α-MSH) and UVB radiation. These results were confirmed by the use of MelanoDerm™ skin model (MEL-312-B, a 3D model in which human melanocytes are incorporated into a well differentiated epidermal tissue): the samples treated with D-tyrosine revealed a reduced amount of melanin and tyrosinase compared to the ones treated with distilled water.

In an additional work [11], the same authors studied the effect of L-tyrosine on TYR by synthetizing four analogues of pentapeptide-18 (L-tyrosyl-D-alanylglycyl-L-phenylalanyl-L-leucine), adding both in position N- and C-terminal L- or D-tyrosine and then studying the effect of these derived peptides on melanin production in Human Melanoma MNT-1 cells, human melanocyte and 3D model of human epidermis. The results highlight that the presence of D-tyrosine, especially in C-terminal position, is able to reduce melanin synthesis and tyrosinase activity induce by both α-MSH and UVA radiation, thus suggesting that the presence of the amino acid could add additional anti-melanogenic effect to cosmetic peptides. However, this additional function is not shown when the addition of D-tyrosine occurs on longer peptides, where the higher number of side chains could interfere negatively with the binding of the peptide to the enzyme and thus on D-tyrosine inhibition efficacy.

Liao et al. studied the effect of ergothioneine (structure shown in Fig. 2), a naturally occurring amino acid found in fungi, oysters and actinobacteria, having antioxidative properties [12]. Ergothioneine is a thiourea derivative of histidine, thus the authors compared the two amino acids as tyrosine inhibitors. The o–diphenolase activity of tyrosine was tested using L-DOPA as substrate at different concentration of histidine, ergothioneine and kojic acid (used as control) on mushroom tyrosinase. Results evidence that L-ERT inhibits TYR in a dose dependent manner with an IC_{50} = 4.47 mM. The Lineweaver–Burk plot of ergothioneine inhibition demonstrates that the reaction velocity increases increasing L-DOPA concentration, suggesting a non-competitive type of inhibition activity. Taken all together, these results indicate that ergothioneine can potentially bind to a substrate site of free enzymes and can also bind to a site distinct from the substrate of an enzyme–substrate complex. On the other hand, histidine instead is not able to inhibit the enzyme.

Luisi et al. studied the antioxidant and tyrosinase inhibitory activity of sulphurated amino acids, with different oxidation state of the sulfur atom, alone or incorporated into small peptides [13]. In particular, the studied compounds were L-cysteine, glutathione and its gamma-oxa analogue, L-cystine, L-ergothioneine and Taurine. Moreover, three additional synthetic compounds containing the non-proteinogenic amino acid tert-Leucine, were synthetized in order to assess their antioxidant and inhibitory activity.

Tyrosinase inhibition activity was measured using the dopachrome method (a schematic representation of the dopachrome formation is reported in Fig. 1).

Final TYR inhibitory activity is expressed as kojic acid equivalents (mgKAE/g sample). All the tested compounds were active against TYR. The results are reported in Table 1.

The effect of cysteine on the o-dihydroxyphenolase activity of PPO has been investigated also by Kahn [14]. Indeed, in the presence of cysteine, an initial delay in dopachrome formation is observed, with higher cysteine concentrations producing longer lag periods. This can be probably attributed to the ability of cysteine to form a conjugate with dopaquinone. The

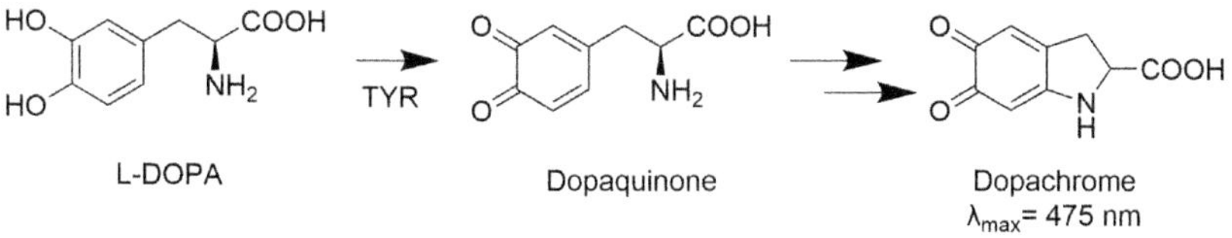

Fig. 1 Schematic representation of the conversion from L-DOPA to dopachrome in melanogenesis.

Table 1 Results of TYR inhibitory activity expressed ad kojic acid equivalents.

Compound:	Structure:	Tyrosinase inhibitory activity (mg KAE/g sample):
L–Cysteine	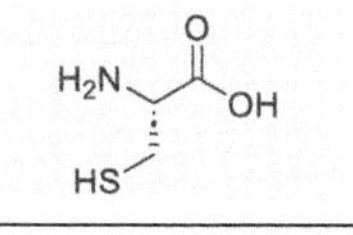	216.40 ± 0.17
L–Cystine	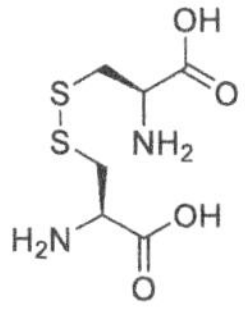	217.16 ± 0.55
L–Ergothioneine	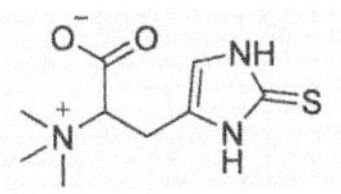	100.97 ± 1.71
Taurine	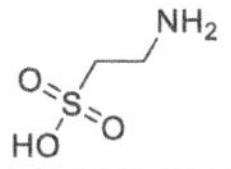	33.87 ± 1.59

results were confirmed by performing experiments on freshly cut fruits like avocado and banana which are prone to fast browning reactions. The application of L-cysteine effectively prevented the discoloration even after 6 h after cutting, suggesting that this amino acid could be potentially use not only in the cosmetic industry but also in the food industry to prevent oxidation reactions in fruits and vegetables.

3. Anti-tyrosinase peptides from natural sources

Naturally occurring TIPs can be found in various biological sources, which include animals, plants, and microorganisms. In recent years, the pursuit of natural ingredients has become a matter of interest especially in the cosmetic field, but also in the pharmaceuticals, giving rise to new active ingredients that can potentially be included in skin care formulations and dermatological applications.

In this paragraph we will elucidate which are the most common peptides from natural sources described in literature, displaying TYR inhibiting activity and antimelanogenic effects on cells. A list of the described compounds is reported, at the end of the paragraph, in Table 2.

3.1 Peptides deriving from mushrooms

The first study reported in the literature describing peptides with anti-tyrosinases activity dates back to 1974, when Madhosingh and Sundberg identified two peptides (named *compound Ia* and *compound Ib* in their paper) from *Agaricus hortensis* mushrooms [15]. Their inhibiting activity was demonstrated by incubation with L-Dihydroxyphenylalanine (L-DOPA) as a substrate, in presence of mushrooms TYR (that has to be inhibited), and they were found to act with different mechanism of action: *compound Ia* is a competitive inhibitor, while *compound Ib* is a non-competitive inhibitor. Amino acids analysis of the extracted peptides revealed that the three most common residues were phenylalanine, aspartic and glutamic acid (ratio 1:1:1). However, the peptide sequences of these two compounds were not described.

3.2 Peptides deriving from bacteria

The first example of a well-defined peptide sequence with anti-tyrosinases activity was reported some years later, in a study based on *Lactobacillus Helveticus* [16]. The authors observed that a cyclopeptide, Cycle[Pro-Tyr-Pro-Val], isolated from this species, was able to inhibit mushroom tyrosinase activity with a half maximal inhibitory concentration (IC_{50}) higher than arbutin that was used as a reference compound.

Some years later, a cyclic peptide capable to inhibit TYR was isolated from *Oscillatoria agardhii* [17]. This cyanobacterium produce cyclic heptapeptide hepatotoxins known as microcystins and one among these, named Oscillapeptin G (whose structure is reported in Fig. 2), were found to be a non-toxic depsipeptide [17,18]. Depsipeptides are cyclic compounds where one or more of their peptide bonds –C(O)NHR–, are substituted by the corresponding ester, –C(O)OR–; principally used as anti-cancer drugs [19,20]. Oscillapeptin G activity against TYR was studied in the presence of L–Tyr and TYR and showed a 0.1 mM IC_{50} [17].

A more recent example reports the dual activity of antioxidant and anti-TYR agent of the tetrapeptide MGRY purified from the microalga *Pavlova lutheri*. TYR-inhibiting properties were demonstrated by testing the peptide as a treatment on B16F10 melanoma cells with α–MSH-induced melanogenesis. Additionally, it showed decreased melanogenesis-related proteins, microphthalmia-associated transcription factor (MITF) and TYR protein expressions.

3.3 Peptides from plant sources

3.3.1 *Pseudostellaria heterophylla*

A study conducted by Morita et al. highlighted the presence of various cyclic peptides isolated from *Pseudostellaria heterophylla* roots, which were tested with the dopachrome method and exhibited mushroom TYR inhibitory activity [21–23].

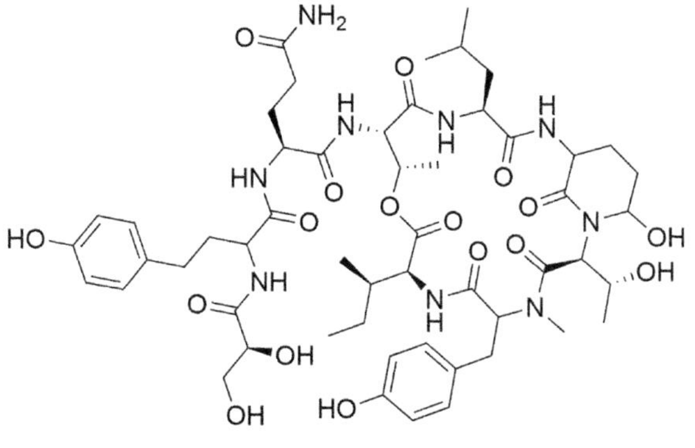

Fig. 2 Oscillapeptin G structure.

These compounds, named pseudostellarins (A,B,C,D,E,F,G) showed lower IC_{50} than the ones observed in the case of the cyclopeptide isolated in the *L. Helveticus* [16], and arbutin, both used as references. Moreover, pseudostellarins C, D and G showed antimelanogenesis effect when used as treatment from B16 melanoma cells, with a $IC_{50} = 171\,\mu M$, $IC_{50} = 49\,\mu M$, and $IC_{50} = 102\,\mu M$, respectively. The importance of investigating the activity in vitro on cell models is due to the fact that mushrooms TYR is present in the cytoplasm, and is different from the mammals TYR present on the membrane of melanin bulks [24].

3.3.2 Rice

In 2009 Ubeid et al. described the activity of some oligopeptides of unknown origin that were able to inhibit both mushrooms and human TYR, showing in particular two highly active sequences: RADSRADC, also called peptide P3, and YRSRKYSSWY, also called peptide P4, with IC_{50} (mushroom TYR) of $123\,\mu M$ and $40\,\mu M$, respectively [25]. Interestingly, P4 is still nowadays the only peptide used as a therapeutic agent for melanin-related skin disorders [26]. Further details on Ubeid and coworkers' studies describing P3 and P4 design and activity is reported below in the *Oligopeptides* paragraph.

Based on the potent activity of P4, and considering rice bran, found in rice kernel and produced as a byproduct of the rice milling process, as a new source of bioactive peptides in a view of new sustainable active ingredients, Ochiai et al. described the activity of rice bran bioactive peptides as TYR inhibitors [27,28]. Among the sequences found in each rice bran protein (RBP), MRSRERSSWY showed the highest grade of homology to P4 (YRSRKYSSWY), and this peptide was designated as TH10. Moreover, in their studies they described analogues of TH10 and P4 that revealed the functional role of tyrosine at various positions in their sequences. They observed that the TH10 variants in which the C-terminal tyrosine residue was replaced, had no effect on TYR, meaning that this position is essential for the tyrosinase inhibitory activity of TH10 and cannot be substituted with other aromatic amino acid residues. Additionally, even by reverting the sequence of TH10, therefore obtaining a tyrosine residue in N-terminal, the effect on TYR was lost. Worth of note, the same correlation was found by replacement of the tyrosine residues in peptide P4. In fact, this sequence contains three tyrosine residues, one in N-terminal, one in C-terminal and one in the middle, and it was observed that only by replacing the one in C-terminal there was a loss of activity, that was indeed not observed by replacing one or two tyrosine in the other positions [27]. Prompted by these indications, the same research group have deepened the possibility to find other TYR inhibitors derived from RBP hydrolysates prepared by mixing chymotrypsin (that cleaves preferentially at the carboxyl sides of large hydrophobic and aromatic amino acids such as tyrosine), and trypsin (in order to obtain short peptide sequences) [28]. The resulted hydrolysate fractions were analyzed, and their amino acid sequences were determined, showing that three of them (namely CT-1, CT-2, and CT-3) contained a tyrosine residue in C-terminal. Their activity was determined both against mono- and diphenolase reaction. In fact, TYR catalyzes two different reactions in the presence of molecular oxygen: the hydroxylation of monophenol (monophenolase activity) and the oxidation of o-diphenol to o-quinone (diphenolase activity) [29], giving rise to DOPA and Dopaquinone respectively, as shown in Fig. 3.

All three CT-1, CT-2 and CT-3 significantly inhibited the monophenolase reaction of tyrosinase; in particular, it was possible to establish that CT-2 had an IC_{50} = 156 µM. Moreover, peptides CT-1 and CT-3 increased melanin production in mouse B16 melanoma cells, suggesting that they do not inhibit mammalian tyrosinase, while CT-2 inhibited melanin production by >50% at 500 µM. Therefore, they conclude that CT-2 (LQPSHY) is a potent peptide to be used as an agent for melanin-related skin disorder treatment.

On the wave of peptide sequences deriving from rice bran, a study conducted by Kubglomsong et al. makes a comparison between different type of rice bran protein in which it was shown that albumin (RBAlb) is the one with higher anti-tyrosinases activity. Moreover, RBAlb was hydrolyzed and this mixture was analyzed to obtain information on the peptide sequences responsible for TYR inhibition and copper-chelating activity [30]. Thirteen peptides composed of 14–50 residues with molecular weights ranging from 1327 to 4819 Da were identified: DASLTRLDLAGANGTDLAYGISLTVAV (sequence n.1), PRCSASSTTPRSTTSAAWRAPSAPTSSTS (sequence n.2),

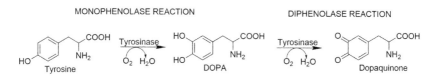

Fig. 3 Tyrosinase monophenolase and diphenolase reaction starting from tyrosine.

QSSKQRRDNGIFFVPQRPYMVLGTLRQQLLYPTW (sequence n.3), DVLEEGNSGDPPLFVSDVLHGAAIEVNEEGTEVAAATVVIMKGRA (sequence n.4), QGWSSSSSEYYGGEGSSSEQGYYGEG (sequence n.5), GWSSSSSEYYGGEGSSSEQGYYGEG (sequence n.6), SSEYYGGEGSSSE-QGYYGEG (sequence n.7), VELEEKGEGAAMTE (sequence n.8), AAAV-AMAAAVAAGEGGAANYLVFVDPPPSGVVCTAYQLSILAAALGSEEK (sequence n.9), SSHAAARRLRRSSSP (sequence n.10), GEPVAAEERDD-GVGVAGE (sequence n.11), SSSEYYGGEGSSSEQGYYGEG (sequence n.12), and LSLIGAVGGIGGSIL (sequence n.13). Remarkably, several peptide sequences (5–7 and 12) had the pattern SSEYYGGEGSSSEQGYYGEG in their peptide sequences, which consists of well-known tyrosinase-inhibitory and metal-chelating amino acids. In particular, serine, glutamic acid, and tyrosine contribute to tyrosinase inhibitory and metal-chelating activities [25,31–37].

3.3.3 Vineyard
A pentapeptide with anti-TYR activity, peptide ECGYF (namely EF-5) was found also in a *Vigna* protein and reported in a study conducted by Shen et al. in which they showed the inhibitory effect of the peptide, but also its radical scavenging effect, and additionally its melanin reducing activity in vitro on A375 melanoma cells. Moreover, the UV–Vis absorption spectra, CD spectra and molecular docking have suggested that EF-5 might induce conformational changes in TYR [38].

3.3.4 Quince
Deng and coworkers reported the activity of peptides derived from Chinese quince (*Chaenomeles speciose*), a plant already in use in Chinese medicine that exhibited excellent anti-inflammatory, antitumor and immunomodulatory properties [39,40]. Hydrolyzed protein fractions were purified, and the amino acid sequence of the purified peptides was identified by MALDI/TOF/TOF mass spectrometry. Their biological properties including antioxidant, copper chelating and tyrosinase inhibitory capacity were studied and the results revealed at least two short sequences with anti-TYR activity: NYRRE (peptide F1-a) and RHAKF (peptide F1-b). The TYR-inhibition IC_{50} of both peptides were calculated and resulted 8.69 mg/mL and 1.15 mg/mL, respectively for F1-a and F1-b. Moreover, as peptide F1-b showed the higher activity in comparison with F1-a, and contain an imidazole ring (His), the authors suggested that this difference in the IC_{50} is due to metal-chelating properties of peptide F1-b [40].

3.3.5 Chia

Peptides deriving from chia seeds of *Salvia hispanica*, and already described for their antioxidant properties, were recently showed to also act as inhibitors of TYR [41,42]. In fact, it was shown by Aguilar-Toalá and Liceaga that fractions hydrolysate from chia seeds containing peptides with molecular mass <3 kDa exhibited inhibitory activity towards different enzymes tested, including TYR, towards which they observed an IC_{50} of 0.66 mg/mL [42].

3.3.6 Walnut

A recent work published by Feng et al. focused on peptides from defatted walnut (*Juglans regia*) meal [43]. This paper represents one among the few examples of well-designed and characterized peptide obtained from natural sources. The authors started from defatted walnut meal protein (DWMP) which was hydrolyzed obtaining defatted walnut meal protein hydrolysates (DWMPHs) and the different fractions obtained were tested to evaluate their possible anti-TYR properties. It was shown that all the four DWMPHs obtained possessed inhibitory activity both for TYR mono- and diphenolase reaction. In particular, it was observed that smaller molecular weight of the ultrafiltration component corresponded to smaller IC_{50} values for the tyrosine monophenolase and diphenolase activity (indicating higher inhibitory properties). DWMPHs-IV displayed the greater TYR inhibitory activity than the other three fractions with IC_{50} values of 3.52 and 2.65 mg/mL for monophenolase and diphenolase, respectively. This fraction was then purified, and the collected fraction were tested again against mushrooms TYR. Among these, fraction F2, was showed to be the most powerful TYR inhibitors, with IC_{50} values of 1.42 and 1.8 mg/mL for monophenolase and diphenolase, respectively. LC-MS/MS analysis on F2 revealed that it was composed of 606 peptide sequences. Considering the large number of peptides to be screened, *in silico* approaches were employed to evaluate the binding ability of TYR to these peptides and to obtain the best inhibitor candidate according to their binding energy based on the affinity scoring function, and peptide FPY exhibited the most stable complex, with the highest binding affinity. Therefore, a synthetic peptide FPY was synthetized and tested, showing excellent TYR inhibitory properties with IC_{50} values of 1.11 and 3.22 mM for monophenolase and diphenolase, respectively.

3.4 Peptides from animal sources

3.4.1 Silk cocoons

Another natural source of TIPs was found in silk cocoons, and in particular in silk sericin hydrolysates recovered from the waste of silk degumming process [32,44]. Despite peptide composition was not described, some key-information were obtained by characterizing the hydrolysates mixture, e.g., that the major amino acid constituent of the hydrolysate is Ser (up to 30%), and the main molecular weight distribution was among 250 and 4000 Da. The tyrosinase inhibiting activity was demonstrated by the decreased conversion of L-DOPA to dopachrome, and the possible mechanism of action proposed was an association with its capacity of chelating elements such as copper and iron by its rich hydroxyl groups. In fact, the correlation between TYR-inhibiting properties and copper-chelating properties is nowadays well known and described [37,45,46].

3.4.2 Fishes

Fish proteins have been also often used as a source of cosmeceutical peptides for the treatment of various skin alterations [47]. One among the most cited studies on this topic is reported by Zhuang et al. and focus on a jellyfish collagen peptide (JPC) rich in Gly, Pro, Ser, Ala, Glu and Asp, with a molecular weight between 400 and 1200 Da. In this study they isolated and purified JCP from jellyfish collagen enzymatic hydrolysate using ion exchange chromatography and gel filtration. The effect on melanogenesis was evaluated by direct inhibition of TYR and melanin content in B16 cells. The observed IC_{50} was 78.2 μg/mL and the content of melanin was decreased to 76.8% at 25 μg/mL and 52.1% at 100 μg/mL compared with the control. JPC was shown to have also copper chelating and antioxidant properties, which is an added value for a melanogenesis regulator, as melanin synthesis is known to be involved in the production of several reactive oxygen species, including $^{\bullet}O2^{-}$, $^{\bullet}OH$ and $NO^{\bullet}$ [46].

JPC case is not unique and there are also other examples of fish collagen-deriving peptides with attributed anti-melanogenic properties, e.g., collagen hydrolysates from *Todarodes pacificus* (squid) skin [48], and collagen peptides from *Chanos chanos* (milkfish) scales [49]. In particular, hydrolysates deriving from squid skin collagen were fractionated and tested to evaluate their hyaluronidase- and TYR-inhibitory activity. The biological assays highlighted that one among the fractions, namely F3, inhibited tyrosinase by 39.65% at 1 mg/mL and was also able to inhibit hyaluronidase. F3 hydrolysate fraction, comprising of 3.4–10 kDa exhibited also strong antioxidant activities and metal chelating activity due to its high content of Ser, Gly, Val and Pro [48]. Indeed, Chen et al. reported the involvement of peptide extracts from milkfish scales (MSCP) in melanogenesis and their result revealed that 500 and 1000 µg/mL MSCP can repress the tyrosinase activity by 44.8% and 55.1%, respectively, in a dose dependent mode with an IC_{50} of 752.4 µg/mL. Moreover, MSCP inhibits melanin synthesis when used as a treatment in melanoma cells (B16) with a IC_{50} of 887.1 µg/mL, and the authors suggested an involvement in the copper chelating properties, as an explanation for the anti-TYR activity [49].

Unfortunately, as observed for several peptides derived from natural sources, especially for those that originate from protein hydrolysis processes, there is not a precise analytical characterization of the peptides described in these examples, therefore, their sequences are not described.

3.4.3 Sea cucumber

Stichopus japonicus (Sea cucumber) is considered a healthy food with anti-cancer, immunoregulatory, and anti-aging properties. Considering that it consists of approximately 70% of collagen [50], that is a gelatin source, in a paper published in 2010 by Wang and coworkers, the putative effect of the sea cucumber body wall hydrolysates was evaluated [51]. Gelatin hydrolysates were collected and after HPLC analysis, they were characterized as low molecular-weight gelatin hydrolysates (LMW-GH) as their molecular weight was from 700 up to 1700 Da. The amino acids compositions of these LMW-GH were described, and the most abundant amino acids resulted Gly, Glu, Asp, and Pro (in order). The potential anti-TYR activity and melanin reduction capability of LMW-GH was evaluated in vitro on B16F10 cells at different concentrations (25, 50 and 100 µg/mL). LMW-GH exhibited excellent inhibitory characteristics in both assays. In particular, concerning the inhibition of TYR, LMW-GH activity was 56.9%, 41.4% and 30.8% at 25, 50, and 100 µg/mL, respectively (expressed as the percentage of the absorbance/mg protein in treated cells of the control).

3.4.4 Milk

In 1997 the activity of the so-called whey proteins, milk proteins that remains soluble in whey after casein precipitation, that include α-lactalbumin, β-lactoglobulin (BLG), serum albumin and IgG was firstly described [52]. In this study the authors depicted the activity of the specific milk proteins on the pigmentation of cultured human melanocytes treated for 72 h with each whey protein and added with L-DOPA after cell lysis, and they showed that β-lactoglobulin decreased TYR activity at a concentration of 1 mg/mL. The mechanism of action proposed is based on the ability of BLG to bind retinol, which is known to inhibit melanin biosynthesis [52,53]. However, Nakajima et al. study considered the whole proteins, instead of specific peptide sequences.

In further studies published some years later, the activity of peptides purified from milk casein and α-lactalbumin was described [54,55]. In particular, peptide YFYPEL was originally isolated from milk casein hydrolysates and several analogues were synthetized in order to study the influence of the deletion of one or more amino acid on the antioxidant activity, founding that the deletion of Y, YF, and YFY causes a loss of activity [54]. Additionally, 42 peptides deriving from α-lactalbumin and BLG hydrolysates with antioxidant properties were described by Hernández-Ledesma and coworkers [55]. Despite in both these cases peptides were only tested for their antioxidant activity and no information on the possible implication in melanogenesis processes were mentioned, they gave rise to further investigation for possible anti-TYR activity. In fact, in a study published in 2020, four types of milk protein-derived peptides deriving from their studies with antioxidant activity (YFYPEL [54], YVEEL, MHIRL and WYSLAMAA [55]) were used as possible leads for anti-TYR peptides, basing on the fact that there could be a correlation between the two activities [56,57]. The authors selected the MHIRL family peptides due to their excellent TYR inhibiting properties, and further studies were conducted to determinate if they could also inhibit melanin production in vitro in different melanocytes (Mel-Ab and B16F10) by interfering with α-MSH binding to MC1R (melanocortin-1 receptor) [58].

In a study conducted by Schurink et al., the activity of several TYR-inhibiting peptides deriving from natural sources, including milk extracts, was described [31]. They started from protein-based octamer peptide libraries made by SPOT synthesis to screen for peptides that show direct interaction with tyrosinase. The authors evaluated the ability of the resulted peptides to bind and inhibit TYR, dividing them into three categories: (I) strong tyrosinase-binding peptides; (II) strong tyrosinase-inhibiting peptides (reported in Table 2); (III) tyrosinase-binding and inhibiting peptides. They observed that the best tyrosinase binding peptides contain an Arg residue and that most tyrosinase-inhibiting peptides contain a Val residue, but also Ala and/or Leu residues (hydrophobic amino acids) seems to be important for the inhibiting activity.

Table 2 List of TYR inhibiting peptides from natural or hybrid sources.

Compound name	Peptide sequence	Natural source (species)	Claimed activity	IC$_{50}$	References
Compound Ia	Unknown	*Agaricus hortensis* (mushrooms)	Competitive inhibition of TYR in presence of DOPA	Unknown	[15]
Compound Ib	Unknown	*Agaricus hortensis* (mushrooms)	Non-competitive inhibition of TYR in presence of DOPA	Unknown	[15]
Compound 1	Cyclo[PYPV] (reported in Fig. 16)	*Lactobacillus Helveticus* (bacteria)	Inhibition of mushrooms TYR	1.5 mM	[16]
Oscillapeptin G	(reported in Fig. 2)	*Oscillatoria agardhii* (bacteria)	Inhibition of mushrooms TYR	0.1 mM	[17]
Pseudostellarins A	Cyclo[GPYLA]	*Pseudostellaria heterophylla* (plants)	Inhibition of mushrooms TYR	131 µM	[21]
Pseudostellarins B	Cyclo[GIGGGPPF]	*Pseudostellaria heterophylla* (plants)	Inhibition of mushrooms TYR	187 µM	[21]
Pseudostellarins C	Cyclo[GTLPSPFL]	*Pseudostellaria heterophylla* (plants)	Inhibition of mushrooms TYR (I) Antimelanogenesis effect on B16 melanoma cells (II)	63 µM (I) 171 µM (II)	[21]

(continued)

Table 2 List of TYR inhibiting peptides from natural or hybrid sources. (*cont'd*)

Compound name	Peptide sequence	Natural source (species)	Claimed activity	IC$_{50}$	References
Pseudostellarins D	Cyclo[GGYPLIL]	*Pseudostellaria heterophylla* (plants)	Inhibition of mushrooms TYR (I) Antimelanogenesis effect on B16 melanoma cells (II)	100 μM (I) 49 μM (II)	[21]
Pseudostellarins E	Cyclo[GPPLGPVIF]	*Pseudostellaria heterophylla* (plants)	Inhibition of mushrooms TYR	175 μM	[21]
Pseudostellarins F	Cyclo[GGYLPPLS]	*Pseudostellaria heterophylla* (plants)	Inhibition of mushrooms TYR	50 μM	[21]
Pseudostellarins G	Cyclo[PFSFGPLA]	*Pseudostellaria heterophylla* (plants)	Inhibition of mushrooms TYR (I) Antimelanogenesis effect on B16 melanoma cells (II)	75 μM (I) 102 μM (II)	[21]
P3	RADSRADC	Unknown	Inhibition of mushrooms TYR	123 μM	[25]
P4	YRSRKYSSWY	Unknown	Inhibition of mushrooms TYR (I) Inhibition of mushrooms TYR monophenolase (II)	40 μM (I) 123 μM (II)	[25,27]

TH10	MRSRERSSWY	*Oryza sativa* (plants)	Inhibition of mushrooms TYR (monophenolase)	102 μM	[27]
CT-1	HGGEGGRPY	*Oryza sativa* (plants)	Inhibition of mushrooms TYR	Unknown	[28]
CT-2	LQPSHY	*Oryza sativa* (plants)	Inhibition of mushrooms TYR (I) Inhibition of melanin production on melanoma cells (II)	156 μM (I) 500 μM (II)	[28]
CT-3	HPTSEV	*Oryza sativa* (plants)	Inhibition of mushrooms TYR	Unknown	[28]
RBAlbH fraction 1	Mixture of peptides composed of the recurring sequence SSEYYGGEGSSSE-QGYYGEG	*Oryza sativa* (plants)	Inhibition of mushrooms TYR (I) Copper-chelating activity (II)	1.31 mg/mL (I) 0.62 mg/mL (II)	[30]
Undefined	APLRVYVE	β-Lactoglobulin, whey protein (animals)	On-membrane inhibition of TYR	Unknown	[31]

(continued)

Table 2 List of TYR inhibiting peptides from natural or hybrid sources. (cont'd)

Compound name	Peptide sequence	Natural source (species)	Claimed activity	IC$_{50}$	References
Undefined	ISLLDAQS	β-Lactoglobulin, whey protein (animals)	On-membrane inhibition of TYR	Unknown	[31]
Undefined	VSLLLVGI	α-Lactalbumin, whey protein (animals)	On-membrane inhibition of TYR	Unknown	[31]
Undefined	MMSFVSLL	α-Lactalbumin, whey protein (animals)	On-membrane inhibition of TYR	Unknown	[31]
Undefined	ASVSVSFG	β-Conglycinin, wheat protein (plant)	On-membrane inhibition of TYR	Unknown	[31]
Undefined	MKTFLILV	Gliadin, wheat protein (plants)	On-membrane inhibition of TYR	Unknown	[31]
Undefined	LILVLLAI	Gliadin, wheat protein (plants)	On-membrane inhibition of TYR	Unknown	[31]
Undefined	SVNVHSSL	Ovalbumin, egg white (animals)	On-membrane inhibition of TYR	Unknown	[31]

Name	Sequence	Source	Mechanism	Value	Reference
Undefined	MVLVNAIV	Ovalbumin, egg white (animals)	On-membrane inhibition of TYR	Unknown	[31]
Undefined	IAIMSALA	Ovalbumin, egg white (animals)	On-membrane inhibition of TYR	Unknown	[31]
Silk sericin hydrolysates	Unknown	Wastewater material from cocoons (animals)	Inhibition of mushrooms TYR (I) Ferrous-ion-chelating activity (II)	8.71 mg/mL (I) 0.128 mg/mL (II)	[30,32]
EF-5	ECGYF	Vigna (plants)	Inhibition of mushrooms TYR	0.46 mM	[38]
F1-a	NYRRE	Chaenomeles speciose (plants)	Inhibition of mushrooms TYR	8.69 mg/mL	[40]
F1-b	RHAKF	Chaenomeles speciose (plants)	Inhibition of mushrooms TYR	1.15 mg/mL	[40]
< 3 kDa	Unknown	Salvia hispanica (plants)	Inhibition of mushrooms TYR	0.66 mg/mL	[42]
DWMPHs-IV	Unknown	Juglans regia (plants)	Inhibition of mushrooms TYR monophenolase (I) Inhibition of mushrooms TYR diphenolase (II)	3.52 mg/mL (I) 2.65 mg/mL (II)	[43]

(continued)

Table 2 List of TYR inhibiting peptides from natural or hybrid sources. (*cont'd*)

Compound name	Peptide sequence	Natural source (species)	Claimed activity	IC$_{50}$	References
F2	Unknown	*Juglans regia* (plants)	Inhibition of mushrooms TYR monophenolase (I) Inhibition of mushrooms TYR diphenolase (II)	1.42 mg/mL (I) 1.8 mg/mL (II)	[43]
Undefined	FPY	*Juglans regia* (plants)	Inhibition of mushrooms TYR monophenolase (I) Inhibition of mushrooms TYR diphenolase (II)	1.11 mM (I) 3.22 mM (II)	[43]
Silk sericin	Unknown	*Bombyx mori cocoons* (animals)	Inhibition of mushrooms TYR	Unknown	[44]
JPC	Unknown	*Rhopilema esculentum* (animals)	Inhibition of melanin production on melanoma cells (I) Copper-chelating activity (II)	78.2 μg/mL (I) 88.7 μg/mL (II)	[46]
MSCP	Unknown	*Chanos chanos* (animals)	Inhibition of mushrooms TYR (I) Inhibition of TYR and melanin synthesis in melanoma cells (II)	752.4 μg/mL (I) 887.1 μg/mL (II)	[49]
LMW-GH	Unknown	*Stichopus japonicus*	Inhibition of mushrooms TYR	Unknown	[51]

β-Lactoglobulin	Protein Data Bank code: 1BEB	Whey Proteins (animals)	Density of cell blot in cultured melanocytes	Unknown	[52]
Undefined	MHIRL	Whey Proteins (animals)	Inhibition of mushrooms TYR	83 μM	[55,58]
Undefined	HIRL	Whey Proteins (animals)	Inhibition of mushrooms TYR	70 μM	[55,58]
Undefined	MGRY	Pavlova lutheri (bacteria)	Reduction of melanin in B16F10 cells Reduction of melanogenesis-related proteins expression	Unknown	[56]
αS-CN hydrolysates	Unknown	Camelus dromedarius (animals)	Inhibition of mushrooms TYR	Unknown	[57]
Phosvitin	Unknown	Egg yolk (animals)	Inhibition of mushrooms TYR (I) Inhibition of TYR and melanin synthesis in melanoma cells (II)	≈ 50 μg/mL (II)	[60]
Compound n.3	ILELPFASGDLLML	Egg white proteins (animals)	Predicted monophenolase and diphenolase inhibition	Unknown	[63]
Compound n.6	GYSLGNWVC–AAK	Egg white proteins (animals)	Predicted monophenolase and diphenolase inhibition	Unknown	[63]

(continued)

Table 2 List of TYR inhibiting peptides from natural or hybrid sources. (cont'd)

Compound name	Peptide sequence	Natural source (species)	Claimed activity	IC$_{50}$	References
Compound n.10	YFGYTGALRCLV	Egg white proteins (animals)	Predicted monophenolase and diphenolase inhibition	Unknown	[63]
Compound n.11	HIATNAVLFFGR	Egg white proteins (animals)	Predicted monophenolase and diphenolase inhibition	Unknown	[63]
Compound n.13	FMMFESQNKDL–LFK	Egg white proteins (animals)	Predicted monophenolase and diphenolase inhibition	Unknown	[63]
Compound n.18	SGALHCLK	Egg white proteins (animals)	Predicted monophenolase and diphenolase inhibition	Unknown	[63]
Compound n.21	YFGYTGALR	Egg white proteins (animals)	Predicted monophenolase and diphenolase inhibition	Unknown	[63]
Pt-5	Unknown	Zebrafish (animals)	Inhibition of mushrooms TYR, TRP-1, TRP-2, and MITF expression in melanoma cells	Unknown	[67]
Leucrocin I	NGVQPKY	*Crocodylus siamensis* (animals)	Inhibition of mushrooms TYR	>200 μM	[69]
TILI-1	NGVQPKC	Modified from Leucrocin I	Inhibition of mushrooms TYR	132 μM	[69]
TILI-2	CNGVQPK	Modified from Leucrocin I	Inhibition of mushrooms TYR	113 μM	[69]

As a last example of peptides deriving from milk protein, Addar et al. reported the effect of trypsin and chymotrypsin on αS-casein (αS-CN) collected from *Camelus dromedarius* milk and its hydrolytic fractions that were evaluated for their antioxidant, TYR and urease inhibitory activities. αS-CN and its hydrolysate fractions classified depending on their different molecular weight were tested at 200 µg/mL, and the results showed that low molecular weight fractions obtained from chymotrypsin hydrolysis had better TYR inhibitory effect than the ones at higher molecular weight (>10 kDa) and αS-CN [57]. The highest activity found for chymotrypsin hydrolysates suggests the importance of hydrophobic and aromatic amino acids at the C-terminus position of the peptide for the TYR inhibitory activity, as reported by Schurink et al. [31].

3.4.5 Honey

Tyrosinase inhibiting peptides of approximately 600 Da were also found in honey, however, their sequences were not described and the information regarding their activity were relatively poor and no further investigated [59].

3.4.6 Egg

In 2012 a Korean research group demonstrated the anti-melanogenesis activity of phosvitin [60]. Phosvitin is a phosphoglycoprotein present in egg yolk with specific amino acid composition comprised of 50% serine, 90% of which are phosphorylated [61] that makes this protein a strong metal chelator agent. Additionally, C-terminal and N-terminal of this protein are composed of hydrophobic amino acids that contribute to TYR active site binding [62]. Phosvitin activity was evaluated by the dopachrome method using mushroom tyrosinase, but also in vitro on B16F10 (melanoma cells) by measuring tyrosinase activity and intracellular melanin content. In these experiments, phosvitin resulted capable of inhibiting mushroom tyrosinase, B16F10 tyrosinase expression and decreasing melanin content, even if in a lower manner in comparison with ascorbic acid, that is one of the gold standards. In particular, it was demonstrated that phosvitin not only interferes with the first phase of eumelanogenesis pathway, but also with the second phase by inhibiting MITF expression.

Indeed, in another study the potential anti-tyrosinases activity of egg white proteins fragments obtained by enzymatic digestion was evaluated [63]. The inhibitory effect was evaluated in terms of monophenolase and diphenolase inhibitory activity. From this study emerged that seven sequences deriving from different egg proteins (including e.g., ovalbumin, ovotransferrin, ovomucoid, ovalbumin-related proteins) could inhibit up to 45.9% monophenolase and up to 48.1% diphenolase, being considered as potential tyrosinase inhibitory peptides. In particular these bioactive peptides were: ILELPFASGDLLML, GYSLGNWVCAAK, YFGYTGALR-CLV, HIATNAVLFFGR, FMMFESQNKDLLFK, SGALHCLK and YFGYTGALR [63].

Phosvitin-derived peptides with anti-TYR activity were also found in zebrafishes' eggs. In particular, in light of the antimicrobial immunomodulating properties and antioxidant activity exhibited by phosvitin-derived peptide Pt5 (encompassing the 55C-terminal amino acids of phosvitin) [64–66], Liu et al. conducted a study on the possible involvement of peptide Pt5 in the melanogenesis processes [67]. Peptide Pt5 was shown to inhibit TYR in vitro (dopachrome method) with inhibitory rates of approximately 3.0%, 7.7%, 12.0%, and 16.0% at 5, 10, 50, and 100 µg/mL, respectively. Additionally, Pt5 also inhibits TYR activity and melanin biosynthesis in melanoma cells (B16F10) by reducing TYR, tyrosinase related protein-1 (TRP-1), tyrosinase related protein-2 (TRP-2), and microphthalmia-associated transcription factor (MITF) levels in cells [67].

3.4.7 Blood

A very recent example of a well characterized peptide deriving from animal source with TYR inhibitory effect, is Leucrocin I (NGVQPKY) originating from *Crocodylus siamensis* (crocodile) white blood cell extract, originally described for its anti-microbial activity [68,69]. In fact, this peptide has a C-terminal tyrosine, a characteristic that can confer tyrosinase inhibitory effect, and this evidence boosted a Thai research group to evaluate its possible implication in the TYR inhibitory processes. Moreover, to understand structure-activity relations, Leucrocin I was rationally modified, and two new peptides were obtained: Tyrosinase Inhibitor Leucrocin I-1 and -2, TILI-1 (NGVQPKC), and TILI-2 (CNGVQPK) respectively. Their efficacy and mechanism were evaluated, surprisingly indicating that the change in the tyrosine of Leucrocin I to cysteine (which led to the TILI-1 sequence) could enhance the inhibitory activity. Additionally, the N-terminal cysteine (TILI-2) showed a higher contribution to the TYR inhibitory activity than the C-terminal one (TILI-1). This effect was depicted both for mushrooms TYR inhibition, and for the reduction in melanin content exhibited when used as treatment in melanoma cells (B16). However, all the tested peptides were relatively weak inhibitors of TYR if compared to other well-known inhibitors, such as kojic acid (IC_{50} = 26 µM) [69].

3.5 Anti-tyrosinase peptides designed from hybrid technologies

Bioactive peptides for different applications can be also designed starting from *in silico* screening of virtual libraries deriving from natural or artificial sources. An example of this application to identify crucial complementary functional groups essential for mushroom TYR inhibition is the study reported by Hsiao et al. [70]. These authors screened compounds reported in a variety of literature references and matched them to find the perfect ligand-based pharmacophore. Among the best 10 molecules found, compound A5 (structure reported in Fig. 4) displayed better results due to the presence of two main region, the first one called Core1, with a tyrosine-like structure and the second one, Core2, similar to a coumarin derivative.

Core1 can act as hydrogen-bond donor, and it can form hydrophobic interactions, while Core2 behave as hydrogen-bond acceptor and can interact through hydrophobic interactions with TYR. Keeping in mind the features of A5, the peptide KFY, with complementary properties, has been identify and tested, exhibiting potent TYR inhibitory activity. Starting from KFY and keeping in mind the constrain of tyrosine in the C-terminal position and the biophysical properties of lysine and phenylalanine, a set of different tripeptides were synthetized. Among them two different peptides CRY and RCY showed considerably high potency in inhibiting tyrosinase activity. The results highlight the necessity of having an arginine residue, both in the N-terminus or in the middle position of the peptide to increase the tyrosine-peptide interaction. CRY particularly showed impressive tyrosinase inhibitory activity, with an IC_{50} value of 6.16 µM, 14-fold lower than kojic acid and 160 times lower than arbutin, a well-known TYR inhibitor. The molecular docking analysis reveal that CRY peptide has features comparable to the model identified from docking studies, forming electrostatic and hydrophobic interactions with glutamic acid and valine residues in the active site of the enzyme. The C-terminal tyrosine instead, is

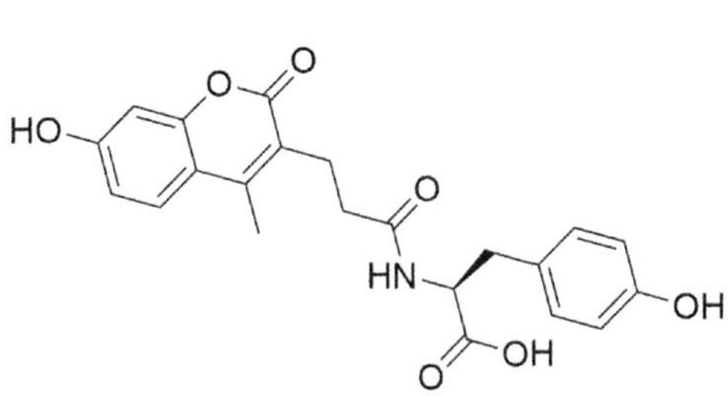

Fig. 4 Structure of A5.

responsible of the establishment of hydrophobic interactions with tryptophan and methionine residues of TYR. Peptide RCY does not display the same interactions of CRY, nor the same fitting inside the enzyme's binding pocket, however its inhibitory potency can be attributed to the presence of cysteine in the C-terminus. Indeed, both tripeptides, possess a cysteine residue close to the copper ions in the active site of mushroom TYR, that thanks to the presence of a thiol group, can act as copper ion chelator.

A second hybrid approach is the use of phage display that represents a potent method for discovering and modifying polypeptides possessing desired functionalities. By merging DNA fragments encoding peptides with genes coding for bacteriophage coat proteins, the resulting fused genes become integrated into phage particles. Subsequently, these particles prominently display the sought-after polypeptides on their exterior surfaces and can be used in screening processes.

Two examples of the application of phage display to the discovery of new peptide TYR inhibitors, are reported by Lee et al. [71], and Nie et al. [72].

In the first example, on the basis of a crystal structure of TYR with a caddie protein in which a short sequence (VSHY) of interaction was shown, the authors exploited the phage display technique to obtain four libraries of tetrapeptides ($CX_1X_2X_3$, $X_3X_2X_1C$, $YX_1X_2X_3$, and $X_3X_2X_1Y$ where C stands for Cys residue and Y stands for non-Cys residue) and screen for new peptide TYR inhibitors. Among these libraries, N-terminal cysteine-containing tetrapeptides showed the most potent tyrosinase-inhibitory abilities, whereas CRVI tetrapeptide exhibited the strongest tyrosinase-inhibitory potency ($IC_{50} = 2.7\,\mu M$), meaning that the N-terminal positioning of the Cys residue is important, but also that the sulfur atom of the thiol group in the cysteine residue is important for copper coordination, as the replacement with Ser negatively affects the potency of the derived peptides [71].

In the second example, indeed, the use of the phage display techniques brought to a novel anti-TYR heptapeptide, whose sequence is IQSHPFF, that was shown to be active both in the mono- and in the diphenolase reactions, with IC_{50} of 1.7 and 4.0 mM, respectively. The authors also reported a kinetic analysis that suggested a competitive and reversible TYR inhibition, possibly occurring through His chelation of copper at the TYR active site, with a inhibition constant (K_i) of 0.765 mM [72].

4. Anti-tyrosinase peptides synthetically produced
4.1 Dipeptides and tripeptides

Girelli et al. [73], with the aim of developing non-toxic inhibitors against mushroom TYR, studied the effect on the enzyme of various glycyl-dipeptides using reverse phase HPLC and UV determination of dopachrome formed. The amount of colored product formed is proportional to the enzyme activity and is chromatographically determined both in absence and in presence of the dipeptides. The results obtained evidence that a non-linear reduction of enzyme activity is present for dipeptides GlyAsp, GlyLys, GlyPhe, and GlyGly, as reported in Table 3.

According to the authors, polyphenol oxidase can bind to both the substrate and the inhibitor forming two different complexes, the enzyme-substrate complex and the enzyme inhibitor complex even though also a ternary inactive, non-productive complex formed by the enzyme and both the substrate and the inhibitor. Comparing the inhibition constant of the different dipeptides, it appeared that the dipeptide GlyAsp has the highest

Table 3 Inhibition constant for the dipeptides GlyAsp, GlyPhe, GlyLys and GlyGly express in mM concentration.

Compound:	Structure:	Inhibition constant (mM):
Glycilaspartic Acid (GlyAsp)		0.66
Glycilphenylalanine (GlyPhe)		32.42
Glycillysine (GlyLys)		33.34
Glycilglycine (GlyGly)		33.20

inhibition power, probably due to the interaction of the peptide with the enzyme through the addition al carboxylic group.

Tseng et al. [74] investigate the inhibitory activity of 400 dipeptides against mushroom TYR. The results evidence that cysteine-containing dipeptides abolish the activity of the enzyme, however also dipeptides PD, DY and YK reduce the activity of tyrosinase by 60–70%. The authors also determine the inhibitory potency against polyphenol oxidase of the dipeptides containing cysteine, as reported in Table 4. The IC_{50} values for all the tested compounds ranged from 2.0 to 22.6 μM, with better results for peptides CS, CY, CF, CC, CI, CM and CQ. Peptides with C-terminal cysteine showed lower inhibition efficacy even though some of them such as IC, NC, QC, FC and VC, have comparable IC_{50} values to the N-terminal cysteine containing dipeptides.

To further elucidate how peptides interact with the enzyme, molecular modeling studies were done using CE, CS, CY, CA, and CW and their reverse forms EC, SC, YC, AC, and WC. Molecular modeling studies were performed on both N-terminal and C-terminal cysteine containing

Table 4 IC_{50} values for dipeptides containing N-terminal cysteine or C-terminal cysteine.

	Compound:	Inhibition constant (μM):
	CS	4.5
	CY	3.1
	CF	2.7
N-terminal cysteine	CC	3.2
	CI	4.0
	CM	4.9
	CQ	3.5
	IC	4.5
	NC	5.3
C-terminal cysteine	QC	5.9
	FC	7.9
	VC	8.3

peptides and according to these studies the cysteine residue, independently of its position in the N- or C-terminal side of the peptides, interacts with the binuclear copper ions inside the active site causing the inhibition of the enzyme.

O-dopaquinone scavengers, such as thiol containing compounds are known to react with dopaquinone leading to the formation of thiol-dopa conjugate. Similarly, cysteine containing dipeptides can form dopaquinone conjugates, limiting the formation of dopachrome. For this reason, the dipeptides reported by the authors can hypothetically act in two different ways: (i) interaction with the active sites of TYR and (ii) formation of conjugates with o-dopaquinone with consequent inhibition of dopachrome formation. To confirm this hypothesis Tseng et al. investigated the use of HPLC analysis to determine the formation of the reported peptide-dopa conjugate but no evidence of their presence was found, thus suggesting that their main mechanism of action is the direct inhibition of the enzyme.

Molecular modeling studies were performed also on tyrosine containing peptides, showing that both YK and DY peptides are able to enter deep inside the enzyme binding pocket and form π-π interactions between the phenol ring of tyrosine ad the imidazole ring of the histidine residue of the enzyme. These peptides were also tested in in vitro experiment showing a substrate like inhibitory behavior.

Finally, cytotoxicity assay on cysteine containing peptide YC and CA, with an IC_{50} of 131.6 μM and 9.6 μM respectively, were performed: melanin levels reduction with no significant cytotoxicity was found suggesting a possible use of these peptides in skin whitening products.

Hsiao et al. [70] used molecular docking and pharmacophore modeling to build a pharmacophore model which was subsequently used to screen a vast library of natural compounds with potential activity as mushroom TYR inhibitor as previously reported in the *Anti-tyrosinase peptides designed from hybrid technologies* paragraph.

Luisi et al., as mentioned above, studied the effect of sulfur containing amino acids and glutathione, its gamma-oxa-analogue H-Glo(Cys-Gly-OH)-OH, two small synthetic tripeptides Z-tLeu-Asp(OtBu)-Sc and Ac-tLeu-Leu-Asp(OtBu)-Sc (whose structures are reported in Fig. 5), on the activity of TYR [13].

In particular, the two peptides Z-tLeu-Asp(OtBu)-Sc and Ac-tLeu-Leu-Asp(OtBu)-Sc contain a semicarbazone moiety and the natural, non-proteinogenic amino acid tert-leucine, which is characterized by higher

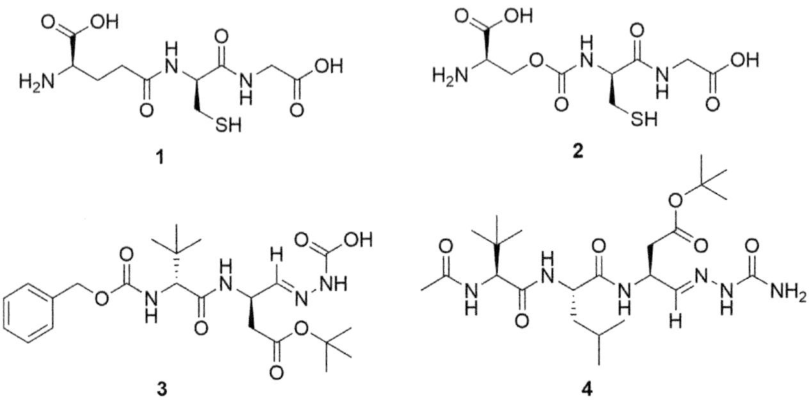

Fig. 5 Structures for glutathione (1), glutathion gamma-oxa-analogue H-Glo(Cys-Gly-OH)-OH (2), peptide Z-tLeu-Asp(OtBu)-Sc (3) and peptide Ac-tLeu-Leu-Asp(OtBu)-Sc (4).

lipophilicity and a bulkier side chain compared to the one of Leucine and Isoleucine. Indeed, this amino acid has been proven to decrease peptide radical formation and increase in vivo stabilization. Results from the dopachrome method are reported in Table 5.

The inhibitory activity of the newly synthetized tripeptides is also accompanied by a good metal reducing power and strong metal chelating activity.

The effect of glutathione on TYR activity has been deeply studied also by Villarama et al. [75]. Glutathione exists in our body in the reduced form GSH which is constantly oxidized and then reduced by the corresponding enzymes. The complex balance among all these species is strongly regulated by internal factors such as hormones and inflammations and external agents like UV lights, chemicals and heat. The mechanism of action of the tripeptide on TYR can be dual: (a) direct inactivation of tyrosinase caused by the binding of the peptide to copper, indeed, TYR is the rate limiting enzyme in the production of melanin; (b) quenching of free radicals and peroxides involved in tyrosinase activation, in particular, GSH can scavenge reactive oxygen species (ROS) generated after UV exposure.

Despite the fact that thiols, and thus GSH, are able to chelate copper ions, present in the active site of TYR, causing the inactivation of the enzyme in a dose dependent manner, TYR is commonly present in an environment rich in cysteine and/or thiol containing compounds without displaying any sign of inhibition. Thus, the reason why glutathione act as TYR inhibitor must be linked to some other phenomena. Indeed, glutathione is involved in TYR transportation, in particular in the inhibition of the transfer of tyrosinase T1 from vesicle to premelanosomes.

Table 5 Tyrosinase inhibitory activity for GSH, its gamma-oxa-analogue and two peptides containing tLeu expressed as milligrams of kojic acid equivalents.

Compound:	Tyrosinase inhibitory activity (mgKAE/g):
GSH	45.60 ± 0.15
H-Glo(Cys-Gly-OH)-OH	216.95 ± 0.17
Z-tLeu-Asp(OtBu)-Sc	163.87 ± 0.90
Ac-tLeu-Leu-Asp(OtBu)-Sc	152.69 ± 1.19

GSH is also involved in the switching process between eumelanin and pheomelanin production. In particular, glutathione interferes with the binding of cysteine to L-dopaquinone causing the induction of cysteinyldopa synthesis and in presence of glutathione-S-transferase it can bind to L-dopaquinone giving glutathionyldopa, a precursor of cystenyldopa causing an increased production of pheomelanin, a yellow-red pigment (as reported in Fig. 6). Thus, glutathione enables the switch from eumelanogenesis to pheaomelanogenesis which results in a lighter skin pigmentation.

Interestingly, as reported by Prota et al., in absence of cysteine or GSH or when there is a decrease in the concentration of these compounds, the hydroxylation of tyrosine followed by dopachrome formation and its transformation into its indole derivative increases, thus increasing eumelanogenesis [76]. The effect of glutathione has also been investigated in a double-blind and placebo-controlled clinical trial of 2014, performed by Watanabe et al. who reported the use of topical lotion containing oxidized glutathione as skin whitening agent [77]. During the 10 weeks of treatment, the skin displayed a progressively lower melanin index, without having marked adverse effects. Although the mechanism of action of GSSG is yet unclear, its efficacy as skin whitening agent could be linked to the presence in the skin of glutathione reductase an enzyme able to rapidly reduce GSSG to GSH.

4.2 Oligopeptides

Mimosine, a non-proteinogenic amino acid reported in Fig. 7, is found in tropical and subtropical plants of genera *Leucaena*. This amino acid has many therapeutic roles, including antitumoral, antiviral and antiproliferative. Therefore, Upadhyay et al. studied as potential TYR inhibitor mimosine-derived tetrapeptides, prepared by solid phase peptide synthesis (SPPS) [78]. The experiments carried out on mushroom tyrosinase highlighted that all the synthetized compounds have better

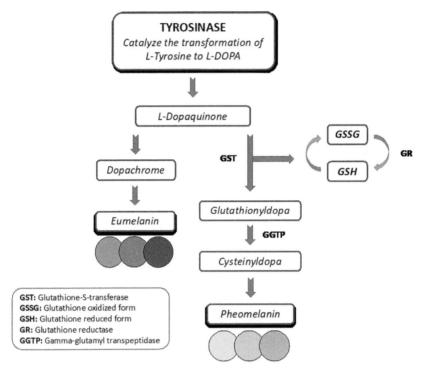

Fig. 6 Scheme of glutathione involvement in melanin synthesis. Highter levels of GHS converted to GSSG led to the synthesis of pheomelanin (red/yellow pigment) in higher percentage than eumelanin (dark-brown/black pigment), resulting in lighter skin tones.

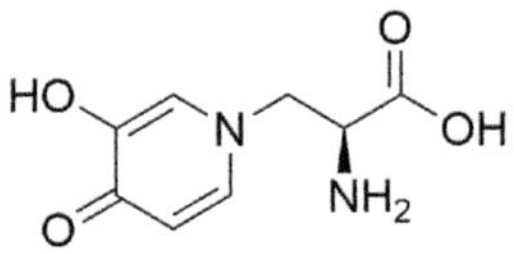

Fig. 7 Structure of mimosine.

inhibitory activity than mimosine itself with an IC_{50} ranging from 5.6 to 36.7 μM. In particular, M–FFY had the best activity amongst the synthetized compounds.

In 2014, Lien and coworkers investigated the effects of acetyl penta-peptides on mushroom tyrosinase as potential inhibitors of melanogenesis [79]. Four different 4n-acetyl-pentapeptides containing Ser, Arg, Phe and Lys were synthetized and tested on B16F10 melanoma cell line. All the compounds were found to be potent TYR inhibitors with IC_{50} below 1 mg/mL, as reported in Table 6.

Table 6 N-Acetyl-pentapeptides inhibitory profile against diphenolase activity of mushroom tyrosinase.

Compound:	IC$_{50}$ (mg/mL):
Ac-P1; Ac-RSRFK	0.75 ± 0.16
Ac-P2; Ac-KSRFR	0.78 ± 0.14
Ac-P3; Ac-KSSFR	0.81 ± 0.02
Ac-P4; Ac-RSRFS	0.29 ± 0.03

Interestingly, the difference in the inhibitory activity of Ac-P1 and Ac-P4 seems to reside in the only different amino acid of the sequence, the serine residue in the C-terminal part of the pentapeptide, suggesting that it plays a critical role on the enzyme activity. Experiments on B16F10 cell line were done using Ac-P4, the most promising compound. The results highlighted that, not only Ac-P4 is not cytotoxic for concentrations ranging from 0.1 to 1.0 mg/mL, but it was also able to reduce the α-MSH induced cellular melanin content in a dose dependent manner, being even more effective than kojic acid. Indeed, IC$_{50}$ for Ac-P4 and kojic acid was 0.09 mM and 1.51 mM, respectively. Taken together, these findings suggest a possible use of Ac-P4 as potential hyperpigmentation reducing agent.

Ubeid et al., after the screening of a proprietary library, identified two promising peptides as tyrosinase inhibitor [25]. The two peptides, P3 and P4 (Seq. P3: RADSRADC as shown in Fig. 8; Seq. P4: YRSRKYSSWY reported in Fig. 9), display enzyme inhibition at very low concentration compared to hydroquinone, with an IC$_{50}$ of 123 μM and 40 μM respectively.

Interestingly, the inhibitory activity of these compounds depends on the substrate and on the presence of cofactors: in presence of L-DOPA as cofactor and L-Tyrosine as substrate, P3 display better inhibitory properties than P4, on the other hand, in presence of only L-tyrosine, P4 gives better results. These findings support the hypothesis of the presence of two different sites on the enzyme, one for the binding of L-DOPA and the other one for the binding of L-tyrosine. Moreover, one of the main advantages of the two peptides is the very low cytotoxicity displayed. While hydroquinone is 100% toxic after 24-hours treatment at 100 μM concentration, P3 does not display cytotoxicity and P4 is only 7% toxic. Furthermore, they both have minimal effects on melanocyte proliferation; thus, they can be potentially used as therapeutic peptides for the treatment of skin

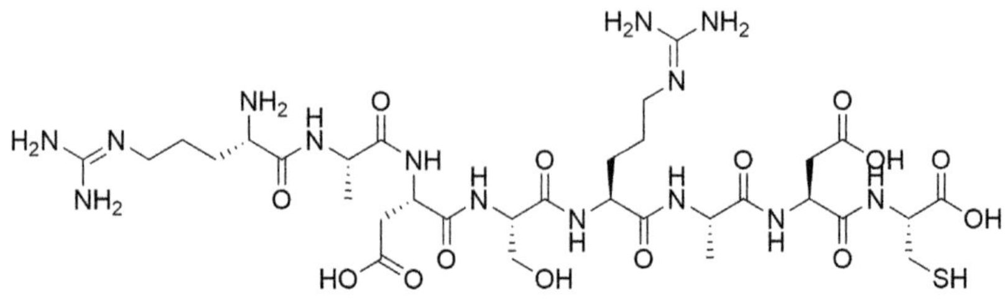

Fig. 8 Structure of peptide P3.

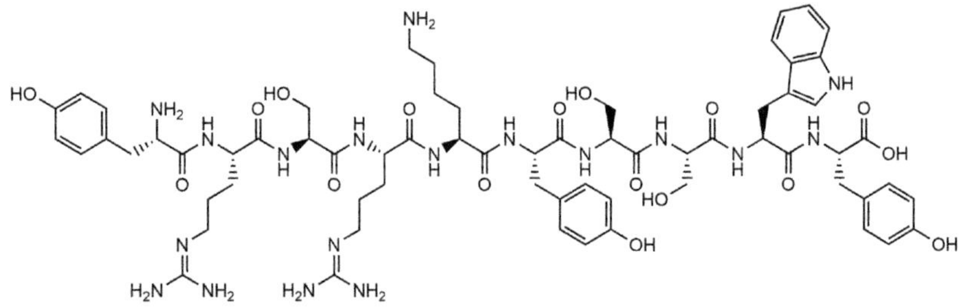

Fig. 9 Structure of peptide P4.

hyperpigmentation. Indeed, P4 is commercialized under the name of Lumixyl ™ a skin-whitening product used to treat melasma [80,81].

In a follow-up study, the same authors described the effect of three octapeptides (Seq. P16: RRWWRRYY, Seq. P17: RRRYWYYR, and Seq. P18: RRYWYWRR, Fig. 10) as tyrosinase inhibitors [82].

The three peptides, which from molecular docking studies, displayed the highest affinity for the TYR catalytic site, have been tested in vitro on human melanocytes, keratinocytes and fibroblast, showing very low cytotoxic effects and revealing a competitive mode of inhibition of the enzyme. The inhibition activity of the peptides was tested both using L-DOPA as substrate and L-tyrosine in presence of L-DOPA as cofactor. As previously reported, these results support the hypothesis of the presence of two catalytic sites as postulated by Hearing et al. [83]. The experiments carried out on primary human melanocytes revealed that the octapeptides are capable of penetrating inside the cells and inhibiting the production of melanin having hypo-pigmenting activity similar to hydroquinone. MTT assays showed minimal toxicity for all the three peptides both after 72 h and 6-days treatments. Interestingly the octapeptides, share the presence of a tryptophan residue that, according to molecular docking simulations, can enter the catalytic site of tyrosinase close to the two copper ions. This

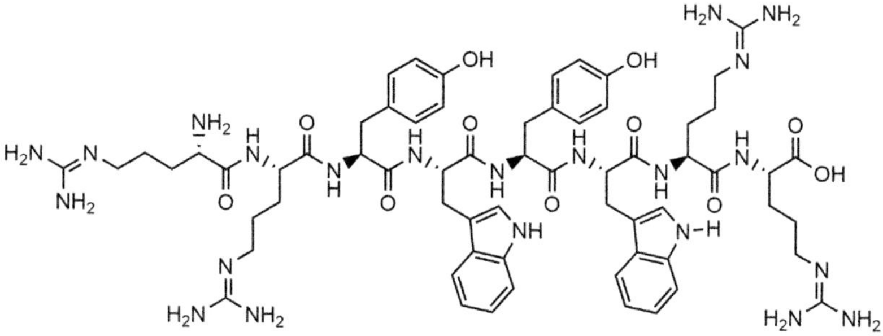

Fig. 10 Structures of peptide P16, P17 and P18.

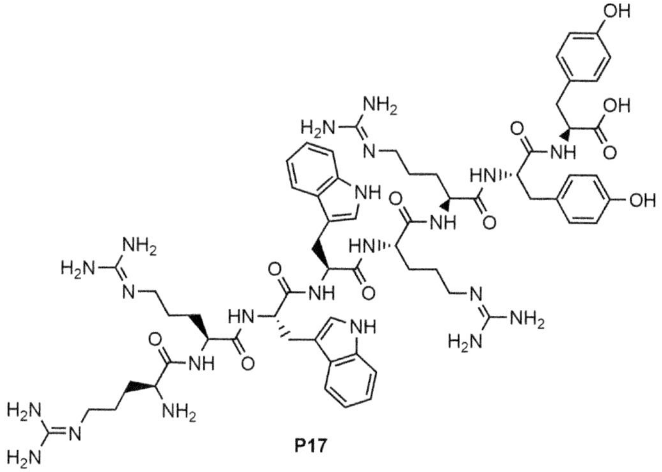

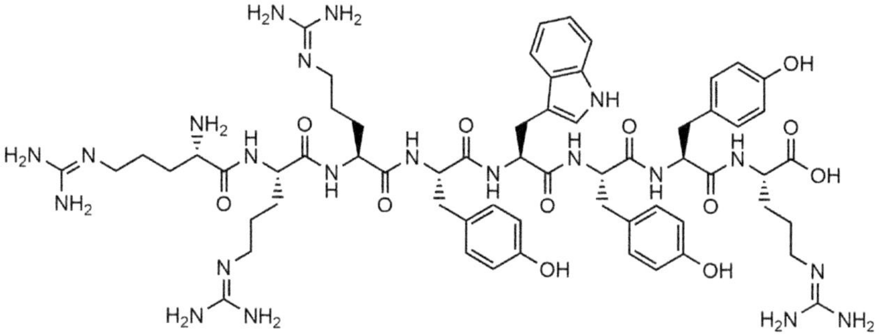

assumption could be validated by the fact that P17, the only octapeptide with one tryptophan residue has lower inhibitory activity respect to the other two compounds, P16 and P18, which contain two tryptophan residues each. Indeed, the indole ring of tryptophan can form a reversible complex with the copper ion of the catalytic site, which can also interact with the imidazole ring of histidine residues via hydrophobic interactions.

4.3 Peptides conjugated with other chemical moieties

4.3.1 Kojic-acid derivatives

In 2009, Noh and coworkers studied the effect of kojic acid–amino acid conjugates as tyrosinase inhibitors [84]. Kojic acid is employed in many countries as skin whitening agents, however the use in cosmeceutical products has been dramatically reduced due to its low stability during storage and high skin irritability. The new compounds show better activity than kojic acid itself, in particular the ones containing amino acids with aromatic side chains like phenylalanine, tyrosine, tryptophan or histidine. Among them KA-F-NH$_2$, displays the best tyrosinase inhibitory activity with an IC$_{50}$ of 14.7 μM. Further kinetic studies on the inhibition of L-DOPA oxidation reveal that KA-F-NH$_2$, can act as a non-competitive inhibitor of mushroom tyrosinase analogously to kojic acid.

The same group also studied the inhibitory activity of tripeptides conjugated with kojic acid [85]. The authors synthetized a library of 22 kojic acid-tripeptides and tested them as inhibitors of TYR comparing the results with the ones previously obtained for kojic acid and kojic acid-tripeptide free acids [86].

Results showed that the inhibitory activity of the conjugated tripeptides were similar to those of the tripeptide-free acids and higher than kojic acid alone. In particular, they investigated the importance of the different residues in the different positions with respect to the kojic acid moiety. Indeed, they find out that peptides having phenylalanine at position KA+1, have the highest inhibitory activity. Substitutions in this position of phenylalanine with hydrophobic amino acids such as tryptophan or tyrosine, create compounds with the best inhibitory activity against TYR, suggesting that this position plays a key role in enzyme inhibition. Fixing the KA+1 and KA+3 positions as phenylalanine and tyrosine the authors studied the effect of the different amino acids in position KA+2 discovering that the amino acid in this position is less relevant for enzyme inhibition. From the combination of all the obtained results the authors selected three different kojic   acid-tripeptides   free   amide   with   an   IC$_{50}$ lower   than   the corresponding kojic acid-tripeptides free acid as reported in Table 7. Interestingly, the FWY peptide conjugated with KA both in the free amid or free acid forms, displays the lower IC_{50}.

Two more peptides conjugated with kojic acid with anti-tyrosinase activity where developed by the group of Singh et al. [87]. The two peptides KA-PS and KA-CDPGYIGSR (renamed KA-CR9) were tested on mushroom tyrosinase displaying a dose-dependent inhibitory activity on the enzyme of 82% and 84% respectively. Due to the lower molecular weight of KA-PS conjugate further studies on mouse B16F10 cells and melan-a cells were performed, highlighting that the compound does not display cytotoxicity and retains the same activity of in vitro experiments.

4.3.2 Other moieties

In 2013, Li et al., studied the potential activity as tyrosinase inhibitors of hydroxypyridinone-L-phenylalanine conjugates [88]. Hydroxypyridinone is a kojic acid bioisoster having NH instead of oxygen at position 1 in the pyranone ring. In their work the authors synthetized six different derivatives and tested them as inhibitors for both monophenolase and diphenolase activity of tyrosinase, as displayed in Fig. 11.

The results revealed that amongst all the synthetized compounds 5d and 5e were the only with an IC_{50} lower than the one of kojic acid alone, with values of 19.2 ± 0.8 and $12.6 \pm 0.3\,\mu M$ respectively. Interestingly the inhibitory activity of the compounds increases with increasing hydrophobicity of the compound itself, the only exception being compound 5f where the too high lipophilicity caused a poor water solubility.

Table 7 IC_{50} value for Kojic Acid, Kojic acid free acid conjugated peptides and Kojic acid free amide conjugated peptides.

	Compound:	IC_{50} (µM):
Kojic Acid	KA	94
	KA-FWY	1.28
Free-acid	KA-FHY	4.55
	KA-FRY	5.92
	KA-FWY-NH_2	2.2
Free-amide	KA-FHY-NH_2	2.36
	KA-FRY-NH_2	3.59

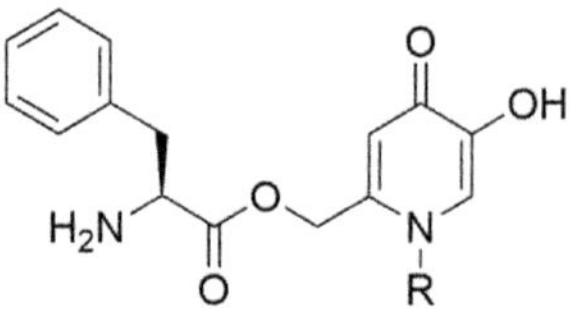

5a: R=CH$_3$
5b: R=C$_2$H$_5$
5c: R=n-C$_4$H$_9$
5d: R=n-C$_6$H$_{13}$
5e: R=n-C$_8$H$_{17}$
5f : R=n-C$_{10}$H$_{21}$

Fig. 11 Structure of hydroxypyridinone-L-phenylalanine conjugates.

Taking into account these findings, the two compounds were also tested on the diphenolase activity of TYR, using L-DOPA as substrate. Even in this case, 5e demonstrated a better inhibitory activity than kojic acid itself, with an IC$_{50}$ of 20 μM. Moreover, from kinetic studies, it was shown that both 5d and 5e molecules were mixed type inhibitors, meaning that they can bind both to the free enzyme and the enzyme-substrate complex. Thus, compound 5e was also tested for its copper chelating ability. 5e can act as bidentate ligand, forming two different complexes with copper ions present in the catalytic site of TYR: CuL$^+$ and CuL$_2$. Thanks to spectrophotometric titration, the stability constant, expressed in log scale were calculated for both CuL$^+$ (log β_1) and CuL$_2$ (log β_2) exhibiting values of 9.294 ± 0.09 and 15.719 ± 0.11, respectively. Indeed, both log β_1 and log β_2 of 5e were higher than the one of kojic acid (log β_1 = 6.6, log β_2 = 11.7) suggesting a higher affinity of the newly synthetized compound for the enzyme's copper ions.

Given the promising results of Li et al., Zhao and coworkers designed and synthetized hydroxypyridinone- amino acid and hydroxypyridinone-dipeptides and tested them as TYR inhibitors [89]. Among the two different series only compound 6e and 12a, whose structures are reported in Fig. 12, had the strongest monophenolase inhibitory activity with an IC$_{50}$ of 1.95 and 2.79 μM respectively.

Keeping in mind these results the authors also investigated the diphenolase activity of the enzyme using compounds 6e and 12a. In these conditions, compound 6e revealed to be a better inhibitor than compound 12a. however, as mentioned above they can be both regarded as reversible mixed type inhibitors of mushroom tyrosinase. In order to further investigate the inhibitory mechanism of the synthetized derivatives copper chelating reducing and copper chelating ability for compound 6e were determined. From the absorbance measurement at 483 nm (λ for cuprous ions determination), it was possible to confirm that compound 6e had a stronger copper reducing capability than kojic acid itself. Moreover,

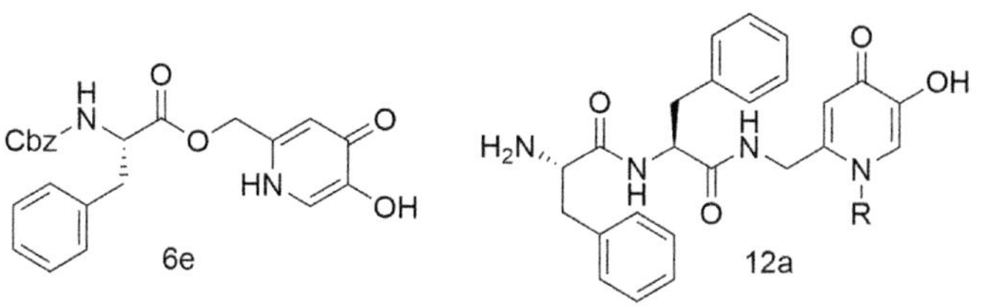

Fig. 12 Structure of hydroxypyridinone-amino acid 6e (Cbz: benzyloxycarbonyl) and hydroxypyridinone-dipeptide 12a.

although the copper chelating ability of compound 6e and kojic acid both increases, increasing the concentration, compound 6e displayed better chelating properties, with values of 91.34 ± 0.28% and 83.34 ± 0.29% at 0.952 mM, at pH 5.0 and 7.4 respectively, while kojic acid at the same values of pH has constants of 82.64 ± 0.25% and 74.35 ± 0.27% at 0.952 mM. Probably compound 6e can entre in the active site of tyrosinase and compete with histidine for the binding of copper ions, causing a loss of enzymatic activity.

Considering that hydroxyphenolic acids and their esters can act as tyrosinase inhibitors and as antioxidant, Noh et al. decided to conjugate protocatechuic acid (PA), a-resocylic acid, gentisic acid (GT) and gallic acid (GA) with aromatic amino acids [90]. The different compounds were tested as inhibitor of TYR using tyrosine as substrate. Amongst them PA-F-NH_2, PA-W-NH_2 and PA-Y-NH_2 display promising results, while the ones conjugated with RA, GT and GA hardly have any inhibitory activity on the enzyme. The IC_{50} values, for both monophenolase and diphenolase activities, of the conjugated compound are listed in Table 8.

All the three compounds can be regarded as reversible competitive-uncompetitive mixed-I type inhibitors. Since PA is a good chelator for copper ions, the mechanism of action of these conjugated molecules can be double. On one hand PA can chelate the copper ions present in the active site of tyrosinase enzyme, on the other hand the aromatic structure of the inhibitor could favor the formation of hydrophobic interactions between the inhibitor itself and the hydrophobic side chains of TYR, thus preventing the entrance of the substrate in the active site.

Recently, peptide fragments derived from β-lactoglobulin, a milk protein, have been reported to be TYR inhibitors [91]. In a work of 2018, Yang and coworkers reported the synthesis of conjugated molecules containing these peptides and caffeic acid (CA), a catechol containing compound with antioxidant activity and which have been proven in a previous

Table 8 IC_{50} value for monophenolase and diphenolase activity of the peptides conjugated with protocatechuic acid.

Compound:	IC_{50} (mM) monophenolase:	IC_{50} (mM) diphenolase:
PA-F-NH$_2$	0.01 ± 0.001	0.64 ± 0.02
PA-W-NH$_2$	0.01 ± 0.002	0.56 ± 0.01
PA-Y-NH$_2$	0.01 ± 0.005	0.66 ± 0.01

study to be able to inhibit α-MSH induced melanin synthesis when in the form of phenethyl ester [92,93]. Using solid phase peptide synthesis, the authors were able to successfully conjugate CA(acetonide)-OH with tetra- and tri-peptides (structure reported in Fig. 13).

The compounds reported in Table 9, were tested against tyrosinase and their inhibitory activity measured using the dopachrome method. To compare the different activities, peptide conjugated with caffeic acids were tested in parallel to the peptides deprived of the caffeic acid moiety.

Results evidence that all the CA-derived peptides were more active against TYR than the unconjugated peptides, suggesting that CA improve the binding of the peptides to the enzyme. Amongst them, CA-MHIR was able to effectively inhibit the enzyme even at a concentration of 50 μM, displaying even better results than Kojic Acid with an IC_{50} of 47.9 μM.

The inhibition mechanism of CA-MHIR (Fig. 14) was investigated revealing a non-competitive type of inhibition. Probably the conjugated peptide can bind outside the active site of the enzyme causing a deformation in the three-dimensional structure of the protein.

Park et al. investigated the effect of caffeic acid and coumaric acid-conjugated peptides in human melanocytes, which in recent studies have been shown to serve as skin-whitening agents inhibiting melanin production more efficiently than KA or arbutin [94]. The group synthetized, using SPPS, peptides containing the -AR- sequence which in previous studies have been proven to be more effective in inhibiting melanogenesis [95]. The peptides were than conjugated to caffeic acid or coumaric acid through a mini-PEG or Gly-Gly-Gly linker to improve the physical properties of the compound and increase its efficacy in inhibiting melanin production, and subsequently tested to measure cell viability, tyrosinase activity and α-MSH-induced melanin content. No signs of cytotoxicity were shown for the different caffeic acid and coumaric acid-conjugated peptides even at 1 mM concentrations. Moreover, the experiment on cells

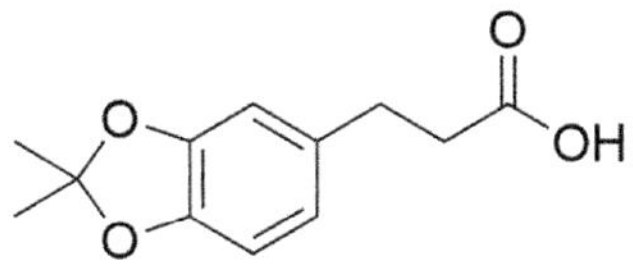

Fig. 13 Structure of caffeic acid acetonide.

Table 9 IC_{50} values for β-lactoglobulin derived peptides with and without caffeic acid moiety.

Compound:	IC_{50} (mM):
CA–MHIR	47.9
CA–HIR	154.8
CA–HIRL	166.2
KA	201.7
HIRL	218.8
MHIR	257.1
HIR	none

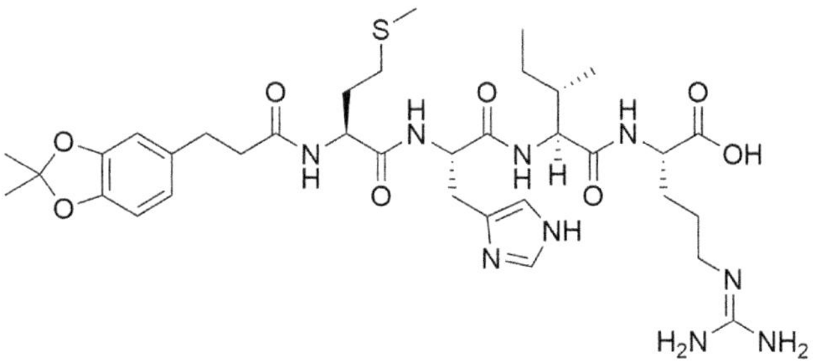

Fig. 14 Structure of CA-MHIR.

treated with α -MSH, to induce melanin production, clearly showed that the conjugated peptides have an inhibitory activity comparable to arbutin, a well-known melanogenesis inhibitor. In particular, four of the synthetized peptides were able to inhibit tyrosinase activity, with coumaric acid-GGG-ARP being stronger than arbutin and thus opening the possibility of use it in cosmetic products to treat or prevent skin pigmentation disorders.

In a study of 2013, Lee at al. synthetized a coumaroyl dipeptide amide, Coumaric acid-LG-NH$_2$, using SPPS and tested it as skin whitening agent [96]. The peptide, which exhibit very low cytotoxicity in MTT assay, demonstrated to be able to inhibit TYR with an IC$_{50}$ value of 182.4 µM. In experiments on B16F1 melanoma cell lines, the measurement of melanin content revealed that the peptide can successfully reduce α-MSH induced melanin production.

Ascorbic acid, called also vitamin C, is characterized by high instability towards oxygen light and heat. Moreover, it cannot penetrate into cell due to its high hydrophilicity. Thus, especially in the cosmetic industry, there has been the necessity to reduce its rapid degradation and, at the same time, increase its permeability. Choi et al. were able to achieve this, by linking ascorbic acid to pentapeptide KTTKS [97]. The compound was tested in in vitro experiment, and its ability to inhibit TYR was measured. The results highlighted that the stabilized ascorbyl pentapeptide (SAP) had comparable inhibitory activity to ascorbic acid at 500 µM and 1000 µM. Moreover, experiments carried out on B16 murine melanoma cells, showed that SAP had a melanogenesis-inhibitory activity more than 10% higher than arbutin. Indeed, SAP displayed more promising results than ascorbic acid in in vivo tyrosinase inhibition at 1 mM concentration.

Recently, a new study on tripeptides conjugated with thiosemicarbazones have been reported as potential TYR inhibitors [98,99]. Ledwoń et al. in their work investigated the inhibitory effect on diphenolase activity of thiosemicarbazone peptide-conjugated using in vitro experiments on mushroom TYR and on B16F0 cell line [100]. A set of 15 different compounds have been investigated, as reported in Table 10.

Thiosemicarbazones (TSC1, TSC2, and TSC3, Fig. 15) have been designed in order to have the free carboxylic group, responsible of peptide conjugation, distant from the thiourea group involved in the binding to copper ions in the active site of the enzyme.

Tripeptides were selected on the basis of the most active sequences reported in the literature, starting for some key requirements, i.e., the presence of a N-terminal phenylalanine residue, and of aromatic and hydrophobic amino acid (Phe, Tyr, Trp). TYR inhibition assay shows that all TSC and tripeptide-conjugates have a dose dependent inhibitory activity, even if some differences are present such as the lowest ability of TSC2, the bulkiest thiosemicarbazone, to inhibit TYR even at 150 mM concentration. Interestingly, the peptides deprived from the thiosemicarbazone moiety are not active against the enzyme. Indeed, TSC1 and

Table 10 Results of melanin production in B16F0 cell line.

Compound:	IC$_{50}$ of melanin production:
Kojic Acid	121.0
TSC1	101.5
TSC2	19.8
TSC3	126.4
Ac-FFY-OH	245.7
Ac-FYY-OH	174.7
Ac-FWY-OH	140.9
TSC1-FFY-OH	138.4
TSC1-FYY-OH	158.5
TSC1-FWY-OH	150.6
TSC2-FFY-OH	146.1
TSC2-FYY-OH	n/a
TSC2-FWY-OH	**89.9**
TSC3-FFY-OH	162.4
TSC3-FYY-OH	144.9
TSC3-FWY-OH	**67.1**

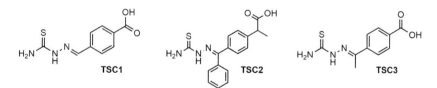

Fig. 15 Structures of TSC1, TSC2, and TSC3.

TSC2 as well as TSC2-FWY-OH and TSC3-FWY-OH are the only compounds displaying stronger potency as melanogenesis inhibitor than kojic acid. On the other hand, melanogenesis inhibition gave different results, highlighting that TSC2 is the most potent melanogenesis inhibitor among all the tested compounds. This observation, coupled to the previous results, suggest that TSC2 probably interfere with melanogenesis by acting on other key enzymes involved in melanin biosynthesis.

5. Peptidomimetics

The use of peptidomimetic compounds has been widely explored in the past years, since these molecules are expected to feature enhanced stability and bioavailability for in vivo applications, as compared to small peptides. Indeed, although peptides present numerous advantages, they also display some challenges which include their stability and availability, since their half-lives are highly influenced by external conditions such as pH, temperatures and interactions with enzymes. Moreover, their availability also depends from their concentration and distributions/accumulation; e.g., peptides with high molecular weight can present solubility problems at high concentrations, and skin permeability problems [4]. Therefore, to reduce some of the issues connected to the use of peptides, peptidomimetics have been studied and synthetized. In a study of 2007, triazole-containing analogues of a naturally occurring tyrosinase inhibitor have been synthetized and tested against the enzyme [101].

L. helveticus, a gram-positive bacteria, can produce Cyclo[Pro-Tyr-Pro-Val] (Fig. 16, [16]) a small cyclic peptide with TYR inhibitory activity.

However, its synthesis is complex due to the ring closure step. For this reason, taking advantage of the *click chemistry* strategies, analogues containing the triazolyl bridge as amide bond isoster, have been synthetized via Cu(I)-alkyne-azide cycloaddition. Interestingly, the 2022 Nobel Prize in Chemistry recognized the development of *click chemistry* and biorthogonal chemistry to

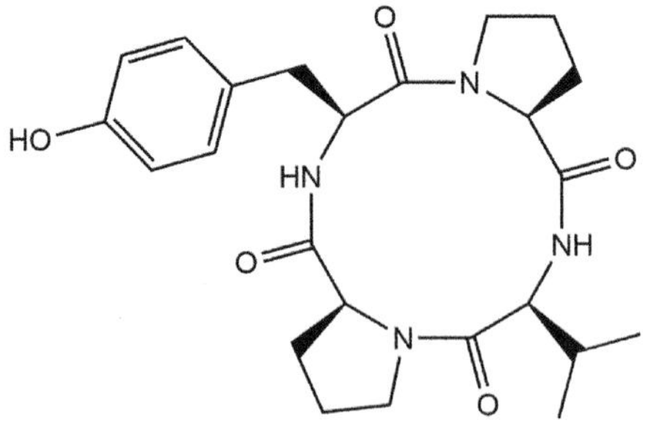

Fig. 16 Structure of cyclo-[Pro-Tyr-Pro-Val].

Table 11 IC$_{50}$ values for the natural compound and the triazolyl-containing peptidomimetics.

Compounds:	IC$_{50}$ (mM):
cyclo-[Pro–Tyr–Pro–Val]	1.5
Triazole analogue 2	0.6
Triazole analogue 3	0.5
Triazole analogue 4	1.6

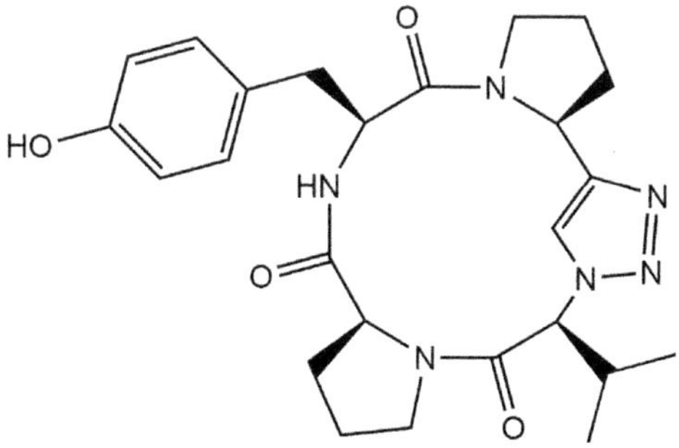

Fig. 17 Structure of the triazole analogue 2.

K. Barry Sharpless, Morten Meldal, and Carolyn R. Bertozzi. The three new peptidomimetics have been tested against mushroom tyrosinase using spectrophotometric assay, displaying the results reported in Table 11.

Surprisingly, not only the analogues retain the inhibitory activity, but triazole analogue 2 (structure reported in Fig. 17) and triazole analogue 3 have better performances compared to the reference compound. Taken together, these results suggest that these triazolyl-containing peptidomimetics, which are readily accessible via click chemistry, could be introduced inside difficult-to-obtain natural compounds, retaining the natural biological activity.

6. Conclusions

This chapter describes the variety of peptides proposed as TIPs, starting from the natural ones, which can be isolated from several sources (i.e., bacteria, mushrooms, plants or animals, as reported in Table 2), taking into account hybrid approaches in which in silico screening strategies are employed, and ending with totally synthetic peptides.

Naturally derived peptides in some cases are considered a sustainable choice in which by-products can be recycled, however, despite the huge variety of peptides deriving from natural sources, a consideration on their safety emerges spontaneously. In fact, in most of the examples the term "peptide" is used to describe protein hydrolysates, which by definition are mixtures of many unknown peptides, but not as single identity. Nevertheless, in some cases peptide characterization from natural extract has been rigorously carried out, leading to the rational purification of peptide sequences whose activity was deeply investigated [25,27,28,38,40,43,58,69].

Indeed, the advantage of fully synthetic approaches is to obtain products that can be easily purified and fully characterized as single molecules that, upon proper biological assays, may result in safe TIPs.

The number of peptides launched in the pharmaceutical market is getting increased through the years, therefore the demand for optimized methodologies to synthetize peptides consequently increases [102,103]. Undoubtedly, the new frontiers in peptide synthesis are the application of greener reagents and solvents to comply nowadays sustainability perspectives [104–106].

References

[1] Á. Sánchez-Ferrer, J. Neptuno Rodríguez-López, F. García-Cánovas, F. García-Carmona, Tyrosinase: a comprehensive review of its mechanism, Biochim. Biophys. Acta (BBA)—Protein Struct. Mol. Enzymol. 1247 (1) (1995) 1–11, https://doi.org/10.1016/0167-4838(94)00204-T

[2] V. Del Marmol, F. Beermann, Tyrosinase and related proteins in mammalian pigmentation, FEBS Lett. 381 (3) (Mar. 1996) 165–168, https://doi.org/10.1016/0014-5793(96)00109-3

[3] X. Lai, H.J. Wichers, M. Soler-Lopez, B.W. Dijkstra, Structure and function of human tyrosinase and tyrosinase-related proteins, Chem. Eur. J. 24 (1) (2018) 47–55, https://doi.org/10.1002/chem.201704410

[4] L.T.N. Ngoc, J.-Y. Moon, Y.-C. Lee, Insights into bioactive peptides in cosmetics, Cosmetics 10 (4) (2023) 111, https://doi.org/10.3390/cosmetics10040111

[5] K. Fosgerau, T. Hoffmann, Peptide therapeutics: Current status and future directions, Drug Discov. Today 20 (1) (2015) 122–128, https://doi.org/10.1016/j.drudis.2014.10.003

[6] F. Errante, P. Ledwoń, R. Latajka, P. Rovero, A.M. Papini, Cosmeceutical peptides in the framework of sustainable wellness economy, Front. Chem. 8 (2020) 572923, https://doi.org/10.3389/fchem.2020.572923

[7] T.-S. Chang, An updated review of tyrosinase inhibitors, IJMS 10 (6) (2009) 2440–2475, https://doi.org/10.3390/ijms10062440

[8] L. Gentilucci, R. De Marco, L. Cerisoli, Chemical modifications designed to improve peptide stability: Incorporation of non-natural amino acids, pseudo-peptide bonds, and cyclization, CPD 16 (28) (2010) 3185–3203, https://doi.org/10.2174/138161210793292555

[9] J. Park, et al., D-tyrosine negatively regulates melanin synthesis by competitively inhibiting tyrosinase activity, Pigment. Cell Melanoma Res. 31 (3) (2018) 374–383, https://doi.org/10.1111/pcmr.12668

[10] A. Slominski, P. Jastreboff, J. Pawelek, L-Tyrosine stimulates induction of tyrosinase activity by MSH and reduces cooperative interactions between MSH receptors in hamster melanoma cells, Biosci. Rep. 9 (5) (1989) 579–586, https://doi.org/10.1007/BF01119801

[11] J. Park, H. Jung, B. Jang, H.-K. Song, I.-O. Han, E.-S. Oh, D-Tyrosine adds an anti-melanogenic effect to cosmetic peptides, Sci. Rep. 10 (1) (2020) 262, https://doi.org/10.1038/s41598-019-57159-3

[12] W.C. Liao, W.H. Wu, P.-C. Tsai, H.-F. Wang, Y.-H. Liu, C.-F. Chan, Kinetics of ergothioneine inhibition of mushroom tyrosinase, Appl. Biochem. Biotechnol. 166 (2) (2012) 259–267, https://doi.org/10.1007/s12010-011-9421-x

[13] G. Luisi, A. Stefanucci, G. Zengin, M. Dimmito, A. Mollica, Anti-oxidant and tyrosinase inhibitory in vitro activity of amino acids and small peptides: new hints for the multifaceted treatment of neurologic and metabolic disfunctions, Antioxidants 8 (1) (2018) 7, https://doi.org/10.3390/antiox8010007

[14] V. Kahn, Effect of proteins, protein hydrolyzates and amino acids on o-dihydrox-yphenolase activity of polyphenol oxidase of mushroom, avocado, and banana, J. Food Sci. 50 (1) (1985) 111–115, https://doi.org/10.1111/j.1365-2621.1985.tb13288.x

[15] C. Madhosingh, L. Sundberg, Purification and properties of tyrosinase inhibitor from mushroom, FEBS Lett. 49 (2) (1974) 156–158, https://doi.org/10.1016/0014-5793(74)80500-4

[16] H. Kawagishi, A. Somoto, J. Kuranari, A. Kimura, S. Chiba, A novel cyclote-trapeptide produced by Lactobacillus helveticus as a tyrosinase inhibitor, Tetrahedron Lett. 34 (21) (1993) 3439–3440, https://doi.org/10.1016/S0040-4039(00)79177-5

[17] T. Sano, K. Kaya, Oscillapeptin G, a tyrosinase inhibitor from toxic Oscillatoria agardhii, J. Nat. Prod. 59 (1) (1996) 90–92, https://doi.org/10.1021/np9600210

[18] K. Fujii, K. Sivonen, E. Naganawa, K. Harada, Non-toxic peptides from toxic cyanobacteria, Oscillatoria agardhii, Tetrahedron 56 (5) (2000) 725–733, https://doi.org/10.1016/S0040-4020(99)01017-0

[19] D.A. Alonzo, T.M. Schmeing, Biosynthesis of depsipeptides, or Depsi: the peptides with varied generations, Protein Sci. 29 (12) (2020) 2316–2347, https://doi.org/10.1002/pro.3979

[20] J. Kitagaki, G. Shi, S. Miyauchi, S. Murakami, Y. Yang, Cyclic depsipeptides as potential cancer therapeutics, Anti-Cancer Drugs 26 (3) (2015) 259–271, https://doi.org/10.1097/CAD.0000000000000183

[21] H. Morita, T. Kayashita, H. Kobata, A. Gonda, K. Takeya, H. Itokawa, Pseudostellarins A–C, new tyrosinase inhibitory cyclic peptides from Pseudostellaria heterophylla, Tetrahedron 50 (23) (1994) 6797–6804, https://doi.org/10.1016/S0040-4020(01)81333-8

[22] H. Morita, T. Kayashita, H. Kobata, A. Gonda, K. Takeya, H. Itokawa, Pseudostellarins D–F, new tyrosinase inhibitory cyclic peptides from Pseudostellaria heterophylla, Tetrahedron 50 (33) (1994) 9975–9982, https://doi.org/10.1016/S0040-4020(01)89612-5

[23] H. Morita, H. Kobata, K. Takeya, H. Itokawa, Pseudostellarin G, a new tyrosinase inhibitory cyclic octapeptide from Pseudostellaria heterophylla, Tetrahedron Lett. 35 (21) (May 1994) 3563–3564, https://doi.org/10.1016/S0040-4039(00)73238-2

[24] X. Lai, H.J. Wichers, M. Soler-Lopez, B.W. Dijkstra, Structure of human tyrosinase related protein 1 reveals a binuclear zinc active site important for melanogenesis, Angew. Chem. Int. Ed. 56 (33) (2017) 9812–9815, https://doi.org/10.1002/anie.201704616

[25] A. Abu Ubeid, L. Zhao, Y. Wang, B.M. Hantash, Short-sequence oligopeptides with inhibitory activity against mushroom and human tyrosinase, J. Investig. Dermatol. 129 (9) (2009) 2242–2249, https://doi.org/10.1038/jid.2009.124

[26] B.M. Hantash, F. Jimenez, A split-face, double-blind, randomized and placebo-controlled pilot evaluation of a novel oligopeptide for the treatment of recalcitrant melasma, J. Drugs Dermatol. 8 (8) (2009) 732–735.

[27] A. Ochiai, et al., New tyrosinase inhibitory decapeptide: molecular insights into the role of tyrosine residues, J. Biosci. Bioeng. 121 (6) (2016) 607–613, https://doi.org/10.1016/j.jbiosc.2015.10.010

[28] A. Ochiai, S. Tanaka, T. Tanaka, M. Taniguchi, Rice bran protein as a potent source of antimelanogenic peptides with tyrosinase inhibitory activity, J. Nat. Prod. 79 (10) (2016) 2545–2551, https://doi.org/10.1021/acs.jnatprod.6b00449

[29] S.G. Burton, Oxidizing enzymes as biocatalysts, Trends Biotechnol. 21 (12) (2003) 543–549, https://doi.org/10.1016/j.tibtech.2003.10.006

[30] S. Kubglomsong, C. Theerakulkait, R.L. Reed, L. Yang, C.S. Maier, J.F. Stevens, Isolation and identification of tyrosinase-inhibitory and copper-chelating peptides from hydrolyzed rice-bran-derived albumin, J. Agric. Food Chem. 66 (31) (2018) 8346–8354, https://doi.org/10.1021/acs.jafc.8b01849

[31] M. Schurink, W.J.H. Van Berkel, H.J. Wichers, C.G. Boeriu, Novel peptides with tyrosinase inhibitory activity, Peptides 28 (3) (2007) 485–495, https://doi.org/10.1016/j.peptides.2006.11.023

[32] J.-H. Wu, Z. Wang, S.-Y. Xu, Enzymatic production of bioactive peptides from sericin recovered from silk industry wastewater, Process. Biochem. 43 (5) (2008) 480–487, https://doi.org/10.1016/j.procbio.2007.11.018

[33] H.M. Ali, A.M. El-Gizawy, R.E.I. El-Bassiouny, M.A. Saleh, The role of various amino acids in enzymatic browning process in potato tubers, and identifying the browning products, Food Chem. 192 (2016) 879–885, https://doi.org/10.1016/j.foodchem.2015.07.100

[34] M.M. Baakdah, A. Tsopmo, Identification of peptides, metal binding and lipid peroxidation activities of HPLC fractions of hydrolyzed oat bran proteins, J. Food Sci. Technol. 53 (9) (2016) 3593–3601, https://doi.org/10.1007/s13197-016-2341-6

[35] R.E. Cian, A.G. Garzón, D.B. Ancona, L.C. Guerrero, S.R. Drago, Chelating properties of peptides from red seaweed Pyropia columbina and its effect on iron bio-accessibility, Plant. Foods Hum. Nutr. 71 (1) (2016) 96–101, https://doi.org/10.1007/s11130-016-0533-x

[36] L. Zhu, J. Chen, X. Tang, Y.L. Xiong, Reducing, radical scavenging, and chelation properties of in vitro digests of alcalase-treated zein hydrolysate, J. Agric. Food Chem. 56 (8) (2008) 2714–2721, https://doi.org/10.1021/jf703697e

[37] V. Kahn, Effect of proteins, protein hydrolyzates and amino acids on o-dihydrox-yphenolase activity of polyphenol oxidase of mushroom, avocado, and banana, J. Food Sci. 50 (1) (1985) 111–115, https://doi.org/10.1111/j.1365-2621.1985.tb13288.x

[38] Z. Shen, Y. Wang, Z. Guo, T. Tan, Y. Zhang, Novel tyrosinase inhibitory peptide with free radical scavenging ability, J. Enzyme Inhib. Med. Chem. 34 (1) (2019) 1633–1640, https://doi.org/10.1080/14756366.2019.1661401

[39] X. Xie, G. Zou, C. Li, Antitumor and immunomodulatory activities of a water-soluble polysaccharide from Chaenomeles speciosa, Carbohydr. Polym. 132 (2015) 323–329, https://doi.org/10.1016/j.carbpol.2015.06.046

[40] Y. Deng, et al., Skin-care functions of peptides prepared from Chinese quince seed protein: sequences analysis, tyrosinase inhibition and molecular docking study, Ind. Crop. Products 148 (2020) 112331, https://doi.org/10.1016/j.indcrop.2020.112331

[41] U. Urbizo-Reyes, M.F. San Martin-González, J. Garcia-Bravo, A. López Malo Vigil, A.M. Liceaga, Physicochemical characteristics of chia seed (*Salvia hispanica*) protein hydrolysates produced using ultrasonication followed by microwave-assisted hydrolysis, Food Hydrocoll. 97 (2019) 105187, https://doi.org/10.1016/j.foodhyd.2019.105187

[42] J.E. Aguilar-Toalá, A.M. Liceaga, Identification of chia seed (*Salvia hispanica* L.) peptides with enzyme inhibition activity towards skin-aging enzymes, Amino Acids 52 (8) (2020) 1149–1159, https://doi.org/10.1007/s00726-020-02879-4

[43] Y. Feng, et al., Separation, identification, and molecular docking of tyrosinase inhibitory peptides from the hydrolysates of defatted walnut (*Juglans regia* L.) meal, Food Chem. 353 (2021) 129471, https://doi.org/10.1016/j.foodchem.2021.129471

[44] N. Kato, S. Sato, A. Yamanaka, H. Yamada, N. Fuwa, M. Nomura, Silk protein, sericin, inhibits lipid peroxidation and tyrosinase activity, Biosci. Biotechnol. Biochem. 62 (1) (1998) 145–147, https://doi.org/10.1271/bbb.62.145

[45] J. Yi, Y. Ding, Dual effects of whey protein isolates on the inhibition of enzymatic browning and clarification of apple juice, Czech J. Food Sci. 32 (6) (2014) 601–609, https://doi.org/10.17221/69/2014-CJFS

[46] Y. Zhuang, L. Sun, X. Zhao, J. Wang, H. Hou, B. Li, Antioxidant and melano-genesis-inhibitory activities of collagen peptide from jellyfish (*Rhopilema esculentum*), J. Sci. Food Agric. 89 (10) (2009) 1722–1727, https://doi.org/10.1002/jsfa.3645

[47] J. Venkatesan, S. Anil, S.-K. Kim, M. Shim, Marine fish proteins and peptides for cosmeceuticals: a review, Mar. Drugs 15 (5) (2017) 143, https://doi.org/10.3390/md15050143

[48] L. Nakchum, S.M. Kim, Preparation of squid skin collagen hydrolysate as an anti-hyaluronidase, antityrosinase, and antioxidant agent, Prep. Biochem. Biotechnol. 46 (2) (2016) 123–130, https://doi.org/10.1080/10826068.2014.995808

[49] Y.-P. Chen, H.-T. Wu, G.-H. Wang, C.-H. Liang, Improvement of skin condition on skin moisture and anti-melanogenesis by collagen peptides from milkfish (*Chanos chanos*) scales, IOP Conf. Ser.: Mater. Sci. Eng. 382 (2018) 022067, https://doi.org/10.1088/1757-899X/382/2/022067

[50] M. Saito, N. Kunisaki, N. Urano, S. Kimura, Collagen as the major edible com-ponent of sea cucumber (*Stichopus japonicus*), J. Food Sci. 67 (4) (2002) 1319–1322, https://doi.org/10.1111/j.1365-2621.2002.tb10281.x

[51] J. Wang, et al., Antioxidation activities of low-molecular-weight gelatin hydrolysate isolated from the sea cucumber *Stichopus japonicus*, J. Ocean. Univ. China 9 (1) (2010) 94–98, https://doi.org/10.1007/s11802-010-0094-9

[52] M. Nakajima, et al., β-Lactoglobulin suppresses melanogenesis in cultured human melanocytes, Pigment. Cell Res. 10 (6) (1997) 410–413, https://doi.org/10.1111/j.1600-0749.1997.tb00700.x

[53] K. Sato, M. Morita, C. Ichikawa, H. Takahashi, M. Toriyama, Depigmenting mechanisms of all-trans retinoic acid and retinol on B16 melanoma cells, Biosci. Biotechnol. Biochem. 72 (10) (2008) 2589–2597, https://doi.org/10.1271/bbb.80279

[54] K. Suetsuna, H. Ukeda, H. Ochi, Isolation and characterization of free radical scavenging activities peptides derived from casein, J. Nutr. Biochem. 11 (3) (2000) 128–131, https://doi.org/10.1016/S0955-2863(99)00083-2

[55] B. Hernández-Ledesma, A. Dávalos, B. Bartolomé, L. Amigo, Preparation of anti-oxidant enzymatic hydrolysates from α-lactalbumin and β-lactoglobulin. Identification of active peptides by HPLC-MS/MS, J. Agric. Food Chem. 53 (3) (2005) 588–593, https://doi.org/10.1021/jf048626m

[56] G.-W. Oh, et al., A novel peptide purified from the fermented microalga *Pavlova lutheri* attenuates oxidative stress and melanogenesis in B16F10 melanoma cells, Process. Biochem. 50 (8) (2015) 1318–1326, https://doi.org/10.1016/j.procbio.2015.05.007

[57] L. Addar, C. Bensouici, S. Si Ahmed Zennia, S. Boudjenah Haroun, A. Mati, Antioxidant, tyrosinase and urease inhibitory activities of camel αS-casein and its hydrolysate fractions, Small Rumin. Res. 173 (2019) 30–35, https://doi.org/10.1016/j.smallrumres.2019.01.015

[58] S. Kong, H.-R. Choi, Y.-J. Kim, Y.-S. Lee, K.-C. Park, S.-Y. Kwak, Milk protein-derived antioxidant tetrapeptides as potential hypopigmenting agents, Antioxidants 9 (11) (2020) 1106, https://doi.org/10.3390/antiox9111106

[59] S. Ates, S. Pekyardimci, C. Cokmus, Partial characterization of a peptide from honey that inhibits mushroom polyphenol oxidase, J. Food Biochem. 25 (2) (2001) 127–137, https://doi.org/10.1111/j.1745-4514.2001.tb00729.x

[60] S. Jung, D.H. Kim, J.H. Son, K. Nam, D.U. Ahn, C. Jo, The functional property of egg yolk phosvitin as a melanogenesis inhibitor, Food Chem. 135 (3) (2012) 993–998, https://doi.org/10.1016/j.foodchem.2012.05.113

[61] R.C. Clark, The primary structure of avian phosvitins, Int. J. Biochem. 17 (9) (1985) 983–988, https://doi.org/10.1016/0020-711X(85)90243-5

[62] B.M. Byrne, A.D. Van Het Schip, J.A.M. Van De Klundert, A.C. Arnberg, M. Gruber, G. Ab, Amino acid sequence of phosvitin derived from the nucleotide sequence of part of the chicken vitellogenin gene, Biochemistry 23 (19) (1984) 4275–4279, https://doi.org/10.1021/bi00314a003

[63] P.-G. Yap, C.-Y. Gan, Chicken egg white—Advancing from food to skin health therapy: Optimization of hydrolysis condition and identification of tyrosinase inhibitor peptides, Foods 9 (9) (2020) 1312, https://doi.org/10.3390/foods9091312

[64] L. Hu, C. Sun, J. Luan, L. Lu, S. Zhang, Zebrafish phosvitin is an antioxidant with non-cytotoxic activity, ABBS 47 (5) (2015) 349–354, https://doi.org/10.1093/abbs/gmv023

[65] S. Wang, Y. Wang, J. Ma, Y. Ding, S. Zhang, Phosvitin plays a critical role in the immunity of zebrafish embryos via acting as a pattern recognition receptor and an antimicrobial effector, J. Biol. Chem. 286 (25) (2011) 22653–22664, https://doi.org/10.1074/jbc.M111.247635

[66] Y. Ding, X. Liu, L. Bu, H. Li, S. Zhang, Antimicrobial–immunomodulatory activities of zebrafish phosvitin-derived peptide Pt5, Peptides 37 (2) (2012) 309–313, https://doi.org/10.1016/j.peptides.2012.07.014

[67] Y.-Y. Liu, et al., Zebrafish phosvitin-derived peptide Pt5 inhibits melanogenesis via cAMP pathway, Fish. Physiol. Biochem. 43 (2) (2017) 517–525, https://doi.org/10.1007/s10695-016-0306-3

[68] S. Pata, et al., Characterization of the novel antibacterial peptide Leucrocin from crocodile (Crocodylus siamensis) white blood cell extracts, Dev. Comp. Immunol. 35 (5) (2011) 545–553, https://doi.org/10.1016/j.dci.2010.12.011

[69] A. Joompang, et al., Evaluation of tyrosinase inhibitory activity and mechanism of Leucrocin I and its modified peptides, J. Biosci. Bioeng. 130 (3) (2020) 239–246, https://doi.org/10.1016/j.jbiosc.2020.04.002

[70] N.-W. Hsiao, et al., Serendipitous discovery of short peptides from natural products as tyrosinase inhibitors, J. Chem. Inf. Model. 54 (11) (2014) 3099–3111, https://doi.org/10.1021/ci500370x

[71] Y.-C. Lee, et al., Phage display–mediated discovery of novel tyrosinase-targeting tetrapeptide inhibitors reveals the significance of N-terminal preference of cysteine residues and their functional sulfur atom, Mol. Pharmacol. 87 (2) (2015) 218–230, https://doi.org/10.1124/mol.114.094185

[72] H. Nie, et al., A novel heptapeptide with tyrosinase inhibitory activity identified from a phage display library, Appl. Biochem. Biotechnol. 181 (1) (2017) 219–232, https://doi.org/10.1007/s12010-016-2208-3

[73] A.M. Girelli, E. Mattei, A. Messina, A.M. Tarola, Inhibition of polyphenol oxidases activity by various dipeptides, J. Agric. Food Chem. 52 (10) (2004) 2741–2745, https://doi.org/10.1021/jf0305276

[74] T.-S. Tseng, et al., Discovery of potent cysteine-containing dipeptide inhibitors against tyrosinase: a comprehensive investigation of 20 × 20 dipeptides in inhibiting dopachrome formation, J. Agric. Food Chem. 63 (27) (2015) 6181–6188, https://doi.org/10.1021/acs.jafc.5b01026

[75] C.D. Villarama, H.I. Maibach, Glutathione as a depigmenting agent: an overview, Intern. J. Cosmetic Sci. 27 (3) (2005) 147–153, https://doi.org/10.1111/j.1467-2494.2005.00235.x

[76] G. Prota, Recent advances in the chemistry of melanogenesis in mammals, J. Investig. Dermatol. 75 (1) (1980) 122–127, https://doi.org/10.1111/1523-1747.ep12521344

[77] F. Watanabe, E. Hashizume, G.P. Chan, A. Kamimura, Skin-whitening and skin-condition-improving effects of topical oxidized glutathione: a double-blind and placebo-controlled clinical trial in healthy women, CCID (2014) 267, https://doi.org/10.2147/CCID.S68424

[78] A. Upadhyay, J. Chompoo, N. Taira, M. Fukuta, S. Gima, S. Tawata, Solid-phase synthesis of mimosine tetrapeptides and their inhibitory activities on neuraminidase and tyrosinase, J. Agric. Food Chem. 59 (24) (2011) 12858–12863, https://doi.org/10.1021/jf203494t

[79] C.-Y. Lien, C.-Y. Chen, S.-T. Lai, C.-F. Chan, Kinetics of mushroom tyrosinase and melanogenesis inhibition by N-acetyl-pentapeptides, Sci. World J. 2014 (2014) 1–9, https://doi.org/10.1155/2014/409783

[80] B. Reddy, T. Jow, B.M. Hantash, Bioactive oligopeptides in dermatology: Part I, Exp. Dermatol. 21 (8) (2012) 563–568, https://doi.org/10.1111/j.1600-0625.2012.01528.x

[81] T. Pillaiyar, V. Namasivayam, M. Manickam, S.-H. Jung, Inhibitors of melanogenesis: an updated review, J. Med. Chem. 61 (17) (2018) 7395–7418, https://doi.org/10.1021/acs.jmedchem.7b00967

[82] A.A. Ubeid, S. Do, C. Nye, B.M. Hantash, Potent low toxicity inhibition of human melanogenesis by novel indole-containing octapeptides, Biochim. Biophys. Acta (BBA)—Gen. Subj. 1820 (10) (2012) 1481–1489, https://doi.org/10.1016/j.bbagen.2012.05.003

[83] V.J. Hearing, T.M. Ekel, Mammalian tyrosinase. A comparison of tyrosine hydroxylation and melanin formation, Biochem. J. 157 (3) (1976) 549–557, https://doi.org/10.1042/bj1570549

[84] J.-M. Noh, S.-Y. Kwak, H.-S. Seo, J.-H. Seo, B.-G. Kim, Y.-S. Lee, Kojic acid–amino acid conjugates as tyrosinase inhibitors, Bioorg. Med. Chem. Lett. 19 (19) (2009) 5586–5589, https://doi.org/10.1016/j.bmcl.2009.08.041

[85] J. Noh, S. Kwak, D. Kim, Y. Lee, Kojic acid–tripeptide amide as a new tyrosinase inhibitor, Biopolymers 88 (2) (2007) 300–307, https://doi.org/10.1002/bip.20670

[86] H. Kim, J. Choi, J.K. Cho, S.Y. Kim, Y.-S. Lee, Solid-phase synthesis of kojic acid-tripeptides and their tyrosinase inhibitory activity, storage stability, and toxicity, Bioorg. Med. Chem. Lett. 14 (11) (2004) 2843–2846, https://doi.org/10.1016/j.bmcl.2004.03.046

[87] B.K. Singh, et al., Kojic acid peptide: a new compound with anti-tyrosinase potential, Ann. Dermatol. 28 (5) (2016) 555, https://doi.org/10.5021/ad.2016.28.5.555

[88] D.-F. Li, et al., Design and synthesis of hydroxypyridinone-L-phenylalanine conjugates as potential tyrosinase inhibitors, J. Agric. Food Chem. 61 (27) (2013) 6597–6603, https://doi.org/10.1021/jf401585f

[89] D.-Y. Zhao, et al., Design and synthesis of novel hydroxypyridinone derivatives as potential tyrosinase inhibitors, Bioorg. Med. Chem. Lett. 26 (13) (2016) 3103–3108, https://doi.org/10.1016/j.bmcl.2016.05.006

[90] J.-M. Noh, Y.-S. Lee, Inhibitory activities of hydroxyphenolic acid–amino acid conjugates on tyrosinase, Food Chem. 125 (3) (2011) 953–957, https://doi.org/10.1016/j.foodchem.2010.09.087

[91] B. Hernández-Ledesma, A. Dávalos, B. Bartolomé, L. Amigo, Preparation of anti-oxidant enzymatic hydrolysates from α-lactalbumin and β-lactoglobulin. Identification of active peptides by HPLC-MS/MS, J. Agric. Food Chem. 53 (3) (2005) 588–593, https://doi.org/10.1021/jf048626m

[92] J.-K. Yang, et al., β-Lactoglobulin peptide fragments conjugated with caffeic acid displaying dual activities for tyrosinase inhibition and antioxidant effect, Bioconjugate Chem. 29 (4) (2018) 1000–1005, https://doi.org/10.1021/acs.bioconjchem.8b00050

[93] J.-Y. Lee, H.-J. Choi, T.-W. Chung, C.-H. Kim, H.-S. Jeong, K.-T. Ha, Caffeic acid phenethyl ester inhibits alpha-melanocyte stimulating hormone-induced melanin synthesis through suppressing transactivation activity of microphthalmia-associated transcription factor, J. Nat. Prod. 76 (8) (2013) 1399–1405, https://doi.org/10.1021/np400129z

[94] K.-Y. Park, J. Kim, Synthesis and biological evaluation of the anti-melanogenesis effect of coumaric and caffeic acid-conjugated peptides in human melanocytes, Front. Pharmacol. 11 (2020) 922, https://doi.org/10.3389/fphar.2020.00922

[95] H.-S. Lee, G.-H. Sin, G.-S. Ryu, G.-Y. Ji, I.S. Cho, H.-Y. Kim, Synthesis of small molecule-peptide conjugates as potential whitening agents, Bull. Korean Chem. Soc. 33 (9) (2012) 3004–3008, https://doi.org/10.5012/BKCS.2012.33.9.3004

[96] H.-S. Lee, et al., Synthesis and evaluation of coumaroyl dipeptide amide as potential whitening agents, Bull. Korean Chem. Soc. 34 (10) (2013) 3017–3021, https://doi.org/10.5012/BKCS.2013.34.10.3017

[97] H.-I. Choi, J.-I. Park, H.-J. Kim, D.-W. Kim, S.-S. Kim, A novel L-ascorbic acid and peptide conjugate with increased stability and collagen biosynthesis, BMB Rep. 42 (11) (2009) 743–746, https://doi.org/10.5483/BMBRep.2009.42.11.743

[98] K. Hałdys, et al., Halogenated aromatic thiosemicarbazones as potent inhibitors of tyrosinase and melanogenesis, Bioorg. Chem. 94 (2020) 103419, https://doi.org/10.1016/j.bioorg.2019.103419

[99] K. Hałdys, et al., Inhibitory properties of aromatic thiosemicarbazones on mushroom tyrosinase: synthesis, kinetic studies, molecular docking and effectiveness in mela-nogenesis inhibition, Bioorg. Chem. 81 (2018) 577–586, https://doi.org/10.1016/j.bioorg.2018.09.003

[100] P. Ledwoń, et al., Tripeptides conjugated with thiosemicarbazones: new inhibitors of tyrosinase for cosmeceutical use, J. Enzyme Inhibition Med. Chem. 38 (1) (2023) 2193676, https://doi.org/10.1080/14756366.2023.2193676

[101] V.D. Bock, D. Speijer, H. Hiemstra, J.H. Van Maarseveen, 1,2,3-Triazoles as peptide bond isosteres: synthesis and biological evaluation of cyclotetrapeptide mimics, Org. Biomol. Chem. 5 (6) (2007) 971, https://doi.org/10.1039/b616751a

[102] A. D'Ercole, et al., An optimized safe process from bench to pilot cGMP production of API eptifibatide using a multigram-scale microwave-assisted solid-phase peptide synthesizer, Org. Process. Res. Dev. 25 (12) (2021) 2754–2771, https://doi.org/10.1021/acs.oprd.1c00368

[103] G. Sabatino, et al., An optimized scalable fully automated solid-phase microwave-assisted cGMP-ready process for the preparation of eptifibatide, Org. Process. Res. Dev. 25 (3) (2021) 552–563, https://doi.org/10.1021/acs.oprd.0c00490

[104] R. Subirós-Funosas, R. Prohens, R. Barbas, A. El-Faham, F. Albericio, Oxyma: an efficient additive for peptide synthesis to replace the benzotriazole-based HOBt and HOAt with a Lower Risk of Explosion[1], Chem. Eur. J. 15 (37) (2009) 9394–9403, https://doi.org/10.1002/chem.200900614

[105] L. Ferrazzano, et al., Sustainability in peptide chemistry: current synthesis and purification technologies and future challenges, Green. Chem. 24 (3) (2022) 975–1020, https://doi.org/10.1039/D1GC04387K

[106] L. Pacini, M. Muthyala, L. Aguiar, R. Zitterbart, P. Rovero, A.M. Papini, Optimization of peptide synthesis time and sustainability using novel eco-friendly binary solvent systems with induction heating on an automated peptide synthesizer, J. Peptide Sci. (2024) e3605, https://doi.org/10.1002/psc.3605

CHAPTER SEVEN

Computational studies of tyrosinase inhibitors

Alessandro Bonardi and Paola Gratteri*

NEUROFARBA Department, Pharmaceutical and Nutraceutical Section, Laboratory of Molecular Modeling Cheminformatics & QSAR, University of Florence, Sesto Fiorentino, Firenze, Italy
*Corresponding author. e-mail address: paola.gratteri@unifi.it

Contents

Abstract

Computational studies have significantly advanced the understanding of tyrosinase (TYR) function, mechanism, and inhibition, accelerating the development of more effective and selective inhibitors. This chapter provides an overview of *in silico* studies on TYR inhibitors, emphasizing key inhibitory chemotypes and the main residues involved in ligand-target interactions. The chapter discusses tools applied in the context of TYR inhibitor development, e.g., structure-based virtual screening, molecular docking, artificial intelligence, and machine learning algorithms.

1. Introduction

The multifunctional membrane-bound type-3 copper-containing enzyme tyrosinase (TYR, EC 1.14.18.1), also named monophenol- or *o*-diphenol oxygen oxidoreductase or polyphenol oxidase, is a glycoprotein located in the membrane of the melanosome, which is crucial for melanin biosynthesis [1].

The Enzymes, Volume 56
ISSN 1874-6047, https://doi.org/10.1016/bs.enz.2024.06.008

Melanogenesis is a physiological process that exerts a protective role against UV radiation in all organisms of the phylogenetic tree [1–3]. On the other hand, abnormal production of melanin is responsible for dermatological problems such as solar lentigo, age spots, freckles, melasma, cancer (e.g., melanoma), and post-inflammatory melanoderma [4].

Moreover, TYRs are involved in three crucial metabolic processes of the insects life including melanization, protective encapsulation, and cuticle sclerotization [5].

Recently, implications of TYR in the Parkinson disease (PD) progression were investigated in several studies and incorporated into the neuromelanin and α-synuclein hypothesis [6]. Generally, neuromelanin, which is abundant in the dopaminergic neurons of the *Substantia Nigra*, exerts a neuroprotective role complexing and removing cytosolic neurotoxic substances, such as free radical species and transition metal (e.g., Fe^{2+}) which catalyze dangerous redox reactions and α-synuclein aggregation (Levy Body) [7,8]. However, during the PD progression, the release of neurotoxins bound to neuromelanin could increase intracellular stress and activate microglia to release neurotoxic cytokines that produce neuroinflammation and neurodegeneration [8]. In this context, TYR directly processes α-synuclein inducing its aggregation in more toxic aggregates and influencing the synthesis of neuromelanin [9]. The role of TYRs in the formation of the Levy Body makes them implicated also in the pathophysiology of senile dementia and Alzheimer disease (AD) [8].

The pharmacological uses of TYR inhibitors could include the prevention of fruit and vegetable browning, treatment of melanoma, hyperpigmentation, and other melanin-related disorders as well as skin whitening agents, applications in antibiotics, antifungals, pesticides, and neurodegenerative disease treatments like PD, AD, and senile dementia [1–9].

Over the last two decades, computational studies on TYRs have become an integral part of understanding the function, mechanism, and inhibition of the enzyme, thereby greatly accelerating the identification and development of more effective and specific TYR inhibitors. This chapter provides a summary of the progresses that has led to the current state of computational studies of TYR inhibitors.

2. Application of *in silico* methods to TYR inhibitors

Over the years, *in silico* or computer-aided approaches, have been widely used in the drug design and development process, significantly reducing time and cost compared to traditional methods. The role of computational methods in the drug discovery process can be categorized into two main strategies: the ligand-based (LB) approach, which attempts to rationalize the biological activity of a series of compounds without information on the three-dimensional structure of the target, and the structure-based (SB) approach, which uses the known 3D structure of the target to study ligand-receptor interactions and to design ligands that enhance these interactions. These approaches are often combined in processing virtual libraries and screening results, as well as in pharmacophore-based virtual screening.

2.1 Ligand-based drug design

LB drug design (LBDD) is a drug design strategy used when the 3D structure of the target is not available. In those cases, drug design relies instead on the measurement of the chemical and physical properties of ligands interacting with a biological target and on the assumption that the biological activities of molecules are related to their chemical and physical features by functional relationships [10,11]. Common tools in LBDD include quantitative structure-activity relationship (QSAR), pharmacophore modeling, and similarity search.

In the early 2000s, QSAR modeling emerged as a valuable tool for the identification of TYR inhibitors (Fig. 1). In a study by Li and Kubo, hydrophobic, electronic, and steric properties were used to characterize a group of 18 benzaldehyde derivatives and investigate their inhibitory behavior against *Sacrophaga neobelliaria* TYR. These compounds were found to inhibit

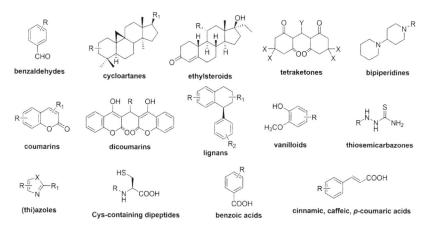

Fig. 1 Some representative classes of tyrosinase inhibitors.

the oxidation of *L*-DOPA by mushroom TYR, likely through the formation of a Schiff base between the aldehyde group of the inhibitors and a primary amino group of the enzyme. The most and least active inhibitors identified in the study were cuminaldehyde and vanillin, respectively [12]. In the following years, Marrero-Ponce et al. successfully applied atom- and bond-based TOMOCOMD-CARDD (Topological Molecular Computational Design-Computer Aided Rational Drug Design) molecular descriptors and linear discriminant analysis (LDA) as statistical classification techniques in the screening of database of heterogeneous compounds to identify molecules with TYR activity. The use of statistical LDA has been shown to well perform in discriminating between TYR inhibitors from inactive compounds. Different datasets were screened either isolated from natural sources [13] or from synthetic origin including triterpenoid cycloartanes [14,15], ethylsteroids [13], tetraketones [16], bipiperidines [17], coumarins and dicoumarins [18,19], lignans [20], vanilloid derivatives [21] as well as new TYR inhibitors [10,22]. In addition to LDA, the authors later improved the performances of the models using non-stochastic and stochastic quadratic indices, and different modeling techniques (quadratic discriminant analysis, QDA; binary logistic regression, BLR; the classification tree—CT—method derived from artificial intelligence) and the 2D TOMOCOMD-CARDD atom-based quadratic molecular descriptors as independent variables in a comparative study of QSAR models for TYR inhibition activity [23]. QSAR models using quantum chemical molecular descriptors were discussed by Pasha et al. [24].

More recently, QSAR studies on TYR inhibitors have been increasingly reported in the literature. In 2018, Tang and collaborators collected 46 TYR inhibitors from the literature and developed a model using a random forest (RF) algorithm. This study provides valuable insights into the complexities of developing QSAR models with heterogeneous activity data and presents a reliable method for detecting and removing outliers from datasets [25]. Unni et al. [26] created a validated QSAR model with a dataset of 69 thiosemicarbazone derivatives using multiple linear regression (MLR) to develop novel antimelanogenic agents. Their study identified novel lead molecules with enhanced inhibitory activity and bioavailability. Moreover, the study pointed out the physicochemical properties essential for TYR inhibition and it performed bioavailability assessments and toxicity predictions for the dataset compounds.

Several machine learning algorithms (MLR; support vector machine; RF; and fully connected neural networks) and descriptors (RDKit, MACCS, ECFP4, and Avalon) were used by Wu et al. [27] in a study on a wide dataset containing mushroom TYR inhibitors. The compounds were successfully clustered into subsets and more predictive QSAR models were identified. Based on 3D pharmacophore models developed with training sets containing thiazole [28] and azole [29] analogs, the authors were able to identify azole derivatives having inhibitory abilities against TYR. Furthermore, 3D pharmacophore models were used to screen an in-house drug-like database, resulting in the identification of three virtual candidates for TYR inhibitors. Molecular fingerprint similarity clustering analysis and genetic algorithms were also applied to cysteine-containing dipeptides. This led to the discovery of three novel natural candidate inhibitors—γ-Glu-Cys, N-Boc-Cys, and acetylcysteine—which exhibit good solubility, gastrointestinal absorption, and potential as topical treatments. [30].

The first 3D QSAR model for TYR inhibitors was developed by Xue et al. [31]. In their 2007 study, they applied comparative molecular field analysis (CoMFA) and comparative molecular similarity indices analysis (CoMSIA) to a dataset consisting of thiosemicarbazone, benzaldehyde, and benzoic acid families against TYR. The CoMFA model provided significant correlations of biological activities with steric and electrostatic fields, while the CoMSIA model correlated biological activities with steric, hydrophobic, and hydrogen bond acceptor fields. These models were used to predict the inhibitory activity of a series of analogs. Later, the same authors extended their investigations to 48 derivatives of benzaldehyde, benzoic acid, and cinnamic acid using both CoMFA and CoMSIA analyses [32]. The 3D studies proposed a potential mechanism for the interaction of benzoic acid derivatives with the TYR active site. Specifically, the formation of a hydrogen bond between the hydroxyl group and the carbonyl oxygen atoms of Y98 stabilizes the position of this residue. This stabilization prevents the amino acid from participating in the interaction between TYR and the open reading frame or "caddie" protein ORF378, a protein that helps transport two copper ions into the active center of TYR. To expand on their 2D QSAR study, which had indicated that electrostatic and lipophilic contributions play a significant role in the activity of benzaldehyde thiosemicarbazone, benzaldehyde, and benzoic acid, Pasha et al. [24] performed CoMFA and CoMSIA analyses. These analyses highlighted the necessary requirements for the investigated set of compounds to achieve good activity.

2.2 Structure-based drug design

SB or receptor-fitting strategies are utilized when the 3D structure of the target biomacromolecule is available. This structure can be determined by methods such as X-ray crystallography, NMR, or cryoEM. Alternatively, a 3D model of the target can be developed using homology modeling from a template with a known 3D structure and high homology to the target, or by employing the learning de novo design capabilities of platforms like AlphaFold [33].

To date, the 3D structure of hTYR is not available. However, its primary sequence shares significant homology with that of the mushroom, with a sequence identity of 24% in the binding site region. This similarity has allowed many researchers to utilize the mushroom's structure for *in silico* investigations. Additionally, other X-ray structures and more recent homology-based models have also been employed [34].

2.3 Homology-built models of tyrosinases

The X-ray structure of TYR from *Streptomyces castaneoglobisporus* (*Sc*TYR) in complex with the ORF378 was the first 3D structure solved in 2006 (PDB ID: 2ZMX, 1.33 Å resolution) and, thanks to its highest sequence homology with the hTYR, the first to be utilized as such for SBDD studies [35]. Since then, the *Sc*TYR structures have been used as templates for the development of homology models of hTYR and *Agaricus bisporus* TYR (*Ab*TYR or mTYR) [36], which has been the most widely used enzyme for the evaluation of TYR activity since 2003 due to its low purification costs [37]. A major breakthrough was achieved in 2011 with the resolution of the three-dimensional structure of *Ab*TYR both in its apo form (PDB ID: 2Y9W, 2.3 Å resolution) and in complex with the inhibitor tropolone (PDB ID: 2Y9X, 2.78 Å resolution) [38]. Structural resolutions of other TYRs and TYR-related proteins (TYRP) have followed, which have undoubtedly contributed to a more reliable understanding of the ligand binding modes within the enzyme and to better explain the *in vitro* results.

In 2013, X-ray structures of *Bacillus megaterium* (*Bm*TYR) in complex with kojic acid (KA) (PDB ID: 3NQ1, 2.3 Å resolution [39] and PDB ID: 5I38, 2.6 Å resolution [40]) were solved, proving to be excellent templates for the development of homology models of human TYR, with which they share 33.5% similarity [41] (Table 1). Three-dimensional coordinates from *S. castaneoglobisporus* and *Ipomea batata* catechol oxidase enzymes were used by Favre et al. to obtain a robust homology-built model of met TYR, i.e., one of the three redox forms formed by the enzyme during the catalytic reaction (met-TYR). The authors carefully included water molecules in their model, based on structural information available from X-ray data from other members of the TYR family [42].

Table 1 Sequence identity percentages for hTYR and ScTYR with respect to IbCO, ScTYR, BmTYR, AbTYR, hTRP-1, and hTYR [12–16].

database code[a]	hTYR	ScTYR
1BUG (from *Ipomea batatas*)	22%	28%
2ZMX (from *Streptomyces castaneoglobisporus*)	35%	100%
2Y9X (from *Agaricus bisporus*)	24%	24%
3NQ1 (from *Bacillus megaterium*)	34%	46%
5M8S (from *Homo sapiens*)	44%	–
AF-P14679-F1 (*H. sapiens*)	100%	–

[a]PDB: Protein Data Bank, www.rcsb.org; AF: AlphaFold, https://alphafold.ebi.ac.uk

A further significant step was taken in 2017 by obtaining the X-ray structures of human TYRP 1 (hTYRP1), both in its apo form (PDB ID: 5M8L), and in complex with several ligands including KA (PDB IDs: 5M8M and 5M8Q), tropolone (PDB IDs: 5M8O and 5M8T), mimosine (PDB IDs: 5M8N and 5M8R), tyrosine (PDB ID: 5M8P), and phenylthiourea (PDB ID: 5M8S) [43].

The advent of artificial intelligence platforms, such as Alphafold by DeepMind [33], has revolutionized the generation of accurate human TYR homology models (https://alphafold.ebi.ac.uk/entry/P14679), by greatly enhancing the precision of studies aimed at designing effective inhibitors for hTYR. In this regard, Alphafold was one of the six software used by Ricci et al. to obtain a pool of six different hTYR homology models based on hTYR similarity to TYRP1 [44]. In their study, the authors identified the structural differences between hTYR-HM and AbTYR that could guide the design of dual-targeting molecules, which were pre-screened on AbTYR as a rapid and cost-effective approach before further testing on hTYR. In particular, the authors found that residues H367 (H263 in AbTYR), I368 (F264 in AbTYR), and V377 (V283 in AbTYR) are conserved or homologs, while a peculiar S380 (A286 in AbTYR) is near the dicopper center of hTYR.

This difference pointed out the importance of designing the copper chelator moiety with an H-bond donor or acceptor function to engage polar contact with crucial residue S380. In addition, the study highlighted the importance of small lipophilic tails in the inhibition of hTYR due to the presence of F347 and I368 at the entrance of the active site Moreover, the presence of an F264 at a greater distance in AbTYR allowed to design of inhibitors with bulkier lipophilic structures [44].

Fig. 2 highlights the amino acids variability in the AbTYR, BmTYR, and HM-hTYR (5M8M as template) active sites.

2.4 Structure-based virtual screening

SB virtual screening (SBVS) is a common *in silico* practice to screen libraries, typically consisting of drug-like compounds, against targets. The procedure utilizes suitable filters (such as ADMET properties, molecular structural features, and calculation precision), along with a scoring function, to evaluate the fitness of the screened compounds for the target, thus helping in the selection of ligands with the highest predicted affinity (hits) [45]. The virtual screening approach, combined with a shape signature computational procedure, was used by Ai and collaborators [46] who explored more than 200,000 commercially available compounds against the three-dimensional hTYR homology-built model (1BUG as modeling template). By using KA and glabridin as queries, they identified a subset of 200 compounds by combining the top 100 hits for each query. The study was complemented with docking studies within the hTYR-HM active site, resulting in the identification of two effective lead structures with antimelanogenic activity (compounds **1**, **2** in Fig. 3).

Traditional Chinese Medicines database@Taiwan containing 61,000 small molecules was used by Tang et al. in an SBVS docking protocol within the hTYR homology model (template PDB: 3NM8 from BmTYR) [47].

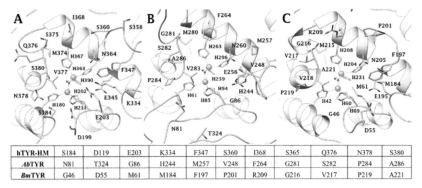

hTYR-HM	S184	D119	E203	K334	F347	S360	I368	S365	Q376	N378	S380
*Ab*TYR	N81	T324	G86	H244	M257	V248	F264	G281	S282	P284	A286
*Bm*TYR	G46	D55	M61	M184	F197	P201	R209	G216	V217	P219	A221

Fig. 2 Residue variability within the TYR active sites of (A) hTYR-HM (5M8M as template), (B) AbTYR, and (C) BmTYR. Copper atoms (CuA and CuB) are represented as gold spheres.

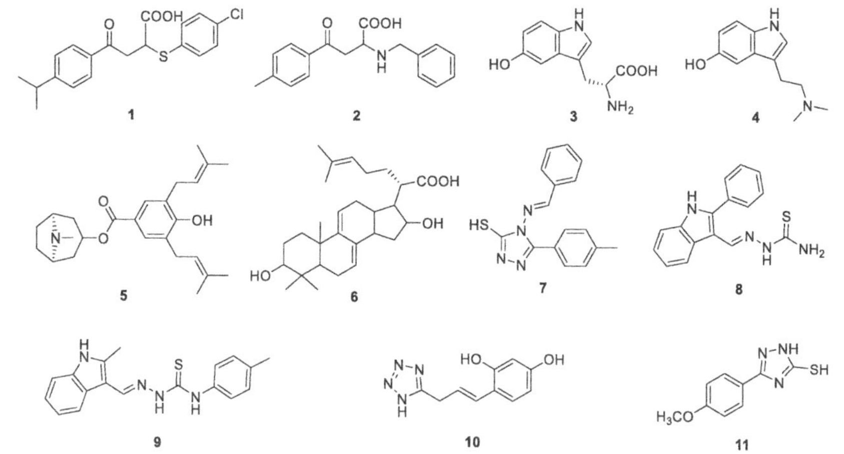

Fig. 3 2D structures of some hits resulting from the application of virtual screening protocols to TYR as the target.

The main finding of the investigation was that 5-hydroxy-L-tryptophan (**3**), bufotenine (**4**), and merresectine C (**5**) emerged as the top three molecules (Fig. 3) [47], with merresectine C being identified as the best potential lead compound. The study also includes molecular dynamics (MDs) simulations to confirm the *in silico* virtual screening results. Shah et al. performed MDs to assess the dynamic behavior of hits obtained from SBVS, evaluating 32 *Poria cocos* phytochemicals retrieved from the TCM System Pharmacology Database. After filtering for drug-likeness scores and toxicity, they identified 7,9-(11)-dehydropachymic acid (**6**) as an effective candidate for TYR inhibition (Fig. 3) [48]. However, the authors did not validate their *in silico* findings with *in vitro* studies [47,48]. Choi et al. [49] employed an enhanced SBVS procedure to address the limitations of docking and to account for conformational changes of receptors upon binding. They used the TYR target structure for SBVS from MD trajectories, exploring 400,000 compounds from the ZINC database. The top 60 molecules inhibited mushroom TYR, identifying tetrazole, triazole, and thiol/thione groups as new TYR inhibitor scaffolds. Moreover, five inhibitors (**7–11**) reduced melanin production by over 30%, with a thiosemicarbazone-containing compound achieving a notable 55% reduction (Fig. 3) [49].

2.5 Docking and molecular dynamic studies on tyrosinases

Common computational tools of SB drug design are docking and MD simulations, which are essential for understanding and predicting how potential drugs interact with their targets at the molecular level. In the field of TYR inhibitor development, docking studies are increasingly popular and widely utilized [50]. There are numerous studies in the literature that report investigations using this approach. TYR has a binuclear copper center in its active site, which is essential for its enzymatic activity. The active site is the primary target for inhibitors, which often bind to the copper ions or nearby residues to inhibit the enzyme activity. When using docking and MDs simulations for designing and predicting the binding modes of TYR inhibitors, it's crucial to consider the form of the enzyme—whether it is in its oxy (E_{oxy}: [Cu(II)–O_2^{2-}–Cu(II)]), deoxy (E_{deoxy}: [Cu(I)–Cu(I)]), or met (E_{met}:[Cu(II)–Cu(II)]) form (Fig. 4) [43].

Additionally, the type of inhibitor plays a significant role in these processes (suicide; competitive: bind to the active site, competing directly with the substrate—e.g., tyrosine or DOPA; uncompetitive: bind only to the enzyme-substrate complex, not the free enzyme; non-competitive; bind to a different part of the enzyme, causing a conformational change that reduces enzyme activity; mixed type). To improve the comparability of their investigation outcomes, researchers use KA, arbutin, or hydroquinone as a standard compound [51]. KA is a natural inhibitor that binds to the active site of TYR where it forms coordination bonds with the copper ions and hydrogen bonds with nearby residues, thereby effectively blocking the access of substrates to the active site. Over the years, docking investigations have been performed on KA and tropolone derivatives (Fig. 5), using as target enzymes co-crystallized with tropolone and KA. Preferred positions for KA substitutions were positions 2, 5, and 6 [54–62], while for tropolone, substituents were located in positions α, β, and γ [63,64].

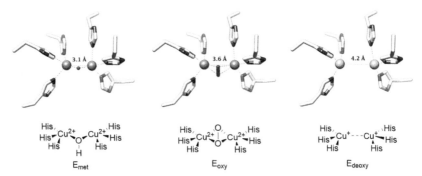

Fig. 4 States of the dicopper center within the oxy- (Cu^{2+}-O_2^{2-}-Cu^{2+}), met- (Cu^{2+}-OH^--Cu^{2+}), and deoxy- (Cu^+-Cu^+) tyrosinase, respectively. Copper atoms (CuA and CuB) are represented as gold spheres.

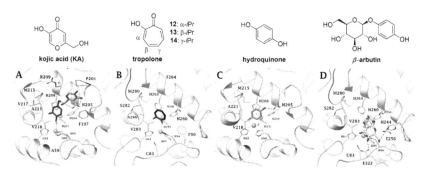

Fig. 5 X-ray solved complexes of (A) KA (magenta in 5I38; orange in 3NQ1), (B) tropolone, and (C) hydroquinone within *Bm*TYR [39,40], *Ab*TYR [38] and *Bm*TYR [52], respectively. Panel (D) shows the *in silico* binding orientations for β-arbutin within *Ab*TYR as reported by ref [53]. Copper atoms (CuA and CuB) are represented as gold spheres.

KA and tropolone derivatives maintained the ability to coordinate the dicopper center. However, introducing very bulky substituents hindered metal coordination, leading the compounds to act as allosteric inhibitors. *In silico* studies on thujaplicins (**12–14**), which are tropolone derivatives, were carried out by Tanuma and colleagues, and they showed that the position of the isopropyl group was crucial for the activity [63,64].

Arbutin, a glycosylated derivative of hydroquinone [52], acts as a competitive inhibitor by interacting with copper ions (CuA and CuB) and by surrounding amino acids through hydrogen bonding and hydrophobic interactions. However, the outcomes of docking analysis by Garcia-Jimenez et al. on both anomers (α and β) of arbutin suggest that the glucosyl moieties are oriented differently, in that they both form polar interactions involving the glucopyranose ring with H244. The discrepancy observed by the authors with previously reported docking of β-arbutin was attributed to the use of the E_{met} instead of the E_{oxy} form of the enzyme, which required rearrangement for the ligand to interact properly with the E_{oxy} form [53]. Hydroquinone exerts the enzyme's inhibition forming interactions with the copper ions and the surrounding residues.

By highlighting key examples from various classes, the chapter aims to provide a comprehensive understanding of the structural diversity and functional mechanisms underlying TYR inhibition.

2.6 Phenols and polyphenols and their derivatives

Simple phenol derivatives (Fig. 6), as well as polyphenols (Figs. 7 and 8) interact with the enzyme, typically involving the copper ions that are

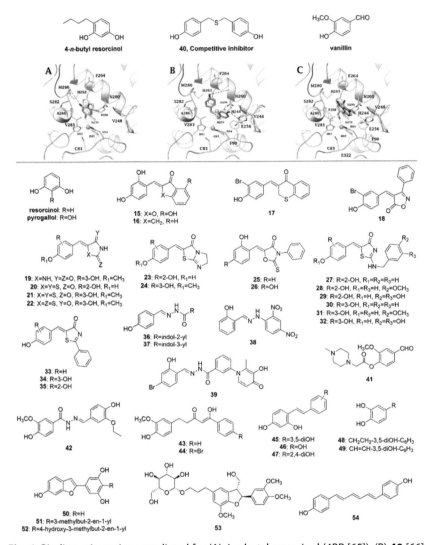

Fig. 6 Binding orientations predicted for (A) 4-*n*-butyl resorcinol (4BR [65]), (B) **40** [66], and (C) vanillin [67] within *Ab*TYR (2Y9X) as reported by the literature. Copper atoms (CuA and CuB) are represented as gold spheres. Chemical structure of some benzylidene phenols, vanilloid, and stilbene analogs.

essential for TYR's catalytic activity, generally acting as competitive TYR inhibitors [52,53,64,67,71–82]. However, other inhibition mechanisms were observed and many phenols, catechols, and resorcinol derivatives can perform a suicide inactivation of TYR [51]. Resorcinol is the lead compound of several TYR inhibitors and acts as a suicide inhibitor. The

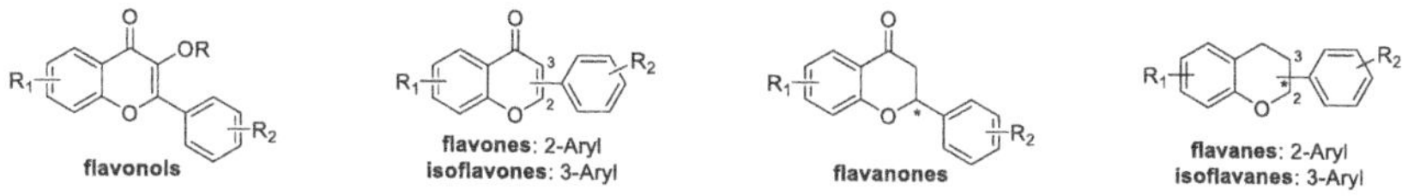

Fig. 7 General structure of flavonoids (flavonols, (iso)flavones, flavanones, and (iso) flavanes).

compound and its derivatives inhibit TYR by interacting with the binuclear copper center and forming hydrogen bonding, and hydrophobic interactions between the aromatic ring and the hydrophobic residues within the active site. Derivatives of resorcinol, with additional functional groups, often exhibit enhanced inhibitory effects due to stronger and more diverse interactions with the enzyme.

4-n-butyl resorcinol (4BR) is considered the most potent inhibitor of hTYR (about 140-fold more active than KA) and has demonstrated clinical efficacy [65]. Research performed by Garcia-Jimenez et al. used docking and dynamic studies to investigate the interaction of 4BR with the active site of TYR enzymes, particularly focusing on two states of the enzyme (E_{met} and E_{oxy}). The studies revealed that the hydroxyl (OH) group of 4BR is positioned toward one of the two copper atoms (CuA) in the enzyme's active site. This positioning is similar to how the amino acid tyrosine interacts with the enzyme [83].

This research underscores the significance of 4BR in inhibiting hTYR and its potential use in therapeutic applications.

Docking and MDs simulation of pyrogallol within 2Y9W were used to evaluate its ability to coordinate with the enzyme's copper ions and form stabilizing hydrogen bonds and hydrophobic interactions within the active site. In particular, the MD trajectory analysis allows for monitoring the distance profiles between the enzyme residues and the functional groups of the ligand [84]. The presence of the OH in position 2 confers non-competitive inhibition features to the compounds.

The majority of benzylidene phenols (**15–40**) (Fig. 6) inhibit the TYR, primarily through coordination with the copper atom in the enzyme's active site, through the formation of hydrogen bonds involving common residues such as histidine, which is part of the copper coordination environment, and other polar residues like asparagine, and serine, and through the formation of hydrophobic interactions. These interactions disrupt the enzyme's catalytic activity, thereby inhibiting melanin production. Although the coordination of copper atoms is often involved in the mechanism of enzyme inhibition, some TYR inhibitors show a different mechanism. This is the case with

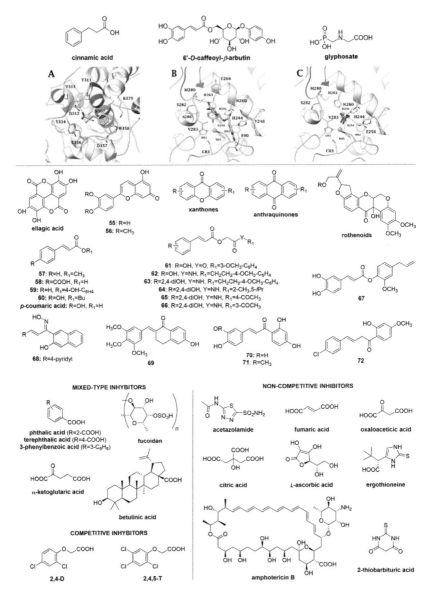

Fig. 8 Binding orientations predicted for (A) cinnamic acid [68], (B) 6′-O-caffeoyl-β-arbutin [69], and (C) glyphosate [70] within AbTYR (2Y9X) as reported by the literature. Copper atoms (CuA and CuB) are represented as gold spheres. Chemical structure of some coumarins, anthocyanins, xanthones, anthraquinones, rotenoids, chalcones, and carboxylates.

derivatives **17** and **18**, where the presence of a bromine atom in *ortho* to the phenolic OH impedes copper ion coordination (Table 2) [85–87,128,129]. This causes the thiochroman-4-one bicycle of **17** to orient in the same area occupied by KA within the *Ab*TYR active site while the phenol OH of **18** engages an H-bond with M280.

Despite being involved in several interactions, *ortho*-methoxy substituted phenols (**19**, **21**, **22**, **24**, **30–32**) and *ortho*-phenols (**25, 39**) are also rarely able to coordinate copper [88–96,130–134]. Instead, derivative **40** coordinates to the dicopper center by the sulfur atom while the phenol groups are in H-bond distance with H244 and N260 [66].

Key residues involved in the binding interactions of the mixed-type inhibitor vanillin with TYR include the copper-binding site (CuA), which can be coordinated by both the phenolic or aldehyde groups [67], and nearby histidine amino acids (e.g., H263, H85, and H259). The coordination through the phenolic OH to the Cu atoms in the active site is also observed for the derivatives **42–44** [97,98,135], while derivative **41** is stabilized within the TYR binding cleft by H-bonds with residues W195 and Q378 [97,98,135].

Surprisingly, even though stilbenes (**45–49**), lignans (**50–53**), and trienes (**54**) possess various phenolic groups, none of them directly coordinate with the metals, and some of them were investigated for binding in possible allosteric sites of TYR [99–103].

Among polyphenols, recent studies (from 2016 to 2021) have explored the role of flavonoids (flavones, flavonols, isoflavones, etc.) as TYR inhibitors [136] and computational docking simulations explained their ability to act as chelators of the di-copper center or, in other cases, as allosteric modulators or as mixed-type inhibitors (Fig. 7).

Coumarin-based derivatives (Fig. 8) have shown potential as antimelanoma agents and TYR inhibitors [137]. MDs simulations assessed the stability of the docking solutions of the derivatives and the persistence of the coordination between the OH (or carbonyl such as for the anthocyanins **55** and **56**) groups and the di-copper center in the TYR active site [104,137,138]. However, exceptions are documented such as, ellagic acid which, in the doubly hydrolyzed form, acts as an allosteric inhibitor [105].

An *in silico* study on cinnamic acid, its methyl ester (**57**), and 4-carboxy-cinnamic acid (**58**) investigate the effect of the pH-dependent increase in TYR activity: lowering the pH favors the protonation state of these inhibitors, allowing them to form more H-bonds with the target active site [139]. Again, the mechanism of interaction with the enzyme may

Table 2 Interactions of the most interesting compounds in Figs. 5–9.

Cmpd	Interaction type	Residues	TYR type	PDB code	Refs.
12–14	Coordination	Cu^{2+}(A), Cu^{2+}(B)	ScTYR	1WX2	[37]
	Coordination	Cu^{2+}(A), Cu^{2+}(B)	hTYR-HM	Template 1WX2	[63]
	H-bond	S280			
15	Coordination	Cu^{2+}(A)	hTYR-HM	Template 5M8Q	[85,86]
	H-bond	E203, S375, S380			
	π–π stacking	F347, H367			
16	Coordination	Cu^{2+}(A)	AbTYR	2Y9X	[87]
	H-bond	N260, M280			
20	Coordination	Cu^{2+}(A), Cu^{2+}(B)	hTYR-HM	Template 5M8Q	[88]
	H-bond	N103, S380			
	π–π stacking	H363, H367			
23, 26	Coordination	Cu^{2+}(A), Cu^{2+}(B)	AbTYR	2Y9X	[89,90]
	π–π stacking	H259, H263			
27	Coordination	Cu^{2+}(A), Cu^{2+}(B)	AbTYR	2Y9X	[91]
	π–π stacking	H259, H263, F264			

			hTYR-HM	Template 5M8Q	[92]
28	Coordination	Cu^{2+}(A), Cu^{2+}(B)			
	H-bond	E203, M374			
	π–π stacking	F347, H367			
29	Coordination	Cu^{2+}(A), Cu^{2+}(B)	*Ab*TYR	2Y9X	[93]
	H-bond	N81, E256			
	π–π stacking	H259, H263			
33	Coordination	Cu^{2+}(A)	*Ab*TYR	2Y9X	[94]
	π–π stacking	H244, H259, H263			
34	Coordination	Cu^{2+}(A), Cu^{2+}(B)	*Ab*TYR	2Y9X	[94]
	π-cation	R268			
35	Coordination	Cu^{2+}(A), Cu^{2+}(B)	*Ab*TYR	2Y9X	[94]
	H-bond	E256			
	π–π stacking	H244, H259, H263			
36	Coordination	Cu^{2+}(A)	*Ab*TYR	2Y9X	[95]
	π–π stacking	H244, H263			

(continued)

Table 2 Interactions of the most interesting compounds in Figs. 5–9. (*cont'd*)

Cmpd	Interaction type	Residues	TYR type	PDB code	Refs.
37	Coordination	Cu^{2+}(A)	*Ab*TYR	2Y9X	[95]
	H-bond	H244, N260			
	π–π stacking	H263			
38	Coordination	Cu^{2+}(A)	*Ab*TYR	2Y9X	[96]
	H-bond	E256			
40	Coordination	Cu^{2+}(A), Cu^{2+}(B)	*Ab*TYR	2Y9X	[66]
	H-bond	H244, N260			
42	Coordination	Cu^{2+}(A), Cu^{2+}(B)	*Ab*TYR	2Y9X	[97]
	H-bond	N260, R268, S282			
	π–π stacking	H263			
43, 44	Coordination	Cu^{2+}(A), Cu^{2+}(B)	*Bm*TYR	3NQ1	[98]
	π–π stacking	H263			
45	H-bond	H263, G281, S282	*Ab*TYR	2Y9X	[99]
	π–π stacking	F264			
	π–cation	R268			

46	H-bond	H244, H263, S282	*Ab*TYR	2Y9X	[99]
47	H-bond	T261, H263, S282	*Ab*TYR	2Y9X	[99]
48, 49	π–π stacking	H263, F264	*Ab*TYR	2Y9X	[100]
50–52	H-bond	S282	*Ab*TYR	2Y9X	[101]
	π–π stacking	F264			
53	H-bond	D312, E335, E356, K379	*Bm*TYR	3NM8	[102]
54	H-bond	H85, M257, T261	*Ab*TYR	2Y9X	[103]
55, 56	Coordination	Cu^{2+}(A), Cu^{2+}(B)	*Ab*TYR	2Y9X	[104]
	H-bond	M280			
Ellagic acid	H-bond	N19, K372	*Ab*TYR	2Y9W	[105]
p-Coumaric acid, 59	Coordination	Cu^{2+}(A)	*Ab*TYR	2Y9X	[106]
	H-bond	H60, H84			
	π–π stacking	H263			
60	H-bond	M280	*Ab*TYR	2Y9X	[106]
	π–π stacking	H263			

(continued)

Table 2 Interactions of the most interesting compounds in Figs. 5–9. (cont'd)

Cmpd	Interaction type	Residues	TYR type	PDB code	Refs.
61	Coordination	Cu^{2+}(A)	*Ab*TYR	2Y9X	[107]
	H-bond	R.268			
	π–π stacking	H84, H259			
63	Coordination	Cu^{2+}(A)	*Ab*TYR	2Y9X	[108]
	π–π stacking	H85, H259, H263			
64	H-bond	M280, E256	*Ab*TYR	2Y9X	[109]
	π–π stacking	H263			
6'-O-caffeoyl-β-arbutin	H-bond	G281, E322	*Ab*TYR	2Y9X	[69]
	π–π stacking	H263			
65	H-bond	H63, H85, N260, E322, T324	*Ab*TYR	2Y9X	[110]
66	H-bond	N81, N260	*Ab*TYR	2Y9X	[111]
	π–π stacking	H259, H263			
67	Coordination	Cu^{2+}(A), Cu^{2+}(B)	*Ab*TYR	2Y9X	[109]
	H-bond	H259, V283			

		Cu^{2+}(A)	AbTYR	2Y9X	[112]
68	Coordination	Cu^{2+}(A)	AbTYR	2Y9X	[112]
	H-bond	N260			
	π–π stacking	H263			
70–72	Coordination	Cu^{2+}(A), Cu^{2+}(B)	AbTYR	2Y9X	[113,114]
	π–π stacking	H263			
Acetazolamide	H-bond	H244, N260	AbTYR	2Y9X	[115]
	π–π stacking	H85			
Fumaric acid	H-bond	H85, H259, H263, V283	AbTYR	2Y9W	[116]
Oxaloacetic acid	H-bond	H61, H85, H259, H263, V283	AbTYR	2Y9W	[117]
Citric acid	H-bond	H61, H259, N260, H263, F264, M280, G281, V283	AbTYR	2Y9W	[118]
Betulinic acid	H-bond	N260	AbTYR	2Y9X	[119]
Glyphosate	Coordination	Cu^{2+}(A), Cu^{2+}(B)	AbTYR	2Y9X	[70]
	Salt bridge	E256			
	π–cation	H85			
2,4-D, 2,4,5–T	Coordination	Cu^{2+}(A), Cu^{2+}(B)	AbTYR	2Y9X	[70]

(continued)

Table 2 Interactions of the most interesting compounds in Figs. 5–9. (cont'd)

Cmpd	Interaction type	Residues	TYR type	PDB code	Refs.
Ergothioneine	Coordination	Cu^{2+}(A), Cu^{2+}(B)	AbTYR	2Y9X	[120]
	H-bond	E256, M280			
Amphotericin B	Coordination	Cu^{2+}(A), Cu^{2+}(B)	hTYR-HM	Template 5M8Q	[121]
	H-bond	D186, H202, H363, N364			
76	Coordination	Cu^{2+}(A), Cu^{2+}(B)	AbTYR	2Y9X	[122]
	H-bond	E256, N260, V283			
	π–π stacking	H263			
82, 83	Coordination	Cu^{2+}(A), Cu^{2+}(B)	ScTYR	2AHK	[123]
	π–π stacking	H263			
HOPNO-AA	Coordination	Cu^{2+}(A), Cu^{2+}(B)	hTYR-HM	Template 5M8Q	[124]
	H-bond	K306, N364, S380			
	π–π stacking	H367			
CRLN	Coordination	Cu^{2+}(A), Cu^{2+}(B)	AbTYR	2Y9X	[125]
	H-bond	N260, R268, E322			

CRVI	Coordination	Cu^{2+}(A), Cu^{2+}(B)	AbTYR	2Y9X	[126]
	H-bond	A246, E256, N260, R268, E322			
PFRMY	Coordination	Cu^{2+}(A), Cu^{2+}(B)	AbTYR	2Y9X	[127]
	H-bond	Y65, N260, H263, E322			
	π–π stacking	H263			
RGFTGM	Coordination	Cu^{2+}(A), Cu^{2+}(B)	AbTYR	2Y9X	[127]
	H-bond	C83, H85, E256, N260, M280, S282			

[106–109] or may not [68,69,140] involve coordination at the binuclear copper center. Indeed, docking studies on cinnamic hydroxyquinone ester (**59**) and 6′-O-caffeoyl-β-arbutin supported the outcomes of the UV–visible spectrophotometric assays, indicating that these molecules do not coordinate with the copper ions (Table 2) [110,141]. Conversely, *p*-coumaric acid, its esters (**60–62**), and caffeic acid esters (**63–67**) chelate the dicopper center within the TYR active site (Table 2) [111,112,142–144].

Among chalcones, compounds **68** coordinate with copper via the pyridine nitrogen, while in derivatives **69–71** the phenolic OH group participates in the coordination at the metal ion [112,113,144–146]. Surprisingly, derivative **72** coordinates the metal via the oxygen atom of the methoxy group [114].

2.7 Carboxylic acids and acidic derivatives

Carboxylic acids and their derivatives typically interact with TYR through a variety of mechanisms, including or not [115,147–150] coordination with copper ions, hydrogen bonding with active site residues, and hydrophobic interactions, involving long alkyl chains or aromatic rings, that help to stabilize the binding. Acidic derivatives (Fig. 8) often contain additional groups (e.g., hydroxyl, nitro) that enhance hydrogen bonding with the enzyme. Mixed-type (i.e., phthalic, as well as cinnamic acid) or non-competitive (i.e., acetazolamide) inhibition mechanisms are observed, also supported by computational investigations [68,70,116–121,151,152]. Computational studies on fumaric acid, oxaloacetic acid, citric acid, L-ascorbic acid, ergothioneine, amphotericin B, and 2-thiobarbituric acid justify their noncompetitive inhibition mechanism, while α-ketoglutaric acid and betulinic acid are confirmed mixed-type inhibitors [68,70,116–121,151,152]. In contrast, glyphosate, and pesticides **2,4-D**, and **2,4,5-T** are competitive inhibitors that were found *in silico* to chelate copper [70].

2.8 Thiosemicarbazones

Molecular mechanical methods, typically molecular docking and a MD simulation, were used to uncover the details and the stability of the interactions between TYR and thiosemicarbazones (TSCs) **73–82** (Fig. 9). In the binding with TYR, TSCs are in deprotonated form with the proton lost by the hydrazine NH group (RNHN$^-$CSNH$_2$) and, as such, they are able to coordinate with the copper ions by the nitrogen and sulfur atoms forming stable complexes [122–124]. On the other hand, functionalizing the terminal NH$_2$ of the thiosemicarbazone/thiourea group with bulky substituents (e.g., **79** and **81**)

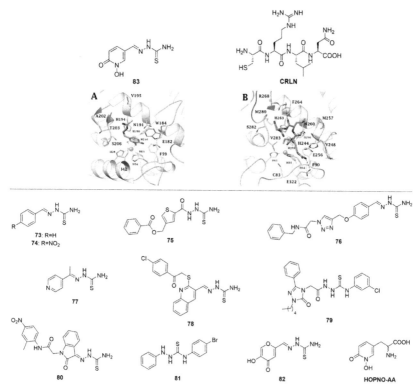

Fig. 9 Binding orientations predicted for (A) **83** [123] and (B) **CRLN** [125] within ScTYR (2AHK) and AbTYR (2Y9X) respectively, as reported by the literature. Copper atoms (CuA and CuB) are represented as gold spheres. Chemical structure of some thiosemicarbazones and HOPNO-AA inhibitor.

prevents metal coordination [153,154]. The thiosemicarbazone moiety has also been used to synthesize ditopic chelators **82** and **83**, which combine the metal-binding abilities of two different chemotypes [123,124]. For example, inhibitor **82** consists of a KA scaffold functionalized at position 2 with the TSC moiety, while compound **83** combines the 2-hydroxypyridine-N-oxide (HOPNO) chelator with the thiosemicarbazone group. In silico studies have demonstrated that metal coordination can also be achieved simultaneously by both pharmacophores [123].

2.9 Peptide inhibitors

Structure-activity relationship studies on 128 peptides behaving as TYR inhibitors were reviewed by Yap and collaborators [155]. Based on the bioactive score in the on line PeptideRanker, water solubility and nontoxicity properties, and docking studies Xue et al. identified three top-scored peptides, DGL, GAR, and SDW, with TYR inhibitory activities better than arbutin [156,157]. The computational investigations showed that DGL, GAR, and SDW occupied the active center of TYR, forming a wide network of interactions with the amino acid residues of the enzyme also including both conventional and non-conventional hydrogen bonds. DGL did not directly interact with the copper atoms, and its inhibition of the catalytic activity of the enzyme is due to the interaction with the six histidine residues involved in the coordination of the Cu^{2+} ions.

Starting from the analysis of structural data and the observation of the contribution provided by the VSHY loop from the caddie protein ORF378 to the interaction with the ScTYR active site, Lee et al. explored libraries containing N- and C-terminal cysteine/tyrosine- tetrapeptides for their TYR-inhibitory abilities [126]. The authors synthesized and tested selected tetrapeptides and gained insights into the binding mode within the TYR by molecular docking simulations. N-terminal cysteine- peptides (e.g., CRVI) showed the best inhibitory profile while C- and N-terminal TYR peptides exhibited less potency. Both hydrogen bondings and electrostatic interactions between charged residues (e.g., arginine) of the tetrapeptide and oppositely charged regions of TYR can occur as well as hydrophobic interactions involving amino acids in the tetrapeptide sequence like phenylalanine, leucine, or isoleucine. Moreover, cysteine can coordinate with copper ions and also the imidazole ring in histidine-containing peptides has a strong affinity for copper, making these peptides effective inhibitors. Aminoacids were embedded by Buitrago et al. [124] in the motif of a well-known catechol able to coordinate the dicopper center, the 2-hydroxypyridine-N-oxide or its corresponding tautomeric form N-hydroxy-2-pyridone (HOPNO) group, and both enantiomers of the HOPNO-aminoacid (HOPNO-AA) were studied in silico (Fig. 9, Table 2).

The gelatin-contained peptide DLGFLARGF was found to block the catalytic activity of TYR through the formation of hydrophobic and hydrogen bonding interactions [158]. In addition to the tri- and tetra-peptides briefly discussed above, other amino acid sequences (pentapeptides, e.g., PFRMY; hexapeptides, e.g., RGFTGM; persipeptides) with TYR inhibitory activity have been reported in the literature [125–127]. All share interactions with the active site of the enzyme, often highlighted by computational studies, involving a sophisticated interplay of coordination with copper ions, hydrogen bonding, hydrophobic interactions, and electrostatic interactions.

Synthetic and from natural sources peptides, as well as TYR inhibitors from natural sources, are the subject of specific chapters in this collection, and reference is made to those chapters for detailed discussion.

3. Conclusion

This chapter provides an overview of the computational studies performed on TYR inhibitors, highlighting the most important inhibitory chemotypes and the key features residues mainly involved in the ligand-target interaction. Among the copper-chelators are KA and tropolone derivatives, phenols, possibly non-*ortho*-substituted, *para*-substituted resorcinol derivatives, certain carboxylic acids, appropriately deprotonated thiosemicarbazones and thioureas, and peptide ligands with N-terminal cysteines and C-terminal tyrosines. Asparagine is a key residue for the catalytic activity, which is highly conserved in the active site of all TYRs (N364/N260/N205/N291 in hTYR/*Ab*TYR/*Bm*TYR/*Sc*TYR). This residue often acts as an anchor for many inhibitors not binding to the metal. Additionally, a histidine, often involved in π–π stacking interactions (H367/H263 in hTYR/*Ab*TYR), and hydrophobic residues (I368/F264 and V377/V283 in hTYR/*Ab*TYR), are conserved among the first/second shell residues of TYR. In hTYR, a serine near the dicopper center (S280/A286 in hTYR/*Ab*TYR) is peculiar in stabilizing the ligand coordination.

In the last two decades, the application of *in silico* techniques for the development of TYR inhibitors has made this field highly prolific, with exponential growth in the discovery of new active molecules. In addition, computer predictions, synthesis, and *in vitro* studies constitute a self-sustaining cycle that becomes increasingly efficient over time due to the growing collection of new data. *In silico* techniques enhance their predictive power with consistent *in vitro* data, which in turn can only be produced after the molecules of interest have been synthesized. The expansion of libraries with real compounds and inhibition data obtained experimentally allows the development of increasingly improved models that help in obtaining more potent and selective molecules. A major limitation in the field of TYR research is the heterogeneity of biological data that makes their comparison difficult, even though the same standards are used as benchmarks in most cases.

References

[1] M.A. Baber, C.M. Crist, N.L. Devolve, J.D. Patrone, Tyrosinase inhibitors: a perspective, Molecules 28 (2023) 5762, https://doi.org/10.3390/molecules28155762

[2] M. Agunbiade, M. Le Roes-Hill, Application of bacterial tyrosinases in organic synthesis, World J. Microbiol. Biotechnol. 38 (2021) 2, https://doi.org/10.1007/s11274-021-03186-0

[3] Y. Yuan, W. Jin, Y. Nazir, C. Fercher, M.A.T. Blaskovich, M.A. Cooper, et al., Tyrosinase inhibitors as potential antibacterial agents, Eur. J. Med. Chem. 187 (2020) 111892, https://doi.org/10.1016/j.ejmech.2019.111892

[4] T. Pillaiyar, M. Manickam, V. Namasivayam, Skin whitening agents: medicinal chemistry perspective of tyrosinase inhibitors, J. Enzyme Inhib. Med. Chem. 32 (2017) 403–425, https://doi.org/10.1080/14756366.2016.1256882

[5] M.M. Al-Rooqi, A. Sadiq, R.J. Obaid, Z. Ashraf, Y. Nazir, R.S. Jassas, et al., Evaluation of 2,3-dihydro-1,5-benzothiazepine derivatives as potential tyrosinase inhibitors: in vitro and in silico studies, ACS Omega 8 (2023) 17195–17208, https://doi.org/10.1021/acsomega.3c01566

[6] T. Nagatsu, A. Nakashima, H. Watanabe, S. Ito, K. Wakamatsu, F.A. Zucca, et al., The role of tyrosine hydroxylase as a key player in neuromelanin synthesis and the association of neuromelanin with Parkinson's disease, J. Neural Transm. (Vienna) 130 (2023) 611–625, https://doi.org/10.1007/s00702-023-02617-6

[7] P. Riederer, C. Monoranu, S. Strobel, T. Iordache, J. Sian-Hülsmann, Iron as the concert master in the pathogenic orchestra playing in sporadic Parkinson's disease, J. Neural Transm. (Vienna) 128 (2021) 1577–1598, https://doi.org/10.1007/s00702-021-02414-z

[8] T. Nagatsu, A. Nakashima, H. Watanabe, S. Ito, K. Wakamatsu, Neuromelanin in Parkinson's disease: tyrosine hydroxylase and tyrosinase, Int. J. Mol. Sci. 23 (2022) 4176, https://doi.org/10.3390/ijms23084176

[9] P. Riederer, D. Berg, N. Casadei, F. Cheng, J. Classen, C. Dresel, et al., α-Synuclein in Parkinson's disease: causal or bystander? J. Neural Transm. (Vienna) 126 (2019) 815–840, https://doi.org/10.1007/s00702-019-02025-9

[10] M.T. Khan, Novel tyrosinase inhibitors from natural resources – their computational studies, Curr. Med. Chem. 19 (14) (2012) 2262–2272, https://doi.org/10.2174/092986712800229041 PMID: 22414108..

[11] D. Giordano, C. Biancaniello, M.A. Argenio, A. Facchiano, Drug design by pharmacophore and virtual screening approach, Pharmaceuticals (Basel) 15 (2022) 646, https://doi.org/10.3390/ph15050646

[12] W. Li, I. Kubo, QSAR and kinetics of the inhibition of benzaldehyde derivatives against Sacrophaga neobelliaria phenoloxidase, Bioorg. Med. Chem. 12 (2004) 701–713, https://doi.org/10.1016/j.bmc.2003.11.014

[13] G.M. Casañola-Martín, Y. Marrero-Ponce, M.T. Khan, A. Ather, S. Sultan, F. Torrens, et al., TOMOCOMD-CARDD descriptors-based virtual screening of tyrosinase inhibitors: evaluation of different classification model combinations using bond-based linear indices, Bioorg. Med. Chem. 15 (2007) 1483–1503, https://doi.org/10.1016/j.bmc.2006.10.067

[14] G.M. Casañola-Martín, M.T. Khan, Y. Marrero-Ponce, A. Ather, M.N. Sultankhodzhaev, F. Torrens, New tyrosinase inhibitors selected by atomic linear indices-based classification models, Bioorg. Med. Chem. Lett. 16 (2006) 324–330, https://doi.org/10.1016/j.bmcl.2005.09.085

[15] Y. Marrero-Ponce, M.T. Khan, G.M. Casañola Martín, A. Ather, M.N. Sultankhodzhaev, F. Torrens, et al., Prediction of tyrosinase inhibition activity using atom-based bilinear indices, ChemMedChem 2 (2007) 449–478, https://doi.org/10.1002/cmdc.200600186

[16] Y. Marrero-Ponce, M.T. Khan, G.M. Casañola-Martín, A. Ather, M.N. Sultankhodzhaev, R. García-Domenech, et al., Bond-based 2D TOMOCOMD-CARDD approach for drug discovery: aiding decision-making in 'in silico' selection of new lead tyrosinase inhibitors, J. Comput. Aided Mol. Des. 21 (2007) 167–188, https://doi.org/10.1007/s10822-006-9094-7

[17] G.M. Casañola-Martín, Y. Marrero-Ponce, M.T. Khan, A. Ather, K.M. Khan, F. Torrens, et al., Dragon method for finding novel tyrosinase inhibitors: biosilico identification and experimental in vitro assays, Eur. J. Med. Chem. 42 (2007) 1370–1381, https://doi.org/10.1016/j.ejmech.2007.01.026

[18] H. Le-Thi-Thu, G.M. Casañola-Martín, Y. Marrero-Ponce, A. Rescigno, L. Saso, V.S. Parmar, et al., Novel coumarin-based tyrosinase inhibitors discovered by OECD principles-validated QSAR approach from an enlarged, balanced database, Mol. Diversity 15 (2011) 507–520, https://doi.org/10.1007/s11030-010-9274-1

[19] G.M. Casañola-Martín, Y. Marrero-Ponce, M. Tareq Hassan Khan, F. Torrens, F. Pérez-Giménez, A. Rescigno, Atom- and bond-based 2D TOMOCOMD-CARDD approach and ligand-based virtual screening for the drug discovery of new tyrosinase inhibitors, J. Biomol. Screening 13 (2008) 1014–1024, https://doi.org/10.1177/1087057108326078

[20] G.M. Casañola-Martin, Y. Marrero-Ponce, M.T. Khan, S.B. Khan, F. Torrens, F. Pérez-Jiménez, et al., Bond-based 2D quadratic fingerprints in QSAR studies: virtual and in vitro tyrosinase inhibitory activity elucidation, Chem. Biol. Drug Des. 76 (2010) 538–545, https://doi.org/10.1111/j.1747-0285.2010.01032.x

[21] A. Rescigno, G.M. Casañola-Martin, E. Sanjust, P. Zucca, Y. Marrero-Ponce, Vanilloid derivatives as tyrosinase inhibitors driven by virtual screening-based QSAR models, Drug Test. Anal. 3 (2011) 176–181, https://doi.org/10.1002/dta.187

[22] Y. Marrero-Ponce, G.M. Casañola-Martín, M.T. Khan, F. Torrens, A. Rescigno, C. Abad, Ligand-based computer-aided discovery of tyrosinase inhibitors. Applications of the TOMOCOMD-CARDD method to the elucidation of new compounds, Curr. Pharm. Des. 16 (24) (2010) 2601, https://doi.org/10.2174/138161210792389216

[23] H. Le-Thi-Thu, G.C. Cardoso, G.M. Casañola-Martin, Y. Marrero-Ponce, A. Puris, F. Torrens, et al., QSAR models for tyrosinase inhibitory activity description applying modern statistical classification techniques: a comparative study, Chemom. Intell. Lab. Syst. 104 (2010) 249–259, https://doi.org/10.1016/j.chemolab.2010.08.016

[24] F.A. Pasha, M. Muddassar, Y. Beg, S.J. Cho, DFT-based de novo QSAR of phenoloxidase inhibitors, Chem. Biol. Drug Des. 71 (2008) 483–493, https://doi.org/10.1111/j.1747-0285.2008.00651.x

[25] H. Tang, F. Cui, L. Liu, Y. Li, Predictive models for tyrosinase inhibitors: challenges from heterogeneous activity data determined by different experimental protocols, Comput. Biol. Chem. 73 (2018 Apr) 79–84, https://doi.org/10.1016/j.compbiolchem.2018.02.007

[26] P.A. Unni, S.S. Lulu, G.G. Pillai, Computational strategies towards developing novel antimelanogenic agents, Life Sci. 250 (2020) 117602, https://doi.org/10.1016/j.lfs.2020.117602

[27] Y. Wu, D. Huo, G. Chen, A. Yan, SAR and QSAR research on tyrosinase inhibitors using machine learning methods, SAR QSAR Environ. Res. 32 (2021) 85–110, https://doi.org/10.1080/1062936X.2020.1862297

[28] S. Ghayas, M. Ali Masood, R. Parveen, M. Aquib, M.A. Farooq, P. Banerjee, et al., 3D QSAR pharmacophore-based virtual screening for the identification of potential inhibitors of tyrosinase, J. Biomol. Struct. Dyn. 38 (2020) 2916–2927, https://doi.org/10.1080/07391102.2019.1647287

[29] B. De, I. Adhikari, A. Nandy, A. Saha, B.B. Goswami, In silico modelling of azole derivatives with tyrosinase inhibition ability: application of the models for activity prediction of new compounds, Comput. Biol. Chem. 74 (2018) 105–114, https://doi.org/10.1016/j.compbiolchem.2018.03.007

[30] X. Li, F. Pan, Z. Yang, F. Gao, J. Li, F. Zhang, et al., Construction of QSAR model based on cysteine-containing dipeptides and screening of natural tyrosinase inhibitors, J. Food Biochem. 46 (2022) e14338, https://doi.org/10.1111/jfbc.14338

[31] C.B. Xue, L. Zhang, W.C. Luo, X.Y. Xie, L. Jiang, T. Xiao, 3D-QSAR and molecular docking studies of benzaldehyde thiosemicarbazone, benzaldehyde, benzoic acid, and their derivatives as phenoloxidase inhibitors, Bioorg. Med. Chem. 15 (2007) 2006–2015, https://doi.org/10.1016/j.bmc.2006.12.038

[32] C.B. Xue, W.C. Luo, Q. Ding, S.Z. Liu, X.X. Gao, Quantitative structure-activity relationship studies of mushroom tyrosinase inhibitors, J. Comput. Aided Mol. Des. 22 (2008) 299–309, https://doi.org/10.1007/s10822-008-9187-6

[33] J. Jumper, R. Evans, A. Pritzel, T. Green, M. Figurnov, O. Ronneberger, et al., Highly accurate protein structure prediction with AlphaFold, Nature 596 (2021) 583–589, https://doi.org/10.1038/s41586-021-03819-2

[34] M.T. Muhammed, E. Aki-Yalcin, Homology modeling in drug discovery: overview, current applications, and future perspectives, Chem. Biol. Drug Des. 93 (2019) 12–20, https://doi.org/10.1111/cbdd.13388

[35] Y. Matoba, T. Kumagai, A. Yamamoto, H. Yoshitsu, M. Sugiyama, Crystallographic evidence that the dinuclear copper center of tyrosinase is flexible during catalysis, J. Biol. Chem. 281 (2006) 8981–8990, https://doi.org/10.1074/jbc.M509785200

[36] S. Takahashi, T. Kamiya, K. Saeki, T. Nezu, S. Takeuchi, R. Takasawa, et al., Structural insights into the hot spot amino acid residues of mushroom tyrosinase for the bindings of thujaplicins, Bioorg. Med. Chem. 18 (2010) 8112–8118, https://doi.org/10.1016/j.bmc.2010.08.056

[37] H.J. Wichers, K. Recourt, M. Hendriks, C.E. Ebbelaar, G. Biancone, F.A. Hoeberichts, et al., Cloning, expression and characterisation of two tyrosinase cDNAs from Agaricus bisporus, Appl. Microbiol. Biotechnol. 61 (2003) 336–341, https://doi.org/10.1007/s00253-002-1194-2

[38] W.T. Ismaya, H.J. Rozeboom, A. Weijn, J.J. Mes, F. Fusetti, H.J. Wichers, et al., Crystal structure of Agaricus bisporus mushroom tyrosinase: identity of the tetramer subunits and interaction with tropolone, Biochemistry 50 (2011) 5477–5486, https://doi.org/10.1021/bi200395t

[39] M. Sendovski, M. Kanteev, V.S. Ben-Yosef, N. Adir, A. Fishman, First structures of an active bacterial tyrosinase reveal copper plasticity, J. Mol. Biol. 405 (2011) 227–237, https://doi.org/10.1016/j.jmb.2010.10.048

[40] B. Deri, M. Kanteev, M. Goldfeder, D. Lecina, V. Guallar, N. Adir, et al., The unravelling of the complex pattern of tyrosinase inhibition, Sci. Rep. 6 (2016) 34993, https://doi.org/10.1038/srep34993

[41] D. Nokinsee, L. Shank, V.S. Lee, P. Nimmanpipug, Estimation of inhibitory effect against tyrosinase activity through homology modeling and molecular docking, Enzyme Res. 2015 (2015) 262364, https://doi.org/10.1155/2015/262364

[42] E. Favre, A. Daina, P.A. Carrupt, A. Nurisso, Modeling the met form of human tyrosinase: a refined and hydrated pocket for antagonist design, Chem. Biol. Drug Des. 84 (2014) 206–215, https://doi.org/10.1111/cbdd.12306

[43] X. Lai, H.J. Wichers, M. Soler-Lopez, B.W. Dijkstra, Structure of human tyrosinase related protein 1 reveals a binuclear zinc active site important for melanogenesis, Angew. Chem. Int. Ed. Engl. 56 (2017) 9812–9815, https://doi.org/10.1002/anie.201704616

[44] F. Ricci, K. Schira, L. Khettabi, L. Lombardo, S. Mirabile, R. Gitto, et al., Computational methods to analyze and predict the binding mode of inhibitors targeting both human and mushroom tyrosinase, Eur. J. Med. Chem. 260 (2023) 115771, https://doi.org/10.1016/j.ejmech.2023.115771

[45] M. Kontoyianni, Docking and virtual screening in drug discovery, Methods Mol. Biol. 1647 (2017) 255–266, https://doi.org/10.1007/978-1-4939-7201-2_18

[46] N. Ai, W.J. Welsh, U. Santhanam, H. Hu, J. Lyga, Novel virtual screening approach for the discovery of human tyrosinase inhibitors, PLoS One 9 (2014) e112788, https://doi.org/10.1371/journal.pone.0112788

[47] H.C. Tang, Y.C. Chen, Identification of tyrosinase inhibitors from traditional Chinese medicines for the management of hyperpigmentation, Springerplus 4 (2015) 184, https://doi.org/10.1186/s40064-015-0956-0

[48] F.H. Shah, Y.S. Eom, S.J. Kim, Evaluation of phytochemicals of *Poria cocos* against tyrosinase protein: a virtual screening, pharmacoinformatics and molecular dynamics study, 3 Biotech. 13 (2023) 199, https://doi.org/10.1007/s13205-023-03626-8

[49] J. Choi, K.E. Choi, S.J. Park, S.Y. Kim, J.G. Jee, Ensemble-based virtual screening led to the discovery of new classes of potent tyrosinase inhibitors, J. Chem. Inf. Model. 56 (2016) 354–367, https://doi.org/10.1021/acs.jcim.5b00484

[50] J.M. Paggi, A. Pandit, R.O. Dror, The art and science of molecular docking, Annu. Rev. Biochem. (2024), https://doi.org/10.1146/annurev-biochem-030222-120000

[51] S. Zolghadri, A. Bahrami, M.T. Hassan Khan, J. Munoz-Munoz, F. Garcia-Molina, F. Garcia-Canovas, et al., A comprehensive review on tyrosinase inhibitors, J. Enzyme Inhib. Med. Chem. 34 (2019) 279–309, https://doi.org/10.1080/14756366. 2018.1545767

[52] E. Solem, F. Tuczek, H. Decker, Tyrosinase versus catechol oxidase: one asparagine makes the difference, Angew. Chem. Int. Ed. Engl. 55 (8) (2016) 2884, https://doi. org/10.1002/anie.201508534

[53] A. Garcia-Jimenez, J.A. Teruel-Puche, P.A. Garcia-Ruiz, A. Saura-Sanmartin, J. Berna, F. Garcia-Canovas, et al., Structural and kinetic considerations on the catalysis of deoxyarbutin by tyrosinase, PLoS One 12 (2017) e0187845, https://doi. org/10.1371/journal.pone.0187845

[54] Z. Peng, G. Wang, Y. He, J.J. Wang, Y. Zhao, Tyrosinase inhibitory mechanism and anti-browning properties of novel kojic acid derivatives bearing aromatic aldehyde moiety, Curr. Res. Food Sci. 6 (2022) 100421, https://doi.org/10.1016/j.crfs.2022. 100421

[55] Z. Peng, G. Wang, J.J. Wang, Y. Zhao, Anti-browning and antibacterial dual functions of novel hydroxypyranone-thiosemicarbazone derivatives as shrimp preservative agents: synthesis, bio-evaluation, mechanism, and application, Food Chem. 419 (2023) 136106, https://doi.org/10.1016/j.foodchem.2023.136106

[56] Y.M. Chen, C. Li, W.J. Zhang, Y. Shi, Z.J. Wen, Q.X. Chen, et al., Kinetic and computational molecular docking simulation study of novel kojic acid derivatives as anti-tyrosinase and antioxidant agents, J. Enzyme Inhib. Med. Chem. 34 (2019) 990–998, https://doi.org/10.1080/14756366.2019.1609467

[57] M. He, M. Fan, W. Yang, Z. Peng, G. Wang, Novel kojic acid-1,2,4-triazine hybrids as anti-tyrosinase agents: synthesis, biological evaluation, mode of action, and anti-browning studies, Food Chem. 419 (2023) 136047, https://doi.org/10.1016/j. foodchem.2023.136047

[58] F. Yousefnejad, A. Iraji, R. Sabourian, A. Moazzam, S. Tasharoie, S. Sara Mirfazli, et al., Ugi bis-amide derivatives as tyrosinase inhibitor; synthesis, biology assessment, and in silico analysis, Chem. Biodivers. 20 (2023) e202200607, https://doi.org/10. 1002/cbdv.202200607

[59] G. Karakaya, A. Türe, A. Ercan, S. Öncül, M.D. Aytemir, Synthesis, computational molecular docking analysis and effectiveness on tyrosinase inhibition of kojic acid derivatives, Bioorg. Chem. 88 (2019) 102950, https://doi.org/10.1016/j.bioorg.2019.102950

[60] D. Rezapour Niri, M.H. Sayahi, S. Behrouz, A. Moazzam, F. Rasekh, N. Tanideh, et al., Design, synthesis, in vitro, and in silico evaluations of kojic acid derivatives linked to amino pyridine moiety as potent tyrosinase inhibitors, Heliyon 9 (2023) e22009, https://doi.org/10.1016/j.heliyon.2023.e22009

[61] L.S. Martins, R.W.A. Gonçalves, J.J.S. Moraes, C.N. Alves, J.R.A. Silva, Computational analysis of triazole-based kojic acid analogs as tyrosinase inhibitors by molecular dynamics and free energy calculations, Molecules 27 (2022) 8141, https://doi.org/10.3390/molecules27238141

[62] E.M. Brasil, L.M. Canavieira, É.T.C. Cardoso, E.O. Silva, J. Lameira, J.L.M. Nascimento, et al., Inhibition of tyrosinase by 4H-chromene analogs: synthesis, kinetic studies, and computational analysis, Chem. Biol. Drug Des. 90 (2017) 804–810, https://doi.org/10.1111/cbdd.13001

[63] A. Yoshimori, T. Oyama, S. Takahashi, H. Abe, T. Kamiya, T. Abe, et al., Structure-activity relationships of the thujaplicins for inhibition of human tyrosinase, Bioorg. Med. Chem. 22 (2014) 6193–6200, https://doi.org/10.1016/j.bmc.2014.08.027

[64] K. Sakuma, M. Ogawa, K. Sugibayashi, K. Yamada, K. Yamamoto, Relationship between tyrosinase inhibitory action and oxidation-reduction potential of cosmetic whitening ingredients and phenol derivatives, Arch. Pharm. Res. 22 (1999) 335–339, https://doi.org/10.1007/BF02979054

[65] M.R. Stratford, C.A. Ramsden, P.A. Riley, Mechanistic studies of the inactivation of tyrosinase by resorcinol, Bioorg. Med. Chem. 21 (2013) 1166–1173, https://doi.org/10.1016/j.bmc.2012.12.031

[66] W.C. Chen, T.S. Tseng, N.W. Hsiao, Y.L. Lin, Z.H. Wen, C.C. Tsai, et al., Discovery of highly potent tyrosinase inhibitor, T1, with significant anti-melanogenesis ability by zebrafish in vivo assay and computational molecular modeling, Sci. Rep. 5 (2015) 7995, https://doi.org/10.1038/srep07995

[67] M.H. Nguyen, H.X. Nguyen, M.T. Nguyen, N.T. Nguyen, Phenolic constituents from the heartwood of *Artocapus altilis* and their tyrosinase inhibitory activity, Nat. Prod. Commun. 7 (2012) 185–186.

[68] S. Hassani, K. Haghbeen, M. Fazli, Non-specific binding sites help to explain mixed inhibition in mushroom tyrosinase activities, Eur. J. Med. Chem. 122 (2016) 138–148, https://doi.org/10.1016/j.ejmech.2016.06.013

[69] D. Xie, W. Fu, T. Yuan, K. Han, Y. Lv, Q. Wang, et al., 6'-O-caffeoylarbutin from Quezui Tea: a highly effective and safe tyrosinase inhibitor, Int. J. Mol. Sci. 25 (2024) 972, https://doi.org/10.3390/ijms25020972

[70] V. Sok, A. Fragoso, Kinetic, spectroscopic and computational docking study of the inhibitory effect of the pesticides 2,4,5-T, 2,4-D and glyphosate on the diphenolase activity of mushroom tyrosinase, Int. J. Biol. Macromol. 118 (2018) 427–434, https://doi.org/10.1016/j.ijbiomac.2018.06.098

[71] S. Chawla, M.A. deLong, M.O. Visscher, R.R. Wickett, P. Manga, R.E. Boissy, Mechanism of tyrosinase inhibition by deoxyArbutin and its second-generation derivatives, Br. J. Dermatol. 159 (2008) 1267–1274, https://doi.org/10.1111/j.1365-2133.2008.08864.x

[72] K. Tasaka, C. Kamei, S. Nakano, Y. Takeuchi, M. Yamato, Effects of certain resorcinol derivatives on the tyrosinase activity and the growth of melanoma cells, Methods Find. Exp. Clin. Pharmacol. 20 (1998) 99–109, https://doi.org/10.1358/mf.1998.20.2.485637

[73] A. Hashimoto, M. Ichihashi, Y. Mishima, The mechanism of depigmentation by hydroquinone: a study on suppression and recovery processes of tyrosinase activity in the pigment cells in vivo and in vitro, Nihon Hifuka Gakkai Zasshi 94 (1984) 797–804.

[74] Y.-R. Chen, R. Y-Y, T.Y. Lin, C.P. Huang, W.C. Tang, S.T. Chen, et al., Identification of an alkylhydroquinone from *Rhus succedanea* as an inhibitor of tyrosinase and melanogenesis, J. Agric. Food Chem. 57 (2009) 2200–2205, https://doi.org/10.1021/jf802617a

[75] A. Sasaki, Y. Yamano, S. Sugimoto, H. Otsuka, K. Matsunami, T. Shinzato, Phenolic compounds from the leaves of *Breynia officinalis* and their tyrosinase and melanogenesis inhibitory activities, J. Nat. Med. 72 (2018) 381–389, https://doi.org/10.1007/s11418-017-1148-8

[76] R.E. Boissy, M. Visscher, M.A. DeLong, Deoxyarbutin: a novel reversible tyrosinase inhibitor with effective in vivo skin lightening potency, Exp. Dermatol. 14 (8) (2005) 601.

[77] S. Chawla, K. Kvalnes, M.A. deLong, R. Wickett, P. Manga, R.E. Boissy, DeoxyArbutin and its derivatives inhibit tyrosinase activity and melanin synthesis without inducing reactive oxygen species or apoptosis, J. Drugs Dermatol. 11 (2012) e28–e34.

[78] L. Kolbe, T. Mann, W. Gerwat, J. Batzer, S. Ahlheit, C. Scherner, et al., 4-n-butylresorcinol, a highly effective tyrosinase inhibitor for the topical treatment of hyperpigmentation, J. Eur. Acad. Dermatol. Venereol. 27 (2013) 19–23, https://doi.org/10.1111/jdv.12051

[79] Z. Ashraf, M. Rafiq, S.Y. Seo, M.M. Babar, N.U. Zaidi, Synthesis, kinetic mechanism and docking studies of vanillin derivatives as inhibitors of mushroom tyrosinase, Bioorg. Med. Chem. 23 (2015) 5870–5880, https://doi.org/10.1016/j.bmc.2015.06.068

[80] S. Shirota, K. Miyazaki, R. Aiyama, M. Ichioka, T. Yokokura, Tyrosinase inhibitors from crude drugs, Biol. Pharm. Bull. 17 (1994) 266–269, https://doi.org/10.1248/bpb.17.266

[81] D. Lin, M. Xiao, J. Zhao, Z. Li, B. Xing, X. Li, et al., An overview of plant phenolic compounds and their importance in human nutrition and management of type 2 diabetes, Molecules 21 (2016) 1374, https://doi.org/10.3390/molecules21101374

[82] W.Y. Huang, Y.Z. Cai, Y. Zhang, Natural phenolic compounds from medicinal herbs and dietary plants: potential use for cancer prevention, Nutr. Cancer 62 (2010) 1–20, https://doi.org/10.1080/01635580903191585

[83] A. Garcia-Jimenez, J.A. Teruel-Puche, C.V. Ortiz-Ruiz, J. Berna, J. Tudela, F. Garcia-Canovas, 4-n-butylresorcinol, a depigmenting agent used in cosmetics, reacts with tyrosinase, IUBMB Life 68 (2016) 663–672, https://doi.org/10.1002/iub.1528

[84] S.L. Xiong, G.T. Lim, S.J. Yin, J. Lee, Y.X. Si, J.M. Yang, et al., The inhibitory effect of pyrogallol on tyrosinase activity and structure: integration study of inhibition kinetics with molecular dynamics simulation, Int. J. Biol. Macromol. 121 (2019) 463–471, https://doi.org/10.1016/j.ijbiomac.2018.10.046

[85] B. Roulier, I. Rush, L.M. Lazinski, B. Pérès, H. Olleik, G. Royal, et al., Resorcinol-based hemiindigoid derivatives as human tyrosinase inhibitors and melanogenesis suppressors in human melanoma cells, Eur. J. Med. Chem. 246 (2023) 114972, https://doi.org/10.1016/j.ejmech.2022.114972

[86] R. Haudecoeur, A. Gouron, C. Dubois, H. Jamet, M. Lightbody, R. Hardré, et al., Investigation of binding-site homology between mushroom and bacterial tyrosinases by using aurones as effectors, Chembiochem 15 (2014) 1325–1333, https://doi.org/10.1002/cbic.201402003

[87] H.J. Jung, S.G. Noh, Y. Park, D. Kang, P. Chun, H.Y. Chung, et al., In vitro and in silico insights into tyrosinase inhibitors with (*E*)-benzylidene-1-indanone derivatives, Comput. Struct. Biotechnol. J. 17 (2019) 1255–1264, https://doi.org/10.1016/j.csbj.2019.07.017

[88] H. Choi, I.Y. Ryu, I. Choi, S. Ullah, H.J. Jung, Y. Park, et al., Novel anti-melanogenic compounds, (*Z*)-5-(substituted benzylidene)-4-thioxothiazolidin-2-one derivatives: in vitro and in silico insights, Molecules 26 (2021) 4963, https://doi.org/10.3390/molecules26164963

[89] H. Choi, I. Young Ryu, I. Choi, S. Ullah, H. Jin Jung, Y. Park, et al., Identification of (*Z*)-2-benzylidene-dihydroimidazothiazolone derivatives as tyrosinase inhibitors: anti-melanogenic effects and in silico studies, Comput. Struct. Biotechnol. J. 20 (2022) 899–912, https://doi.org/10.1016/j.csbj.2022.02.007

[90] I. Choi, Y. Park, I.Y. Ryu, H.J. Jung, S. Ullah, H. Choi, et al., In silico and in vitro insights into tyrosinase inhibitors with a 2-thioxooxazoline-4-one template, Comput. Struct. Biotechnol. J. 19 (2020) 37–50, https://doi.org/10.1016/j.csbj.2020.12.001

[91] J. Lee, Y.J. Park, H.J. Jung, S. Ullah, D. Yoon, Y. Jeong, et al., Design and synthesis of (*Z*)-2-(benzylamino)-5-benzylidenethiazol-4(5H)-one derivatives as tyrosinase inhibitors and their anti-melanogenic and antioxidant effects, Molecules 28 (2023) 848, https://doi.org/10.3390/molecules28020848

[92] Y. Jung Park, H. Jin Jung, H. Jin Kim, H. Soo Park, J. Lee, D. Yoon, et al., Thiazol-4(5H)-one analogs as potent tyrosinase inhibitors: synthesis, tyrosinase inhibition, antimelanogenic effect, antioxidant activity, and in silico docking simulation, Bioorg. Med. Chem. 98 (2024) 117578, https://doi.org/10.1016/j.bmc.2023.117578

[93] M.K. Kang, D. Yoon, H.J. Jung, S. Ullah, J. Lee, H.S. Park, et al., Identification and molecular mechanism of novel 5-alkenyl-2-benzylaminothiazol-4(5H)-one analogs as anti-melanogenic and antioxidant agents, Bioorg. Chem. 140 (2023) 106763, https://doi.org/10.1016/j.bioorg.2023.106763

[94] D. Yoon, M.K. Kang, H.J. Jung, S. Ullah, J. Lee, Y. Jeong, et al., Design, synthesis, in vitro, and in silico insights of 5-(substituted benzylidene)-2-phenylthiazol-4(5H)-one derivatives: a novel class of anti-melanogenic compounds, Molecules 28 (2023) 3293, https://doi.org/10.3390/molecules28083293

[95] A. Iraji, N. Sheikhi, M. Attarroshan, G. Reaz Sharifi Ardani, M. Kabiri, A. Naghibi Bafghi, et al., Design, synthesis, spectroscopic characterization, in vitro tyrosinase inhibition, antioxidant evaluation, in silico and kinetic studies of substituted indolecarbohydrazides, Bioorg. Chem. 129 (2022) 106140, https://doi.org/10.1016/j.bioorg.2022.106140

[96] M. Alijanianzadeh, A.A. Saboury, M.R. Ganjali, H. Hadi-Alijanvand, A.A. Moosavi-Movahedi, Inhibition of mushroom tyrosinase by a newly synthesized ligand: inhibition kinetics and computational simulations, J. Biomol. Struct. Dyn. 30 (2012) 448–459, https://doi.org/10.1080/07391102.2012.682210

[97] A. Iraji, T. Adelpour, N. Edraki, M. Khoshneviszadeh, R. Miri, M. Khoshneviszadeh, Synthesis, biological evaluation and molecular docking analysis of vaniline-benzylidenehydrazine hybrids as potent tyrosinase inhibitors, BMC Chem. 14 (2020) 28, https://doi.org/10.1186/s13065-020-00679-1

[98] G. Rocchitta, C. Rozzo, M. Pisano, D. Fabbri, M.A. Dettori, P. Ruzza, et al., Inhibitory effect of curcumin-inspired derivatives on tyrosinase activity and melanogenesis, Molecules 27 (2022) 7942, https://doi.org/10.3390/molecules27227942

[99] K. Parndaeng, T. Pitakbut, C. Wattanapiromsakul, J.S. Hwang, W. Udomuksorn, S. Dej-Adisai, Chemical constituents from streblus taxoides wood with their antibacterial and antityrosinase activities plus in silico study, Antibiotics (Basel) 12 (2023) 319, https://doi.org/10.3390/antibiotics12020319

[100] E. Chaita, G. Lambrinidis, C. Cheimonidi, A. Agalou, D. Beis, I. Trougakos, et al., Anti-melanogenic properties of Greek plants. A novel depigmenting agent from *Morus alba* wood, Molecules 22 (2017) 514, https://doi.org/10.3390/molecules22040514

[101] S. Sari, B. Barut, A. Özel, A. Kuruüzüm-Uz, D. Şöhretoğlu, Tyrosinase and α-glucosidase inhibitory potential of compounds isolated from *Quercus coccifera* bark: in vitro and in silico perspectives, Bioorg. Chem. 86 (2019) 296–304, https://doi.org/10.1016/j.bioorg.2019.02.015

[102] D. Şöhretoğlu, S. Sari, B. Barut, A. Özel, Tyrosinase inhibition by a rare neolignan: inhibition kinetics and mechanistic insights through in vitro and in silico studies, Comput. Biol. Chem. 76 (2018) 61–66, https://doi.org/10.1016/j.compbiolchem.2018.06.003

[103] Y.M. Ha, H.J. Lee, D. Park, H.O. Jeong, J.Y. Park, Y.J. Park, et al., Molecular docking studies of (1*E*,3*E*,5*E*)-1,6-bis(substituted phenyl)hexa-1,3,5-triene and 1,4-bis(substituted trans-styryl)benzene analogs as novel tyrosinase inhibitors, Biol. Pharm. Bull. 36 (2013) 55–65, https://doi.org/10.1248/bpb.b12-00605

[104] P. Correia, H. Oliveira, P. Araújo, N.F. Brás, A.R. Pereira, J. Moreira, et al., The role of anthocyanins, deoxyanthocyanins and pyranoanthocyanins on the modulation of tyrosinase activity: an in vitro and in silico approach, Int. J. Mol. Sci. 22 (2021) 6192, https://doi.org/10.3390/ijms22126192

[105] Q. Huang, W.M. Chai, Z.Y. Ma, W.L. Deng, Q.M. Wei, S. Song, et al., Antityrosinase mechanism of ellagic acid in vitro and its effect on mouse melanoma cells, J. Food Biochem. 43 (2019) e12996, https://doi.org/10.1111/jfbc.12996

[106] M.T. Varela, M. Ferrarini, V.G. Mercaldi, B.D.S. Sufi, G. Padovani, L.I.S. Nazato, et al., Coumaric acid derivatives as tyrosinase inhibitors: efficacy studies through in silico, in vitro and ex vivo approaches, Bioorg. Chem. 103 (2020) 104108, https://doi.org/10.1016/j.bioorg.2020.104108

[107] Y. Nazir, A. Saeed, M. Rafiq, S. Afzal, A. Ali, M. Latif, et al., Hydroxyl substituted benzoic acid/cinnamic acid derivatives: tyrosinase inhibitory kinetics, anti-melanogenic activity and molecular docking studies, Bioorg. Med. Chem. Lett. 30 (2020) 126722, https://doi.org/10.1016/j.bmcl.2019.126722

[108] Y. Nazir, H. Rafique, N. Kausar, Q. Abbas, Z. Ashraf, P. Rachtanapun, et al., Methoxy-substituted tyramine derivatives synthesis, computational studies and tyrosinase inhibitory kinetics, Molecules 26 (2021) 2477, https://doi.org/10.3390/molecules26092477

[109] J. Li, X. Min, X. Zheng, S. Wang, X. Xu, J. Peng, Synthesis, anti-tyrosinase activity, and spectroscopic inhibition mechanism of cinnamic acid-eugenol esters, Molecules 28 (2023) 5969, https://doi.org/10.3390/molecules28165969

[110] M. Rafiq, Y. Nazir, Z. Ashraf, H. Rafique, S. Afzal, A. Mumtaz, et al., Synthesis, computational studies, tyrosinase inhibitory kinetics and antimelanogenic activity of hydroxy substituted 2-[(4-acetylphenyl)amino]-2-oxoethyl derivatives, J. Enzyme Inhib. Med. Chem. 34 (1) (2019) 11, https://doi.org/10.1080/14756366.2019.1654468

[111] Y. Nazir, H. Rafique, S. Roshan, S. Shamas, Z. Ashraf, M. Rafiq, et al., Molecular docking, synthesis, and tyrosinase inhibition activity of acetophenone amide: potential inhibitor of melanogenesis, Biomed. Res. Int. 2022 (2022) 1040693, https://doi.org/10.1155/2022/1040693

[112] S. Radhakrishnan, R. Shimmon, C. Conn, A. Baker, Integrated kinetic studies and computational analysis on naphthyl chalcones as mushroom tyrosinase inhibitors, Bioorg. Med. Chem. Lett. 25 (2015) 4085–4091, https://doi.org/10.1016/j.bmcl.2015.08.033

[113] W. Pan, I. Giovanardi, T. Sagynova, A. Cariola, V. Bresciani, M. Masetti, et al., Potent antioxidant and anti-tyrosinase activity of butein and homobutein probed by molecular kinetic and mechanistic studies, Antioxidants (Basel) 12 (2023) 1763, https://doi.org/10.3390/antiox12091763

[114] A. Iraji, Z. Panahi, N. Edraki, M. Khoshneviszadeh, M. Khoshneviszadeh, Design, synthesis, in vitro and in silico studies of novel Schiff base derivatives of 2-hydroxy-4-methoxybenzamide as tyrosinase inhibitors, Drug Dev. Res. 82 (2021) 533–542, https://doi.org/10.1002/ddr.21771

[115] A. Bonardi, A. Nocentini, S. Bua, J. Combs, C. Lomelino, J. Andring, et al., Sulfonamide inhibitors of human carbonic anhydrases designed through a three-tails approach: improving ligand/isoform matching and selectivity of action, J. Med. Chem. 63 (2020) 7422–7444, https://doi.org/10.1021/acs.jmedchem.0c00733

[116] Gou L, J. Lee, J.M. Yang, Y.D. Park, H.M. Zhou, Y. Zhan, et al., Inhibition of tyrosinase by fumaric acid: integration of inhibition kinetics with computational docking simulations, Int. J. Biol. Macromol. 105 (2017) 1663–1669, https://doi.org/10.1016/j.ijbiomac.2016.12.013

[117] Gou L, J. Lee, H. Hao, Y.D. Park, Y. Zhan, Z.R. Lü, The effect of oxaloacetic acid on tyrosinase activity and structure: integration of inhibition kinetics with docking simulation, Int. J. Biol. Macromol. 101 (2017 Aug) 59–66, https://doi.org/10.1016/j.ijbiomac.2017.03.073

[118] Gou L, J. Lee, J.M. Yang, Y.D. Park, H.M. Zhou, Y. Zhan, et al., The effect of alpha-ketoglutaric acid on tyrosinase activity and conformation: Kinetics and molecular dynamics simulation study, Int. J. Biol. Macromol. 105 (Pt 3) (2017) 1654–1662, https://doi.org/10.1016/j.ijbiomac.2016.12.015

[119] R. Biswas, J. Chanda, A. Kar, P.K. Mukherjee, Tyrosinase inhibitory mechanism of betulinic acid from Dillenia indica, Food Chem. 232 (2017) 689–696, https://doi.org/10.1016/j.foodchem.2017.04.008

[120] H.M. Liu, W. Tang, X.Y. Wang, J.J. Jiang, Y. Zhang, Q.L. Liu, et al., Experimental and theoretical studies on inhibition against tyrosinase activity and melanin biosynthesis by antioxidant ergothioneine, Biochem. Biophys. Res. Commun. 682 (2023) 163–173, https://doi.org/10.1016/j.bbrc.2023.10.007

[121] P. Mahalapbutr, S. Sabuakham, S. Nasoontorn, T. Rungrotmongkol, A. Silsirivanit, U. Suriya, Discovery of amphotericin B, an antifungal drug as tyrosinase inhibitor with potent anti-melanogenic activity, Int. J. Biol. Macromol. 246 (2023) 125587, https://doi.org/10.1016/j.ijbiomac.2023.125587

[122] H. Hosseinpoor, S. Moghadam Farid, A. Iraji, S. Askari, N. Edraki, S. Hosseini, et al., Anti-melanogenesis and anti-tyrosinase properties of aryl-substituted acetamides of phenoxy methyl triazole conjugated with thiosemicarbazide: Design, synthesis and biological evaluations, Bioorg. Chem. 114 (2021) 104979, https://doi.org/10.1016/j.bioorg.2021.104979

[123] E. Buitrago, C. Faure, L. Challali, E. Bergantino, A. Boumendjel, L. Bubacco, et al., Ditopic chelators of dicopper centers for enhanced tyrosinases inhibition, Chemistry 27 (2021) 4384–4393, https://doi.org/10.1002/chem.202004695

[124] E. Buitrago, C. Faure, M. Carotti, E. Bergantino, R. Hardré, M. Maresca, et al., Exploiting HOPNO-dicopper center interaction to development of inhibitors for human tyrosinase, Eur. J. Med. Chem. 248 (2023) 115090, https://doi.org/10.1016/j.ejmech.2023.115090

[125] A. Joompang, P. Anwised, S. Klaynongsruang, L. Taemaitree, A. Wanthong, K. Choowongkomon, et al., Rational design of an N-terminal cysteine-containing tetrapeptide that inhibits tyrosinase and evaluation of its mechanism of action, Curr. Res. Food Sci. 7 (2023) 100598, https://doi.org/10.1016/j.crfs.2023.100598

[126] Y.C. Lee, N.W. Hsiao, T.S. Tseng, W.C. Chen, H.H. Lin, S.J. Leu, et al., Phage display-mediated discovery of novel tyrosinase-targeting tetrapeptide inhibitors reveals the significance of N-terminal preference of cysteine residues and their functional sulfur atom, Mol. Pharmacol. 87 (2015) 218–230, https://doi.org/10.1124/mol.114.094185

[127] Y. Song, J. Li, H. Tian, H. Xiang, S. Chen, L. Li, et al., Copper chelating peptides derived from tilapia (*Oreochromis niloticus*) skin as tyrosinase inhibitor: biological evaluation, in silico investigation and in vivo effects, Food Res. Int. 163 (2023 Jan) 112307, https://doi.org/10.1016/j.foodres.2022.112307

[128] E. Bang, S.G. Noh, S. Ha, H.J. Jung, D.H. Kim, A.K. Lee, et al., Evaluation of the novel synthetic tyrosinase inhibitor (Z)-3-(3-bromo-4-hydroxybenzylidene)thio-chroman-4-one (MHY1498) in vitro and in silico, Molecules 23 (2018) 3307, https://doi.org/10.3390/molecules23123307

[129] S.J. Kim, J. Yang, S. Lee, C. Park, D. Kang, J. Akter, et al., The tyrosinase inhibitory effects of isoxazolone derivatives with a (Z)-β-phenyl-α, β-unsaturated carbonyl scaffold, Bioorg. Med. Chem. 26 (2018) 3882–3889, https://doi.org/10.1016/j.bmc.2018.05.047

[130] Y.M. Ha, J.A. Kim, Y.J. Park, D. Park, J.M. Kim, K.W. Chung, et al., Analogs of 5-(substituted benzylidene)hydantoin as inhibitors of tyrosinase and melanin formation, Biochim. Biophys. Acta 1810 (2011) 612–619, https://doi.org/10.1016/j.bbagen.2011.03.001

[131] E. Bang, E.K. Lee, S.G. Noh, H.J. Jung, K.M. Moon, M.H. Park, et al., In vitro and in vivo evidence of tyrosinase inhibitory activity of a synthesized (Z)-5-(3-hydroxy-4-methoxybenzylidene)-2-thioxothiazolidin-4-one (5-HMT), Exp. Dermatol. 28 (2019) 734–737, https://doi.org/10.1111/exd.13863

[132] H.J. Jung, D.C. Choi, S.G. Noh, H. Choi, I. Choi, I.Y. Ryu, et al., New benzi-midazothiazolone derivatives as tyrosinase inhibitors with potential anti-melano-genesis and reactive oxygen species scavenging activities, Antioxidants (Basel) 10 (2021) 1078, https://doi.org/10.3390/antiox10071078

[133] B. Hassani, F. Zare, L. Emami, M. Khoshneviszadeh, R. Fazel, N. Kave, et al., Synthesis of 3-hydroxypyridin-4-one derivatives bearing benzyl hydrazide substitu-tions towards anti-tyrosinase and free radical scavenging activities, RSC Adv. 13 (2023) 32433–32443, https://doi.org/10.1039/d3ra06490e

[134] A. Bagheri, S. Moradi, A. Iraji, M. Mahdavi, Structure-based development of 3,5-dihydroxybenzoyl-hydrazineylidene as tyrosinase inhibitor; in vitro and in silico study, Sci. Rep. 14 (2024) 1540, https://doi.org/10.1038/s41598-024-52022-6

[135] M. Hassan, Z. Ashraf, Q. Abbas, H. Raza, S.Y. Seo, Exploration of novel human tyrosinase inhibitors by molecular modeling, docking and simulation studies, Interdiscip. Sci. 10 (2018) 68–80, https://doi.org/10.1007/s12539-016-0171-x

[136] H.A.S. El-Nashar, M.I.G. El-Din, L. Hritcu, O.A. Eldahshan, Insights on the inhibitory power of flavonoids on tyrosinase activity: a survey from 2016 to 2021, Molecules 26 (2021) 7546, https://doi.org/10.3390/molecules26247546

[137] J.A. Nunes, R.S.A. Araújo, F.N.D. Silva, J. Cytarska, K.Z. Łączkowski, S.H. Cardoso, et al., Coumarin-based compounds as inhibitors of tyrosinase/tyrosine hydroxylase: synthesis, kinetic studies, and in silico approaches, Int. J. Mol. Sci. 24 (2023) 5216, https://doi.org/10.3390/ijms24065216

[138] S. Masuri, B. Era, F. Pintus, E. Cadoni, M.G. Cabiddu, A. Fais, et al., Hydroxylated coumarin-based thiosemicarbazones as dual antityrosinase and antioxidant agents, Int. J. Mol. Sci. 24 (2023) 1678, https://doi.org/10.3390/ijms24021678

[139] H. Jiang, L. Zhou, Y. Wang, G. Liu, S. Peng, W. Yu, et al., Inhibition of cinnamic acid and its derivatives on polyphenol oxidase: effect of inhibitor carboxyl group and system pH, Int. J. Biol. Macromol. 259 (2024) 129285, https://doi.org/10.1016/j.ijbiomac.2024.129285

[140] D. Xie, K. Han, Q. Jiang, S. Xie, J. Zhou, Y. Zhang, et al., Design, synthesis, and inhibitory activity of hydroquinone ester derivatives against mushroom tyrosinase, RSC Adv. 14 (2024) 6085–6095, https://doi.org/10.1039/d4ra00007b

[141] Q. Abbas, Z. Ashraf, M. Hassan, H. Nadeem, M. Latif, S. Afzal, et al., Development of highly potent melanogenesis inhibitor by in vitro, in vivo and computational studies, Drug Des. Devel Ther. 11 (2017) 2029–2046, https://doi.org/10.2147/DDDT.S137550

[142] Neelam, S. Ahlawat, A. Shankar, A. Lather, A. Khatkar, K.K. Sharma, Bioevaluation and molecular docking analysis of novel phenylpropanoid derivatives as potent food preservative and anti-microbials, 3 Biotech. 11 (2021) 70, https://doi.org/10.1007/s13205-020-02636-0

[143] S. Sari, B. Barut, A. Özel, D. Şöhretoğlu, Tyrosinase inhibitory effects of *Vinca major* and its secondary metabolites: enzyme kinetics and in silico inhibition model of the metabolites validated by pharmacophore modelling, Bioorg. Chem. 92 (2019) 103259, https://doi.org/10.1016/j.bioorg.2019.103259

[144] Y. Sıcak, H. Kekeçmuhammed, A. Karaküçük-İyidoğan, T. Taşkın-Tok, E.E. Oruç-Emre, M. Öztürk, Chalcones bearing nitrogen-containing heterocyclics as multi-targeted inhibitors: design, synthesis, biological evaluation and molecular docking studies, J. Mol. Recognit. 36 (2023) e3020, https://doi.org/10.1002/jmr.3020

[145] H.J. Jung, S.G. Noh, I.Y. Ryu, C. Park, J.Y. Lee, P. Chun, et al., (E)-1-(Furan-2-yl)-(substituted phenyl)prop-2-en-1-one derivatives as tyrosinase inhibitors and melanogenesis inhibition: an in vitro and in silico study, Molecules 25 (2020) 5460, https://doi.org/10.3390/molecules25225460

[146] S. Ranjbar, M.M. Kamarei, M. Khoshneviszadeh, H. Hosseinpoor, N. Edraki, M. Khoshneviszadeh, Benzylidene-6-hydroxy-3,4-dihydronaphthalenone chalco-noids as potent tyrosinase inhibitors, Res. Pharm. Sci. 16 (2021) 425–435, https://doi.org/10.4103/1735-5362.319580

[147] S.J. Yin, Y.X. Si, Y.F. Chen, G.Y. Qian, Z.R. Lü, S. Oh, et al., Mixed-type inhibition of tyrosinase from *Agaricus bisporus* by terephthalic acid: computational simulations and kinetics, Protein J. 30 (2011) 273–280, https://doi.org/10.1007/s10930-011-9329-x

[148] S.J. Yin, Y.X. Si, G.Y. Qian, Inhibitory effect of phthalic acid on tyrosinase: the mixed-type inhibition and docking simulations, Enzyme Res. 2011 (2011) 294724, https://doi.org/10.4061/2011/294724

[149] Q. Abbas, H. Raza, M. Hassan, A.R. Phull, S.J. Kim, S.Y. Seo, Acetazolamide inhibits the level of tyrosinase and melanin: an enzyme kinetic, in vitro, in vivo, and in silico studies, Chem. Biodivers. 14 (2017), https://doi.org/10.1002/cbdv.201700117

[150] Z.J. Wang, Y.X. Si, S. Oh, J.M. Yang, S.J. Yin, Y.D. Park, et al., The effect of fucoidan on tyrosinase: computational molecular dynamics integrating inhibition kinetics, J. Biomol. Struct. Dyn. 30 (2012) 460–473, https://doi.org/10.1080/07391102.2012.682211

[151] E.D. Kaya, A. Türkhan, F. Gür, B. Gür, A novel method for explaining the product inhibition mechanisms via molecular docking: inhibition studies for tyrosinase from *Agaricus bisporus*, J. Biomol. Struct. Dyn. 40 (2022) 7926–7939, https://doi.org/10.1080/07391102.2021.1905069

[152] L.M.A.F. Khraisat, S. Sabuncuoğlu, G. Girgin, O. Unsal Tan, Synthesis and tyrosinase inhibitory activity of novel benzimidazole/thiazolidin-4-one hybrid derivatives, Chem. Biodivers. 21 (2024) e202301489, https://doi.org/10.1002/cbdv.202301489

[153] E. Gultekin, O. Bekircan, Y. Kolcuoğlu, A. Akdemir, Synthesis of new 1,2,4-tria-zole-(thio)semicarbazide hybrid molecules: their tyrosinase inhibitor activities and molecular docking analysis, Arch. Pharm. (Weinh. 354 (2021) e2100058, https://doi.org/10.1002/ardp.202100058

[154] D. Sousa-Pereira, O.A. Chaves, C.M. Dos Reis, M.C.C. de Oliveira, C.M.R. Sant'Anna, J.C. Netto-Ferreira, et al., Synthesis and biological evaluation of N-aryl-2-phenyl-hydrazinecarbothioamides: experimental and theoretical analysis on tyrosinase inhibition and interaction with HSA, Bioorg. Chem. 81 (2018) 79–87, https://doi.org/10.1016/j.bioorg.2018.07.035

[155] P.G. Yap, C.Y. Gan, Tyrosinase inhibitory peptides: structure-activity relationship study on peptide chemical properties, terminal preferences and intracellular regulation of melanogenesis signaling pathways, Biochim. Biophys. Acta Gen. Subj. 1868 (2024) 130503, https://doi.org/10.1016/j.bbagen.2023.130503

[156] W. Xue, X. Liu, W. Zhao, Z. Yu, Identification and molecular mechanism of novel tyrosinase inhibitory peptides from collagen, J. Food Sci. 87 (2022) 2744–2756, https://doi.org/10.1111/1750-3841.16160

[157] Z. Yu, H. Lv, M. Zhou, P. Fu, W. Zhao, Identification and molecular docking of tyrosinase inhibitory peptides from allophycocyanin in *Spirulina platensis*, J. Sci. Food Agric. 104 (2024) 3648–3653, https://doi.org/10.1002/jsfa.13249

[158] W. Wang, H. Lin, W. Shen, X. Qin, J. Gao, W. Cao, et al., Optimization of a novel tyrosinase inhibitory peptide from *Atrina pectinata* mantle and its molecular inhibitory mechanism, Foods 12 (2023) 3884, https://doi.org/10.3390/foods12213884

CHAPTER EIGHT

Bacterial tyrosinases and their inhibitors

Ali Irfan[a], Yousef A. Bin Jardan[b],*, Laila Rubab[c], Huma Hameed[d], Ameer Fawad Zahoor[a],*, and Claudiu T. Supuran[e],*

[a]Department of Chemistry, Government College University Faisalabad, Faisalabad, Pakistan
[b]Department of Pharmaceutics, College of Pharmacy, King Saud University, Riyadh, Saudi Arabia
[c]Department of Chemistry, Sargodha Campus, The University of Lahore, Sargodha, Pakistan
[d]Faculty of Pharmaceutical Sciences, University of Central Punjab (UCP), Lahore, Pakistan
[e]Department of NEUROFARBA—Section of Pharmaceutical and Nutraceutical Sciences, University of Florence, Sesto Fiorentino, Florence, Italy
*Corresponding authors. e-mail address: ybinjardan@ksu.edu.sa; fawad.zahoor@gcuf.edu.pk; claudiu.supuran@unifi.it

Contents

Abstract

Bacterial tyrosinase is a copper-containing metalloenzyme with diverse physio-chemical properties, that have been identified in various bacterial strains, including actinobacteria and proteobacteria. Tyrosinases are responsible for the rate-limiting catalytic steps in melanin biosynthesis and enzymatic browning. The physiological role of bacterial tyrosinases in melanin biosynthesis has been harnessed for the production of coloring and

ISSN 1874-6047, https://doi.org/10.1016/bs.enz.2024.06.003
231

dyeing agents. Additionally, bacterial tyrosinases have the capability of cross-linking activity, demonstrated material functionalization applications, and applications in food processing with varying substrate specificities and stability features. These characteristics make bacterial tyrosinases a valuable alternative to well-studied mushroom tyrosinases. The key feature of substrate specificity of bacterial tyrosinase has been exploited to engineer biosensors that have the ability to detect the minimal amount of different phenolic compounds. Today, the world is facing the challenge of multi-drugs resistance in various diseases, especially antibiotic resistance, skin cancer, enzymatic browning of fruits and vegetables, and melanogenesis. To address these challenges, medicinal scientists are developing novel chemotherapeutic agents by inhibiting bacterial tyrosinases. To serve this purpose, heterocyclic compounds are of particular interest due to their vast spectrum of biological activities and their potential as effective tyrosinase inhibitors. In this chapter, a plethora of research explores applications of bacterial tyrosinases in different fields, such as the production of dyes and pigments, catalytic applications in organic synthesis, bioremediation, food and feed applications, biosensors, wool fiber coating and the rationalized synthesis, and structure-activity relationship of bacterial tyrosinase inhibitors.

1. Introduction

Tyrosinase, and catechol oxidase are groups of binuclear copper–containing enzymes that are widely distributed in various species, including bacteria, fungi, animals and plants. These enzymes play an integral role in the melanin biosynthetic pathway and other phenolic compounds. Tyrosinases catalyzed the oxidation of phenols such as in the melanin synthetic pathway, l-tyrosine to l-Dopa, which further transforms into quinones, and after complex multi-steps, quinones polymerize to form pigments [1–4]. In recent years, bacterial tyrosinases exhibited wide variety of application in various industrial fields, biotechnology, organic synthesis, bioremediation, food and feed processing, and biomedical fields. These tyrosinase demonstrated excellent structural characteristics, such as stability, have wide substrate specificities, and are easy to produce in large quantities on commercial scales which make these tyrosinase attractive to utilized in different fields in comparison to other tyrosinases derived other microorganisms [5–7].

The first crystal of structure of a tyrosinase from *Streptomyces castaneoglobisporus* provided significant insights into mechanism of phenol hydroxylation This structure revealed the enzyme complexed with its accessory caddie protein. The bacterial tyrosinase predominantly demonstrated α-helical structural features and six histidine residues coordinated with two copper ions to form the enzyme's active site. A recently published another structure of bacterial tyrosinase isolated from *Bacillus megaterium,* showed similar folding pattern with that *Streptomyces castaneoglobisporus* [8,9,10].

1.1 Types of bacterial tyrosinases

Bacterial tyrosinases have been divided into five types based on domain organization and the potential requirement of caddie protein for enzyme activity [7,11].

1.1.1 Type I bacterial tyrosinases

The necessity of a secondary helper protein (caddie protein) for secretion, correct folding, assembly of the cooper atoms, and enzyme activity is common to type-1 tyrosinases, such as enzymes from *S. castaneoglobisporus* and *S. antibioticus* [7,12,13].

1.1.2 Type II bacterial tyrosinases

Type II bacterial tyrosinases are present in *B. megaterium,* and are small, monomeric enzymes containing only the catalytic domain, that don't require additional helper proteins; they are possibly secreted [7,9].

1.1.3 Type III bacterial tyrosinases

The tyrosinase enzyme isolated from the *Verrucomicrobium spinosum* bacterial strain is designated as a Type III bacterial tyrosinase which has a C-terminal domain similar to that of fungal tyrosinase whose removal led to 100-fold higher activity [7,14]. The concept that the C-terminal domain plays a role in keeping the tyrosinase enzyme in an inactive state in the cells of fungi and plants, is also supported by the bacterial tyrosinase type III enzyme [7,15–17].

1.1.4 Bacterial tyrosinase type IV enzyme

Bacillus thuringiensis (14 kDa) and *Sterptomyces nigrifaciens* produced bacterial tyrosinase type IV enzyme. The smallest enzyme among all five types of bacterial tyrosinases, is the type IV bacterial tyrosinase enzyme [7,18,19].

1.1.5 Bacterial tyrosinase type V enzyme

The bacterial tyrosinase enzyme is unique in nature because it lacks the typical sequence characteristics of bacterial tyrosinase enzymes, demonstrates features usually associated with laccase bacterial enzyme and displays minimum activity on the tyrosine substrate. For example, a membrane-bounded bacterial tyrosinase enzyme of *Marinomonas mediterranea* is active on the standard laccase substrate ABTS with NCBI ID: AAF75831.2, and a tyrosinase enzyme from the same bacterial strain shows classical substrate specificity and is activated on substrate SDS with NCBI ID: AAV49996.1 has been reported. On the basis of the aforementioned five types of bacterial tyrosinases, a comprehensive summary of bacterial tyrosinases, including their characteristics, sources, and any relevant information from different bacterial strains [20], is outlined in Table 1.

2. Catalytic mechanism of bacterial tyrosinases

In bacterial tyrosinases, the type 3 binuclear copper center is composed of two copper atoms, namely CuA and CuB, which are coordinated by conserved histidine residues, forming enzyme's active site. This is a common feature that is an integral part of all types of tyrosinases except types IV and V [38]. In the bacterial tyrosinases, the CuA binding motif is characterized by the sequence H-Xn-H-X8-H, while the CuB motif follows the sequence H-X3-H-Xn-H, where Xn denotes a variable number of unspecified amino acids [39]. Currently, the structures of only a few bacterial tyrosinases have been determined, some of whose PDB IDs are provided in Table 1. All these structures demonstrate the coordination of two copper atoms, highlighting the role of caddie protein in the formation, coordination, and stabilization of these cooper atoms in tyrosinases [21–37]. On the basis of structural insights, it is predicted that the enzyme active site can exit in one of three oxidative states (oxy, met, and deoxy) during the oxidation of substrate as displayed in Figs. 1 and 2.

- In the oxy state, the two copper ions in the enzyme are bridged by a dioxygen molecule, forming a peroxide linkage. This configuration facilitates the deprotonation of mono-phenolic substrate, ultimately leading to its orthro hydroxylation.
- Upon the reaction with substrate, the enzyme transitions into the met state.
- The met form is then converted into deoxy form, during the oxidation of the diphenolic compound, a process involving a two–electron reduction.

When the deoxy form subsequently binds dioxygen, the enzyme is restored to oxy state [38–40]. For the enzyme to perform a monophenol monooxygenase type reaction, it must be in oxy state. Conversely, both the met and deoxy states are essential for diphenolase activity. Notably, the met form of tyrosinase can not hydroxylate the monophenol, underscoring the specificity of each oxidative state in enzyme's functions as depicted in Figs. 1 and 2 [39,40].

Table 1 List of bacterial tyrosinase cited in the literature.

Bacterial strain	Tyrosinase type	Features of tyrosinases	References
Streptomyces lincolnensis	I	This bacterial strain tyrosinase played role in lincomycin biosynthesis. PDB ID: 8IL0	[21,22]
Streptomyces antibioticus	I	This bacterial strain is well studied and structural information are confirmed through EPR NMR, and mutagenesis	[23–25]
Streptomyces glauescens	I	The information on this bacterial strain tyrosinase enzyme is comprehensively studied and reported	[26,27]
Bacillus megaterium	II	The *Bacillus megaterium* tyrosinase was crystalized as dimer and was confirmed to be active in its crystalline form at a resolution of 2.0–2.3 Å. PDB ID: 3NM8	[9,28–30]
Pseudomonas aeruginosa	II	Tyrosinase PvdP is dimeric enzyme of bacterial strain *Pseudomonas aeruginosa*.PDB ID: 6RRR	[31]
Ralstonia solanacearum	III	This bacterial strain tyrosinase monophenolase activity is significantly higher than diphenolase activity	[32,33]
Bacillus thuringiensis	II and IV	Tyrosinase of *Bacillus thuringiensis* is monomer and act as potential herbicide	[34,35]
Marinomonas mediterranea	I and V	In this bacterial strain, both tyrosinase and laccase having tyrosinase like activities are present which behave like type 1 and type v tyrosinase	[36,37]

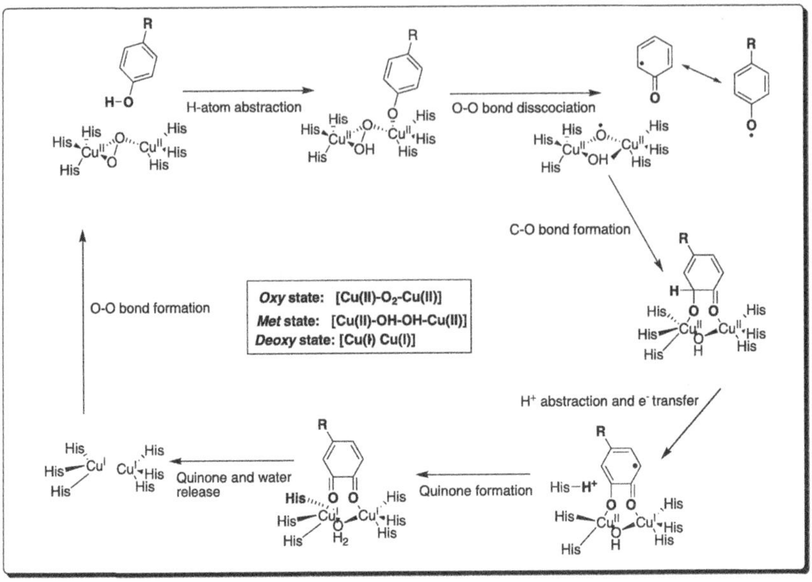

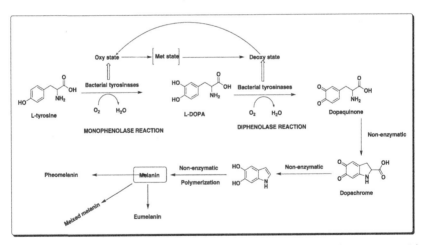

Fig. 1 Proposed catalytic mechanism of tyrosinases with oxidation states.

Fig. 2 Biosynthetic strategy of melanin synthesis by using bacterial tyrosinase with oxidative states.

3. Biosynthetic strategy of melanin synthesis by bacterial tyrosinases

Melanin is a high molecular weight biopolymer formed through the oxidative polymerization of phenolic or indolic substrates. This melanin pigment serves as a protective agent across a wide range of organisms, manifesting predominantly in black or brown hues [41]. Commercially available melanin is obtained in 100 mg or 1 g from cuttlefish, so this method is not sustainable and economical. The alternative way of producing melanin is through chemical-based synthetic methodologies, melanin can be synthesized with the oxidation of tyrosine by utilizing hydrogen peroxide as an oxidizing agent. Sigma Aldrich synthesized melanin through chemical protocols and offered melanin in various small quantities, such as 100 mg, 250 mg, and 1 g packs. In contrast, melanin production by using bacterial tyrosinases is a cost effective, high-yield, and sustainable alternative. The schematic methodology of melanin production by using bacterial tyrosinases is outlined in Fig. 2.

The highest melanin production was reported by Guo and coworkers was 28.8 g per liter, which was obtained with the overexpression of bacterial tyrosinase in *S. kathirae SC-1* bacterial strain. This bacterial biosynthetic pathway of melanin production revealed a high yield, and is eco-friendly with reduced hazardous environmental impacts in comparison with the traditional extraction approach and chemicals-based synthetic protocols [42]. In the melanin biosynthetic pathway, the first monophenol substrate, L-tyrosine underwent hydroxylation to form the diphenol product L-Dopa by utilizing bacterial tyrosinase. The reaction is called monophenolase or cresolase reaction. In this cresolase reaction, the tyrosinase enzyme is in its oxy state. In the next step, the tyrosinase enzyme catalyzes the diphenol L-Dopa substrate to dopaquinones, the reaction is termed as diphenolase or catecholase. After the monphenolase activity, the tyrosinase enzyme convert into its met state which changed to deoxy state after diphenolase reaction. After this, the tyrosinase enzyme converts into its oxy state. A series of complex transformations such as oxidation-reduction, cyclization, and polymerization, convert dopaquinone into melanin pigments, which can be brown, black, or yellow. These sequential enzymatic and non-enzymatic reactions depicted in Fig. 2 [43–45].

Tyrosine is the first precursor involved in the general process for melanogenesis in bacteria, which may be highlighted. In order to manufacture DOPA type melanin, tyrosine serves as a main substrate for two enzymes such as tyrosinase and laccase Fig. 3. In many microbes like fungi and bacteria,

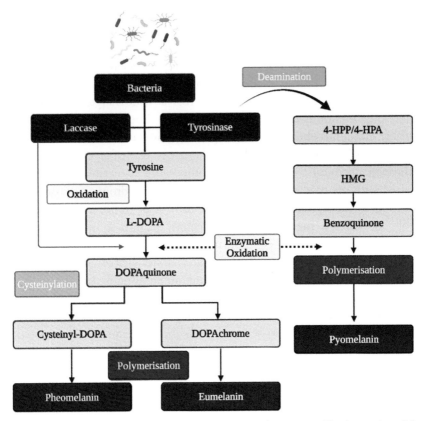

Fig. 3 Bacteria's general melanogenesis pathway. Where L-3,4-dihydroxy-phenylala-nine (L-DOPA), hydroxy-phenyl pyruvate (HPP), hydroxy-phenyl acetate (HPA), 2,5-dihydroxyphenylacetate (HMG).

mammals, and other species, L-3,4-dihydroxyphenylalanine (L-DOPA) is the starting point of the melanin synthesis pathway. The primary molecules of DOPA melanin are L-DOPA, catecholamines, L-tyrosine, and derivatives produced by tyrosinases and laccases. However, in certain instances, the initial oxidation of DOPA production is solely the result of laccase CNLAC1 [46]. These molecules oxidize to form dihydroxy-indoles (DHI), which polymerize into melanin [47]. L-DOPA is produced by oxidation of tyrosine mainly due to involvement of amino group; otherwise, chemicals like homogentisate (2,5-dihydroxyphenylacetate; HMG) are produced when molecules without amino groups participate in the reaction. L-DOPA and HMG are the two dihydroxylated aromatic chemicals that result from tyrosine oxidation in bacterial melanogenesis process. After that, all of these substances oxidize on their own or with the help of catalysts, producing the quinones, which are the direct precursors of melanin (i.e., dopaquinone produced by enzymatic oxidation of DOPA quinone and benzoquinone produced by enzymatic oxidation of HMG). These erratic precursors auto-polymerize to produce the distinct melanins such as pheomelanin, eumelanin, and pyromelanin, respectively [48].

4. Tyrosinase inhibition mechanism

The enzymatic browning of vegetables and fruits, and melanin pigments formation can be inhibited without exposure to UV. This inhibition can be carried out by the inhibition of tyrosinase, the removal of melanin with corneal ablation, and the inhibition of melanocytes metabolism and proliferation [49,50]. The existing standard topical treatments available for the treatment of hyperpigmentation disorders are, melasma, bleaching with hydroquinones, retinoids anti-inflammatory therapy, and use of tyrosinase inhibitors derived from both synthetic and natural sources. These tyrosinase inhibitors can inhibit browning of fruits and vegetables, inhibit the formation of melanin pigment and can treat the skin cancer by the inhibition of monophenolase or diphenolase activities or even both of these activities which have been discovered till date as depicted in Fig. 4 [51,52].

4.1 Inhibiting bacterial melanogenesis at the transcriptional level

Targeting the genetic machinery that produces melanin in bacteria is one way to inhibit bacterial melanogenesis at the transcriptional level. Numerous species, including bacteria, synthesize the pigment melanin, which has a number of functions including defense against UV rays, oxidative stress, and antibiotics.

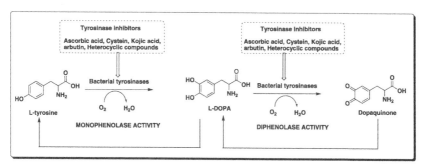

Fig. 4 Inhibition of tyrosinase catalyzed monophenolase and diphenolase activities.

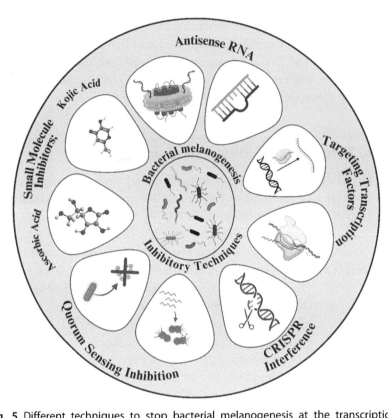

Fig. 5 Different techniques to stop bacterial melanogenesis at the transcriptional level.

The synthesis of melanin in some pathogenic bacteria can enhance their virulence and resistance to drugs and host immunological responses [45]. One strategy is to discover and target the regulatory genes or transcription factors involved in regulating the expression of melanin biosynthesis genes in order to suppress bacterial melanogenesis at the transcriptional level. Bacteria can have their melanin production lowered by interfering with these genes' transcriptional activity. Few tactics to stop bacterial melanogenesis at the transcriptional level are mentioned below and in Fig. 5.

1. **Targeting Transcription Factors:** Create inhibitors that can stop the function of transcription factors linked to genes expressions involved in melanin production. This can entail searching small chemical libraries for substances that obstruct transcription factor binding or alter protein-DNA interactions [53].

2. **Antisense RNA:** Create antisense RNA molecules that precisely target and impede the expression of important genes needed for the manufacture of melanin. By base-pairing with the target mRNA, these antisense RNAs can cause translation inhibition or mRNA destruction [54].

3. **CRISPR Interference (CRISPRi):** Specifically suppress the transcription of genes involved in melanin manufacture by utilizing CRISPR-based technologies. RNA polymerase binding and transcription initiation can be inhibited by CRISPR interference by using catalytically inactive Cas proteins (dCas) to target the promoter regions of these genes [55].

4. **Small Molecule Inhibitors (e.g., tyrosinase inhibitors):** Look through compound libraries for tiny compounds such as kojic acid and ascorbic acid that can stop the action of tyrosinase and other enzymes involved in the manufacture of melanin. These inhibitors can work at the transcriptional level by downregulating important enzymes involved in the process that produces melanin [56,57].

5. **Quorum Sensing Inhibition:** Quorum sensing is a cell-to-cell communication system that controls the production of melanin in certain bacteria. Melanin biosynthesis can be indirectly inhibited by creating substances that disrupt the regulatory networks that govern its expression, hence interfering with quorum sensing signals [58].

It is feasible to create novel antimicrobial methods that can reduce virulence and increase the susceptibility of pathogenic bacteria to host immune responses and drugs by focusing on bacterial melanogenesis at the transcriptional level. However, when developing transcriptional inhibitors for bacterial melanogenesis, it's crucial to take resistance development and other off-target consequences into account [59].

5. Bacterial tyrosinase inhibitors

The medical and economic impact of tyrosinase inhibitors can be seen in the increasing number of compounds being synthesized or derived from different natural sources to determine their chemotherapeutic efficacy against different origins of tyrosinases [60]. In humans, the higher level of melanin pigment causes a variety of disorders such as ocular retinitis pigmentosa and cutaneous hyperpigmentation-related disorders of solar lentigo, melasma, naevi, etc. [61]. It may also be involved in neurodegenerative disorders such as Parkinson's disease development, due to excessive degradation of L-Dopa levels, caused by a higher level of melanin synthesis [62]. Cosmetology researchers are focusing on the development of such a tyrosinase inhibitors that can act as skin-whitening agents and also treat skin cancer by inhibiting melanin formation [63,64]. Tyrosinase inhibitors plays a significant role in the food industry in controlling browning of fruits and vegetables by inhibiting the formation melanin [51,65,66].

5.1 Aurones as bacterial tyrosinase inhibitors

Aurones are an emerging class of tyrosinase inhibitors that belong to the naturally occurring flavonoids family, and are distributed throughout the plant kingdom [67–69]. Haudecoeur and colleagues already reported aurones, also called benzylidenebenzofuran-3(2H)-ones, against human melanocytes to inhibit melanin biosynthesis [70].

According to the reported synthetic protocol [71,72], the different structural features containing aurone ((Z)-2-(2,4-dihydroxybenzylidene)-4-hydroxybenzofuran-3(2H)-one) 3 was achieved via aldol condensation of 4-hydroxybenzofuran-3(2H)-one 1 with 2,4-dihydroxybenzaldehyde 2 under the basic condition of potassium hydroxide in methanol at 65 °C as demonstrated in Scheme 1.

In this study, Haudecoeur and colleagues synthesized 24 aurone derivatives and screened them against bacterial tyrosinase of *Streptomyces antibioticus* strain (TyB3). Among all the synthesized compounds, the aurone compound 3 displayed higher tyrosinase inhibitory therapeutic potential with an IC_{50} value of $4 \pm 1\,\mu M$. The structure–activity relationship showed that the phenyl ring containing hydroxy moiety at *orthro* and *para* position enhanced the bacterial tyrosinase inhibition in comparison with other synthesized scaffolds as shown in Fig. 6.

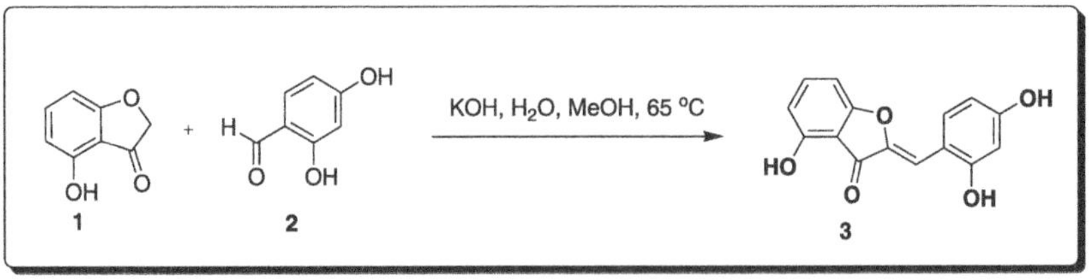

Scheme 1 Synthesis of ((Z)-2-(2,4-dihydroxybenzylidene)-4-hydroxybenzofuran-3(2H)-one).

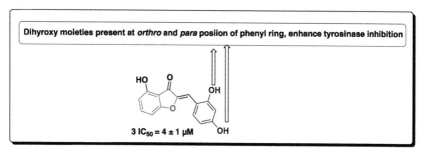

Fig. 6 Aurone compound **3** tyrosinase inhibitory activity.

5.2 Common reference drugs as bacterial tyrosinase inhibitors

In this study, Nokinsee et al. [73], selected four common standard reference drugs such as tropolone, ascorbic acid, arbutin, and kojic acid to determine the highly potent bacterial tyrosinase inhibitor as displayed in Fig. 7. For this purpose, the protein template of the *B. megaterium* tyrosinase (PDB ID: 3NQ1) enzyme was selected on the basis of its 33.5% highest known identity to the human tyrosinase enzyme. Therefore, this bacterial tyrosinase enzyme acted as a template for the construction of a human model. In 2011, the crystal structure of *Bacillus megaterium* was reported [9] and catalytic mechanistic studies of melanogenesis were carried out on the 3D-model constructed on the basis of X-ray data.

Kojic acid is well known due to its tyrosinase inhibitory effects, which explained by the chelation of copper at the active site of the enzyme, as reported by Chen et al. [74]. A nonphenolic reference tyrosinase inhibitor is arbutin, which is chemically glycosylated benzoquinone and is known for tyrosinase inhibitory activities, because in the body, arbutin transforms into phenolic agent, hydroquinone, which is famous for inhibiting melanin synthesis [75]. Previous data revealed that all three reference tyrosinase inhibitors (kojic acid, tropolone, and arbutin) formed chelation at the active site of tyrosinase enzyme to inhibit tyrosinase activity, while ascorbic acid acts as a reducing agent to inhibit melanin formation by reducing dopaquinone to L-Dopa. The outcome of this study is the generation of a homology model of human tyrosinase by using bacterial tyrosinase template due to its greater similarity to human than mushroom tyrosinase. The molecular docking and MD simulation studies indicated that E195 and H208 are important residues in the structure of bacterial tyrosinase because E195 residue contribute to binding to inhibitor, while H208 forms a pi interaction in bacterial tyrosinase [73,76]. The ascorbic acid showed good

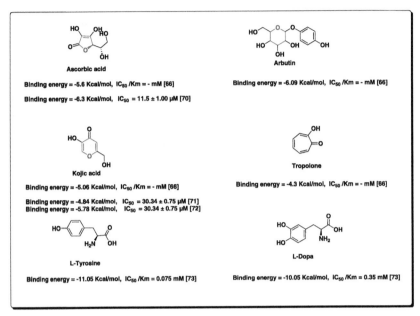

Fig. 7 Standard reference bacterial tyrosinase inhibitors and substrate inhibitory activity data.

binding affinity of −5.6 Kcal/mol [66] and −6.3 Kcal/mol with the active site of bacterial tyrosinase and showed excellent bacterial tyrosinase inhibition activity with an IC_{50} value of 11.5 ± 1.00 μM [28]. In the molecular docking study, the arbutin reference drug showed a binding affinity score of −6.09 Kcal/mol which indicated good bacterial tyrosinase inhibition potential of arbutin [73], as depicted in Fig. 7. Kojic acid exhibited strong binding affinity scores of −5.06 Kcal/mol with the active site of the bacterial tyrosinase enzyme, which showed interactions of kojic acid with the active site tyrosinase, which validated the in vitro results [77,78]. The tropolone showed comparatively lower binding affinity scores of −4.3 Kcal/mol with the active site of bacterial tyrosinase than the ascorbic acid as well as kojic acid [73]. Similarly, L-tyrosine and L-Dopa showed good molecular docking scores as displayed in Fig. 7 [73].

Complex chemicals called melanins are necessary for a number of defense processes in microbes. The structural characteristics of the three main forms of melanin such as eumelanin, pheomelanin, and pyromelanin. When combined, these compounds strengthen intricate defense mechanisms in microbes against oxidative stress [79]. The complex interaction between

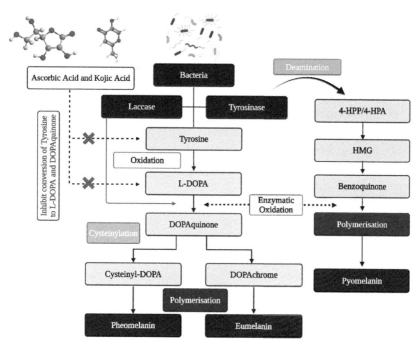

Fig. 8 Mechanism of tyrosinase inhibitory activity in bacteria's melanogenesis pathway.

melanogenesis in bacteria and protective adaptation is delicately balanced, as seen in Fig. 8. Tyrosine is first converted to L–Dopa in the manufacture of dopaquinone, and dopaquinone is then used as a substrate to increase the amount of melanin produced by converting into cysteiny L–Dopa and Dopachrome that further after polymerization converted to Pheomelanin and Eumelanin. Ascorbic acid, kojic acid, hydroquinone, and arbutin are examples of tyrosinase inhibitors that interfere with the conversion of tyrosine to L–DOPA and block the all sequential pathways involved in melanin production subsequently. This disturbance reduced the activity of L–Dopa in the manufacture of melanin. Tyrosine inhibitors also disrupt the catalytic process that converts L–Dopa into an intermediate molecule called Dopaquinone, which raises the tyrosinase enzyme's activity [80]. Tyrosinase inhibitors primarily work by interfering with the cyteinylation, enzymatic oxidation, and polymerization steps involved in the production of three distinct types of melanin: eumelanin, pheomelanin, and pyomelanin. This reduces the amount of melanin produced. But substances like hydroquinone, arbutin, kojic acid, and ascorbic acid prevent melanosomes from forming, which reduces the quantity of all three forms of melanin by bacteria.

5.3 Synthesis of furan-1,3,4-oxadiazole molecules as bacterial tyrosinase inhibitors

Irfan et al. reported the synthesis of highly biologically active furan-1,3,4-oxadiazole engrafted with S-alkylated-N-phenyl acetamides by utilizing ultrasonic-assisted synthetic methodology as presented in Scheme 2 [28,81–83]. The first step of this approach is the synthesis benzofuran-2-carboxylate as reported by Kowalewska et al. [84]. The benzofuran-2-carboxylate was obtained in good yield by treating 5-bromo-2-hydroxybenzaldehyde 4 and ethyl chloroacetate in specified synthetic conditions. In the next step, benzofuran-2-carboxylate was refluxed with hydrazine hydrate to form benzofuran-2-carbohydrazide which react with CS2/KOH to synthesize bromobenzofuran-1,3,4-oxadiazole-2 thiol 5 [85,86]. The other fragment of targeted compounds was prepared by treating aromatic amines 6 with bromoacetyl bromide to yield substituted bromoace-tanilides 7a–j in good to excellent yield [87]. The furan-1,3,4-oxadiazole engrafted with S-alkylated-N-phenyl acetamides 8a–j were furnished in 53–79% yield by the ultrasonic-assisted coupling reaction of 7a–j percussors with bro-mobenzofuran-1,3,4-oxadiazole-2 thiol as demonstrated in Scheme 2.

5.3.1 Bacterial tyrosinase inhibitory activity and SAR of Bromobenzofuran-1,3,4-oxadiazoles

The bromobenzofuran-1,3,4-oxadiazoles were screened against bacterial tyrosinase enzyme which was isolated from bacterial strain (*Bacillus subtilis* NCTC 10,400) [28]. The bacterial tyrosinase inhibitory activity of bromobenzofuran-1,3,4-oxadiazoles 8a–j was outlined in below mentioned Fig. 9.

In the synthesized bromobenzofuran-1,3,4-oxadiazole compounds 8a–j, SAR indicated that nature and the position of substituents on the phenyl ring of acetamide significantly affect the bacterial tyrosinase inhibitory activity. The results of bacterial tyrosinase inhibition showed that the greater the electronegative effect of substituents (the electron withdrawing

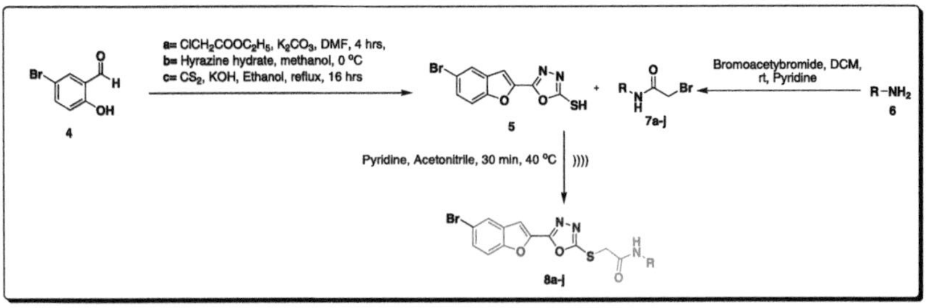

Scheme 2 Synthesis of bromobenzofuran-1,3,4-oxdiazole as bacterial tyrosinase inhibitors.

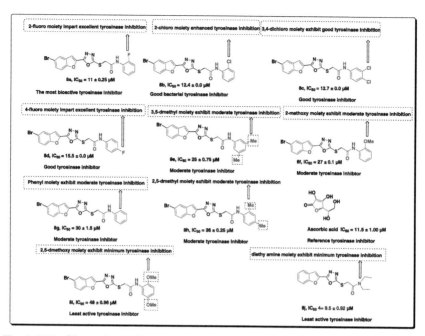

Fig. 9 SAR of bromobenzofuran-1,3,4-oxadiazoles as bacterial tyrosinase inhibitors.

effect of substituents), the higher bacterial tyrosinase inhibition, and vice versa. The bromobenzofuran-1,3,4-oxadiazole compounds containing the highest electronegative fluorine (F) and chlorine (Cl) substituents at 2nd (*orthro*) position on the phenyl ring of acetamide fragment, exhibited the excellent bacterial tyrosinase inhibitory potency, higher than tyrosinase inhibitory activity of reference drug ascorbic acid, while 3rd, 4TH and 5TH positions of acetamide phenyl ring are less bioactive than 2nd positions. The bromobenzofuran-1,3,4-oxadiazole molecule **8a** containing fluorine moiety at *orthro* position, showed the excellent and the highest tyrosinase inhibition potential (IC$_{50}$ = 11 ± 0.25 μM) among the screened derivatives in comparison reference drug ascorbic acid (IC$_{50}$ = 11.5 ± 1.00 μM). The bromobenzofuran-based oxadiazoles **8a–d**, displayed better tyrosinase inhibition due to presence of halo groups while the bromobenzofuran-based oxadiazoles **8e–j** containing electron donating (EDG) alkyl (ethyl, methyl and methoxy) substituents on the phenyl ring of S-linked acetamide attached to bromobenzofuran-1,3,4-oxadiazole. The 2-chloro moiety containing benzofuran-oxadiazole compound **8b** with an IC$_{50}$ value of 12.4 ± 0.0 μM, 3,4-dichloro substituents containing benzofuran-oxadiazole compound **8c** with an IC$_{50}$ value of 12.7 $\pm$ 0.0 μM and 4-fluoro moiety containing benzofuran-oxadiazole **8d** with an IC$_{50}$ value of 15.5 $\pm$ 0.0 μM demonstrated good bacterial tyrosinase inhibition potential due to EWD effect of 2-chloro, 3,4-dichloro and 4-fluoro substituents as shown in Fig. 9. The benzofuran-oxadiazole compounds exhibited moderate bacterial tyrosinase inhibition potency due to presence of electron donating groups such as 3,4-dimethyl group containing molecule **8e** with an IC$_{50}$ value of 25 $\pm$ 0.75 μM, 2-methoxy group containing **8f** with an IC$_{50}$ value of 27 $\pm$ 1.00 μM, phenyl containing **8g** with an IC$_{50}$ value of 30 $\pm$ 1.50 μM and 2,4-dimethyl substituent containing **8h** with an IC$_{50}$ value of 36 $\pm$ 0.25 μM respectively as depicted in Fig. 9. The molecule containing 2,5-dimethoxy (EDG) substituent **8i** (IC$_{50}$ = 48 $\pm$ 0.96 μM) and diethyl group containing **8j** (IC$_{50}$ = 49.5 $\pm$ 0.92 μM) were exhibited least tyrosinase inhibitory activity as shown in Fig. 9. The results of this study indicated that lower the IC$_{50}$ value of bromobenzofuran-oxadiazole compounds, least would be bacterial tyrosinase the tyrosinase inhibition activity and vice versa. SAR study revealed that bacterial tyrosinase activity increased in following order: diethyl < 2,5-dimethoxy < 2,4-dimethyl < phenyl < 2-methoxy < 3,4-dimethyl < 4-fluoro < 3,4-dichloro < 2-chloro < 2-flouro [28].

5.3.2 Structural assessment of Bacillus megaterium bacterial tyrosinase (PDB ID: 3NM8)

Bacillus megaterium bacterial tyrosinase is composed of a couple of chains which consist of 303 residues [9]. The overall structure of this bacterial tyrosinase comprises 34% α-helices, 12% β-sheets, and 53% coil; its R-value is 0.273 and has a resolution of 2.00 Å. The unit cell crystal dimensions were determined by using X-ray diffraction, in which 100.0% deposits were found in the permitted part of the Ramachandran graph, while 98% of protein dregs were found in the preferential section [9,28,88].

5.4 Synthesis of benzofuran-1,2,4-triazole based coumarins as bacterial tyrosinase inhibitors

Kausar et al. [89] synthesized a series of novel coumarin containing benofuran-1,2,4-triazoles and evaluated their tyrosinase inhibitory activity. Firstly, resorcinol **9** reacted with ethyl acetoacetate via the Pechmann condensation reaction to prepare 7-hydroxy-4-methylcoumarin **10**, catalyzed by sulfuric acid. Furthermore, the derivative 4-methyl-2-oxo-2*H*-chromen-7-yl carbonobromidate **11** was obtained by the treating 2-bromoacetylbromide with precursor **10** by using pyridine as a catalyst.

In next step, to synthesize other fragment of targeted compounds, substituted 2-hydoxybenzaldehydes **12a–h**, ethyl chloroacetate, and K_2CO_3, react to yield various ethyl benzofuran-2-carboxylates **13a–h**. Moreover, the derivatives **13a–h** reacted with hydrazine hydrate in the presence of MeOH to afford substituted benzofuran-2-carbohydrazides **14a–h**. Thiosemicarbazides were obtained by further reaction of derivative **14a–h** with arylisothiocyanates. The cyclization of substituted 2-(benzofuran-2-carbonyl)hydrazine-1-sulfinimidamides in aqueous NaOH solution generated variety of benzofuran-1,2,4-triazoles **(15a–h)**. Finally, substituted 5-(benzofuran-2-yl)-4H-1,2, 4-triazole-3-thiols **15a–h** reacted with precursor **11** in the presence of K_2CO_3 and DCM (dichloromethane) and stirred the reaction mixture to afford desired coumarin moiety containing benzofuran–1,2,4-triazole hybrids **16a–h** in good 62–75% yield (Scheme 3).

5.4.1 Bromobenzofuran-1,2,4-triazoles containing coumarins as bacterial tyrosinase inhibitory activity and SAR

In this study, a series of novel coumarins containing benzofuran-1,2, 4-triazole structural hybrids **16a–h** displayed promising inhibitory chemotherapeutic potential against the tyrosinase enzyme, which was derived from *Bacillus subtilis* NCTC 10400. The 5-bromobenzofuran moiety containing derivative **16e** (4-methyl-2-oxo-2H-chromen-7-yl-2-((5-(5-bromobenzofuran-2-yl)-4-(4-methoxyphenyl)-4H-1,2,4-triazol-3-yl)thio) acetate) and 5-chlorobenzofuran moiety containing scaffold **16f** (4-methyl-2-oxo-2H-chromen-7-yl-2-((5-(5-chlorobenzofuran-2-yl)-4-phenyl-4H-1,2,4-triazol-3-yl)thio)acetate) exhibited the most powerful anti-tyrosinase

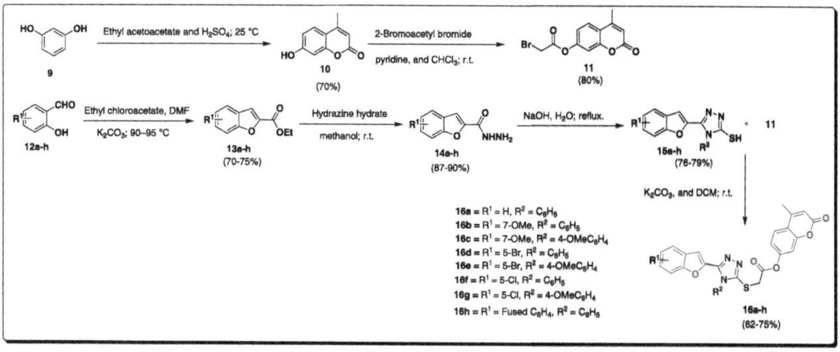

Scheme 3 Synthesis of benzofuran-1,2,4-triazole based coumarins as bacterial tyrosinase inhibitors.

activity with IC_{50} ranges from $0.339 \pm 0.25\,\mu M$ to $3.148 \pm 0.23\,\mu M$ as compared to standard reference inhibitors ascorbic acid (IC_{50} = $11.5 \pm 1.00\,\mu M$) and kojic acid (IC_{50} = $30.34 \pm 0.75\,\mu M$), respectively. The SAR analysis showed that bromo (Br) and chloro substitution (Cl) substitutions at the 5th position of benzofuran ring and methoxy phenyl and phenyl rings at the 4th position of 1,2,4-triazole ring of compounds **16e** and **16f** increased the tyrosinase inhibition activity. In the same way, structural motif **16d** also demonstrated the remarkably higher tyrosinase inhibition (IC_{50} = $5.893 \pm 0.09\,\mu M$) due to the Br group on the aromatic ring of benzofuran and phenyl ring at the 4th position of triazole. The benzofuran-triazole molecule **16a**, having an IC_{50} value of $8.138 \pm 0.13\,\mu M$, possessed moderate activity due to the non-substituted group on the benzofuran ring and phenyl moiety at 4th position of triazole ring. For the above structural insights, it is confirmed that halo atoms, in this case bromine and chlorine at the 5th position of benzofuran ring impart significant effect to enhance the chemotherapeutic efficacy of benzofuran-1,2,4-triazole engrafted with s-alkylated coumarin compounds. Furthermore, the introduction of methoxy moiety at the 7th position of benzofuran ring as in the case of derivative **16b** (IC_{50} = $14.062 \pm 0.17\,\mu M$) and **16c** (IC_{50} = $12.987 \pm 0.32\,\mu M$) displayed the least tyrosinase activity as depicted in Fig. 10. [89].

5.5 Synthesis of tosyl piperazine engrafted dithiocarbamates as bacterial tyrosinase inhibitors

Zahoor and coworkers [77] reported the ultrasonic assisted synthetic strategy to synthesize a series of 1-tosyl piperazine-based dithiocarbamate molecules **20a–j** in good to excellent yield. These synthesized scaffolds were tested for their in vitro bacterial tyrosinase inhibitory activity [90]. This synthetic strategy began to carryout reaction between anhydrous piperazine **17** and p-toluenesulfonyl chloride in dichloromethane (DCM) to achieve a scaffold 1-tosyl piperazine **18**, which further treated substituted N-phenyl bromoacetamides **19a–j** in carbon disulfide (CS2) and sodium acetate (NaOAc) under ultrasonic conditions, leading to the formation of the final targets 1-tosyl piperazine-dithiocarbamate hybrids **20a–j** in a 75–90% yield. The percussor-substituted N-phenyl bromoacetamides **19a–j** was afforded by the reaction of substituted aryl amines with bromoacetyl bromide in the presence basic pyridine and DCM under mild conditions [77,90] (Schemes 4 and 5).

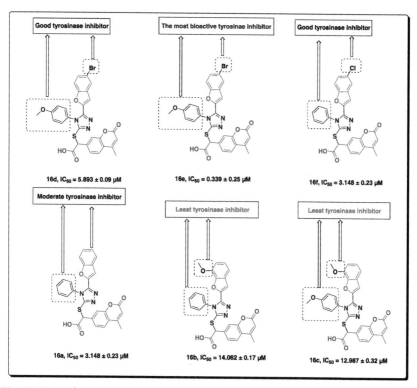

Fig. 10 Benzofuran-1,2,4-triazole based coumarin derivatives as bacterial tyrosinase inhibitor.

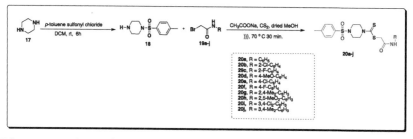

Scheme 4 Synthesis of tosyl piperazine based dithiocarbamates.

5.5.1 Tosyl piperazine based dithiocarbamates as bacterial tyrosinase inhibitory activity and SAR

The piperazine-based dithiocarbamate derivatives **20a–j** were afforded in good yield and screened for in vitro tyrosinase (derived from *Bacillus subtilis* NA2) inhibitory activity. In this study, the most active derivative

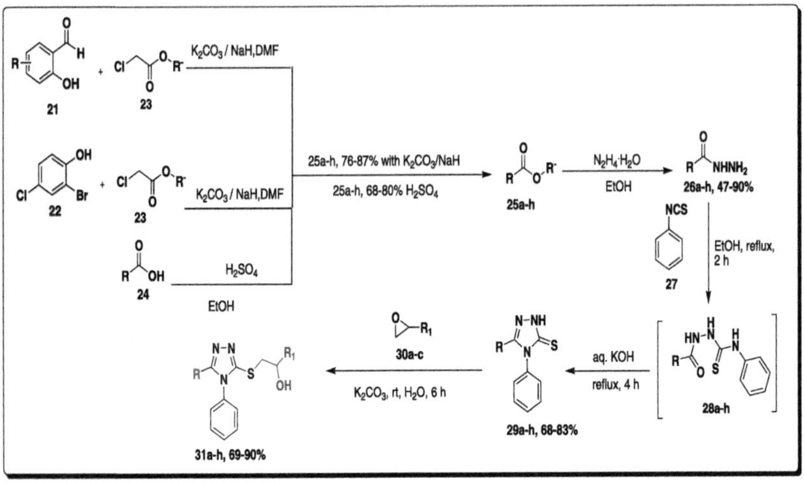

Scheme 5 Synthesis of 1,2,4-triazole tethered β-hydroxy sulfides.

was 2-((4-methoxyphenyl)amino)-2-oxoethyl4-tosylpiperazine-1-carbo-dithioate **20d**, which displayed the highest and the most significant tyrosinase inhibition potential with an IC_{50} value of 6.88 ± 0.11 µM less than IC_{50} values of reference drugs kojic acid (IC_{50} = 30.34 ± 0.75 µM) and ascorbic acid (IC_{50} = 11.5 ± 1.00 µM), respectively. The lower the IC_{50} value of any compound, the greater its chemotherapeutic efficacy, and vice versa. The synthesized piperazine molecules were investigated for SAR study, to determine the effect of nature and position of moieties attached to the phenyl ring of acetamide fragment, which either increased or decreased the therapeutic potential of piperazine-based dithiocarbamate structural motifs **20a–j**.

The SAR analysis of **20a–j** compounds investigated that electron with-drawing or electron donating groups attached to phenyl ring of acetamide results in either an increase or decrease of bacterial tyrosinase inhibition potential, but in general, EDG groups enhanced the tyrosinase inhibition. The introduction of electron-donating methoxy group on *p*-position of phenyl ring of the compound **20d**, increased its bacterial tyrosinase inhibition activity. On the other hand, the hybrid **20a** with an IC_{50} value of 34.8 µM displayed less activity un-substituent Ph-ring. In a similar way, various deri-vatives such as **20c**, **29e**, and **20g**, having IC_{50} values 8.1 µM, 11.11 µM, and 7.24 µM demonstrated good activities, due to attachment of F, Cl and Me substituents on Ph-group, respectively as depicted in Fig. 11 [77].

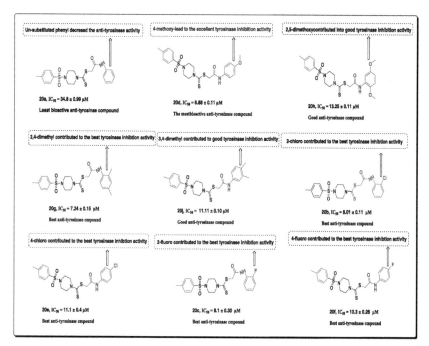

Fig. 11 Anti-tyrosinase tosyl piperazine based dithiocarbamates.

5.6 Synthesis of benzofuran-1,2,4-triazole based coumarins as bacterial tyrosinase inhibitors

Saeed et al. described the synthesis of 1,2,4-triazole-containing β-hydroxy sulfide hybrids, which were evaluated for bacterial tyrosinase inhibitory activity. In this synthetic approach, the esters **25** were prepared by the treatment of different substituted aromatic aldehydes **21**, phenol **22**, and methyl/ ethyl chloroacetate **23** by using potassium carbonate as a basic agent in DMF. On the other hand, various carboxylic acids **24** underwent esterification along with H_2SO_4, were also resulting in the provision of corresponding esters **25**. In next step, various esters **5a–h** reacted with hydrazine monohydrate using ethanol, leading to the formation of corresponding hydrazides **26a–h** which further refluxed with aryl isothiocyanate **27** in ethanol to get key intermediate (thiosemicarbazide) **28a–h**. This intermediate, further treated with KOH solution and converted into the desired 1,2,4-triazoles **29a–h**. Targeted 1,2, 4-triazole-based β-hydroxysulfides **31a–h** were prepared in excellent yield (69–90%) by ring opening of epoxides **(30a–c)** with triazoles **29a–h**, in the presence of potassium carbonate [78].

5.6.1 1,2,4-triazole-based β-hydroxysulfides as bacterial tyrosinase inhibitory activity and SAR

A series of newly reported 1,2,4-triazole-tethered β-hydroxy sulfides were evaluated for their bacterial tyrosinase enzyme inhibition therapeutic potential. Among all the 1,2,4-triazole-tethered β-hydroxy sulfide target molecules **31a–h**, the promising structural motifs **31a**, **31c**, **31d**, and **31f** demonstrated excellent bacterial tyrosinase activity with IC_{50} values of 7.67 ± 1.00 μM, 4.52 ± 0.09 μM, 6.60 ± 1.25 μM, and 5.93 ± 0.50 μM, significantly higher than reference standard drugs kojic acid (IC_{50} = 30.34 ± 0.75 μM), and ascorbic acid (IC_{50} = 11.5 ± 1.00 μM), respectively. The SAR highlighted that the hybrid **31c** bearing a benzofuran moiety was found to have the most potent bacterial tyrosinase activity. The benzofuran moiety of scaffold **31c** is replaced by aryl group of **31a** with an IC_{50} value of 7.67 ± 1.00 μM, and naphthofuran **31d**, (IC_{50} = 6.60 ± 1.25 μM). The benzofuran and propan-2-ol moieties of compound **31c**, were exchanged with the naphthofuran and 3-phenoxypropan-2-ol moieties in compound **31f**, (IC_{50} = 5.93 ± 0.50 μM), which imparted a significant effect in decreasing tyrosinase inhibitory activity. Moreover, derivative **31e** (IC_{50} = 51.40 ± 0.16) with 5-bromobenzofuran and 3-(naphthalen-2-yloxy) propan-2-ol substituents displayed the least bacterial tyrosinase inhibitory activity. The introduction of electron-donating groups on the aryl ring, i.e., 4-CH_3 of **31b** (IC_{50} = 14.43 ± 1.50 μM), showed less tyrosinase activity. The substitution of bromo and chloro in the 2- and 4-positions of the phenyl ring (attached to 1,2,4-triazole) of derivative **31g**, led to a decrease in the tyrosinase inhibition activity with an IC_{50} value of 13.50 ± 0.50 μM. The anti-tyrosinase effect of **31h** (IC_{50} = 13.41 ± 1.15 μM) bearing acefylline was found to be almost similar in activity to that of the compound **31g** (IC_{50} = 13.50 ± 0.50 μM), as depicted in Fig. 12 [78].

6. Conclusion and future perspective of bacterial tyrosinase

Numerous bacterial tyrosinases have been identified in various bacterial strains, and patent applications for four bacterial tyrosinases have been filed. In recent years, the surge in the availability of bacterial genome sequences has significantly facilitated production of novel bacterial tyrosinases. This advancement in the genome data has accelerated the identification of new bacterial tyrosinases with unique properties and promising applications in various fields.

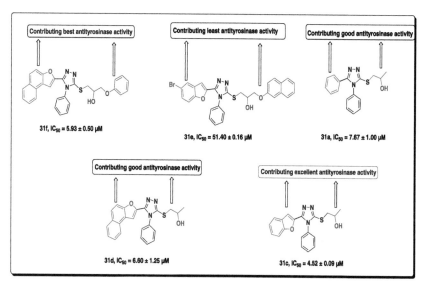

Fig. 12 1,2,4-triazole-based β-hydroxysulfides as bacterial tyrosinase inhibitors.

The future of bacterial tyrosinases appears promising, because of their approximately 38% structural resemblance to human tyrosinases compared to the highly studied mushroom tyrosinase. This structural similarity suggests that bacterial tyrosinase inhibitors have applications in various industries, like in the cosmetic industry, where they can be applied in the development of skin care products for skin whitening and melanogenesis aimed at regulating melanin and other related pigments. In the food industry, bacterial tyrosinase inhibitors can be used as food preservatives and additives to mitigate the economic damage caused by the enzymatic browning of fruits and vegetables. The bacterial tyrosinase inhibitors are potent antibacterial that have the potential to become future antibiotics. In organic synthesis, bacterial tyrosinases can be used as catalytic agents for producing pigments (melanin), dyes, and medicinally valuable compounds such as L–Dopa, which is used in the treatment of Parkinson's disease, a neurodegenerative disorder. Bacterial tyrosinases demonstrated versatile applications, including as a biosensor to detect phenolic compounds, as catalyst in wool coating, bio-based material applications, diversified applications in food and feed processes, and applications in the synthesis of antibiotics. Bacterial tyrosinases can be produced easily in recombinant form by using host models such as *E. coli*, due to which the production of bacterial tyrosinases can be possible in larger quantities, making protein-engineering studies more straightforward and time-efficient. So, it can be concluded that this capability of bacterial tyrosinases opens up new potential for various future applications.

References

[1] S. Zolghadri, A. Bahrami, M.T. Hassan-Khan, J. Munoz-Munoz, F. Garcia-Molina, F. Garcia-Canovas, et al., A comprehensive review on tyrosinase inhibitors, J. Enzyme Inhib. Med. Chem. 34 (1) (2019) 279–309.

[2] H. Claus, H. Decker, Bacterial tyrosinases, Syst. Appl. Microbiol. 29 (1) (2006) 3–14.

[3] Y. Yuan, W. Jin, Y. Nazir, C. Fercher, M.A.T. Blaskovich, M.A. Cooper, et al., Tyrosinase inhibitors as potential antibacterial agents, Eur. J. Med. Chem. 187 (2020) 111892.

[4] E.I. Solomon, U.M. Sundaram, T.E. Machonkin, Multicopper oxidases and oxygenases, Chem. Rev. 96 (7) (1996) 2563–2605.

[5] M. Agunbiade, M. Le Roes-Hill, Application of bacterial tyrosinases in organic synthesis, World J. Microbiol. Biotechnol. 38 (1) (2021) 2.

[6] J.C. García-Borrón, F. Solano, Molecular anatomy of tyrosinase and its related proteins: beyond the Histidine-bound metal catalytic center, Pigment. Cell Res. 15 (3) (2002) 162–173.

[7] G. Faccioa, K. Kruus, M. Saloheimo, L. Thöny-Meyer, Bacterial tyrosinases and their applications, Process. Biochem. 47 (2012) 1749–1760.

[8] Y. Matoba, T. Kumagai, A. Yamamoto, H. Yoshitsu, M. Sugiyama, Crystallographic evidence that the dinuclear copper center of tyrosinase is flexible during catalysis, J. Biol. Chem. 281 (2006) 8981–8990.

[9] M. Sendovski, M. Kanteev, V.S. Ben-Yosef, N. Adir, A. Fishman, First structures of an active bacterial tyrosinase reveal copper plasticity, J. Mol. Biol. 405 (2011) 227–237.

[10] A. Dolashki, W. Voelter, A. Gushterova, J. Van-Beeumen, B. Devreese, B. Tchorbanov, Isolation and characterization of novel tyrosinase from Laceyella sacchari, Protein Pept. Lett. 19 (5) (2012) 538–543.

[11] M. Fairhead, L. Thöny-Meyer, Bacterial tyrosinases: old enzymes with new relevance to biotechnology, N. Biotechnol. 29 (2012) 183–191.

[12] A.M. Betancourt, V. Bernan, W. Herber, E. Katz, Analysis of tyrosinase synthesis in Streptomyces antibioticus, Microbiology 138 (1992) 787–794.

[13] P.Y. Kohashi, T. Kumagai, Y. Matoba, A. Yamamoto, M. Maruyama, M. Sugiyama, An efficient method for the overexpression and purification of active tyrosinase from Streptomyces castaneoglobisporus, Protein Exp. Purif. 34 (2004) 202–207.

[14] M. Fairhead, L. Thöny-Meyer, Role of the C-terminal extension in a bacterial tyrosinase, FEBS J. 277 (2010) 2083–2095.

[15] Y. Kawamura-Konishi, S. Maekawa, M. Tsuji, H. Goto, C-terminal processing of tyrosinase is responsible for activation of Pholiota microspora proenzyme, Appl. Microbiol. Biotechnol. 90 (2010) 227–234.

[16] N. Fujieda, M. Murata, S. Yabuta, T. Ikeda, C. Shimokawa, Y. Nakamura, et al., Multifunctions of MelB, a fungal tyrosinase from Aspergillus oryzae, ChemBioChem 13 (2012) 193–201.

[17] W.H. Flurkey, J.K. Inlow, Proteolytic processing of polyphenol oxidase from plants and fungi, J. Inorg. Biochem. 102 (2008) 2160–2170.

[18] A.M. Nambudiri, J.V. Bhat, P.V. Rao, Enzymic conversion of p-coumarate into caffeate by Streptomyces nigrifaciens, Biochem. J. 128 (1972) 63.

[19] N. Liu, T. Zhang, Y.J. Wang, Y.P. Huang, J.H. Ou, P. Shen, A heat inducible tyrosinase with distinct properties from Bacillus thuringiensis, Lett. Appl. Microbiol. 39 (2004) 407–412.

[20] E. Fernandez, A. Sanchez-Amat, F. Solano, Location and catalytic characteristics of a multipotent bacterial polyphenol oxidase, Pigment. Cell Res. 12 (1999) 331–339.

[21] J. Michalik, W. Emilianowicz-Czerska, L. Switalski, K. Raczyńska-Bojanowska, Monophenol monooxygenase and lincomysin biosynthesis in Streptomyces lincolnensis, Antimicrob. Agents Chemother. 8 (5) (1975) 526–531.

[22] Y. Dai, Y. Cheng, W. Ding, H. Qiao, D. Zhang, G. Zhong, et al., Structural basis of low-molecular-weight thiol glycosylation in lincomycin a biosynthesis, ACS Chem. Biol. 18 (6) (2023) 1271–1277.

[23] H.C. Tseng, C.K. Lin, B.J. Hsu, W.M. Leu, Y.H. Lee, S.J. Chiou, et al., The melanin operon of Streptomyces antibioticus: expression and use as a marker in gram-negative bacteria, Gene 86 (1) (1990) 123–128.

[24] K. Han, J. Hong, H.C. Lim, C.H. Kim, Y. Park, J.M. Cho, Tyrosinase production in recombinant E. coli containing trp promoter and ubiquitin sequence, Ann. N. Y. Acad. Sci. 721 (1994) 30–42.

[25] L. Bubacco, M. Van Gastel, E.J. Groenen, E. Vijgenboom, G.W. Canters, Spectroscopic characterization of the electronic changes in the active site of Streptomyces antibioticus tyrosinase upon binding of transition state analogue inhibitors, J. Biol. Chem. 278 (9) (2003) 7381–7389.

[26] G. Hintermann, M. Zatchej, R. Hütter, Cloning and expression of the genetically unstable tyrosinase structural gene from Streptomyces glaucescens, Mol. Gen. Genet. 200 (3) (1985) 422–432.

[27] M. Huber, K. Lerch, Identification of two histidines as copper ligands in Streptomyces glaucescens tyrosinase, Biochemistry 27 (1988) 5610–5615.

[28] A. Irfan, A.F. Zahoor, S. Kamal, M. Hassan, A. Kloczkowski, Ultrasonic-assisted synthesis of benzofuran appended oxadiazole molecules as tyrosinase inhibitors: mechanistic approach through enzyme inhibition, molecular docking, chemoinformatics, ADMET and drug-likeness studies, Int. J. Mol. Sci. 23 (2022) 10979.

[29] Y. Xu, X. Liang, C.G. Hyun, Isolation, characterization, genome annotation, and evaluation of tyrosinase inhibitory activity in secondary metabolites of Paenibacillus sp. JNUCC32: a comprehensive analysis through molecular docking and molecular dynamics simulation, Int. J. Mol. Sci. 25 (4) (2024) 2213 12.

[30] J.P. Wibowo, F.A. Batista, N. Van-Oosterwijk, M.R. Groves, F.J. Dekker, W.J. Quax, A novel mechanism of inhibition by phenylthiourea on PvdP, a tyrosinase synthesizing pyoverdine of Pseudomonas aeruginosa, Int. J. Biol. Macromol. 146 (2020) 212–221.

[31] D. Hernández-Romero, A. Sanchez-Amat, F. Solano, A tyrosinase with an abnormally high tyrosine hydroxylase/dopa oxidase ratio, FEBS J. 273 (2) (2006) 257–270.

[32] D. Hernández-Romero, F. Solano, A. Sanchez-Amat, Polyphenol oxidase activity expression in Ralstonia solanacearum, Appl. Environ. Microbiol. 71 (11) (2005) 6808–6815.

[33] N. Liu, T. Zhang, Y.J. Wang, Y.P. Huang, J.H. Ou, P. Shen, A heat inducible tyrosinase with distinct properties from Bacillus thuringiensis, Lett. Appl. Microbiol. 39 (5) (2004) 407–412.

[34] L. Ruan, W. He, J. He, M. Sun, Z. Yu, Cloning and expression of mel gene from Bacillus thuringiensis in Escherichia coli, Antonie Van. Leeuwenhoek 87 (4) (2005) 283–288.

[35] D. López-Serrano, F. Solano, A. Sanchez-Amat, Identification of an operon involved in tyrosinase activity and melanin synthesis in Marinomonas mediterranea, Gene 342 (1) (2004) 179–187.

[36] A. Sanchez-Amat, P. Lucas-Elío, E. Fernández, J.C. García-Borrón, F. Solano, Molecular cloning and functional characterization of a unique multipotent polyphenol oxidase from Marinomonas mediterranea, Biochim. Biophys. Acta 1547 (1) (2001) 104–116.

[37] M. Huber, K. Lerch, Identification of two histidines as copper ligands in Streptomyces glaucescens tyrosinase, Biochemistry 27 (1988) 5610–5615.

[38] G. Faccio, M. Richter, L. Thony-Meyer, Bioprospecting for microbial tyrosinases, Microbial Catalysts, vol. 2, Nova Science Publishers, New York, 2019, pp. 1–16.

[39] P. Agarwal, M. Singh, J. Singh, R.P. Singh, Microbial tyrosinases: a novel enzyme, structural features, and applications, in: P. Shukla (Ed.), Applied microbiology and bioengineering, Academic Press, 2019, pp. 3–19.

[40] J.N. Hamann, B. Herzigkeit, R. Jurgeleit, F. Tuczek, Small-molecule models of tyrosinase: from ligand hydroxylation to catalytic monooxygenation of external substrates, Coord. Chem. Rev. 334 (2017) 54–66.

[41] E. Franciscon, M. Grossman, J.A. Paschoal, F.G. Reyes, L. Durrant, Decolorization and biodegradation of reactive sulfonated azo dyes by a newly isolated *Brevibacterium* sp. Strain VN-15. SpringerPlus 1 (2012) 37.

[42] J. Guo, Z. Rao, T. Yang, Z. Man, M. Xu, X. Zhang, et al., Cloning and identification of a novel tyrosinase and its overexpression in *Streptomyces kathirae* SC-1 for enhancing melanin production, FEMS Microbiol. Lett. 362 (8) (2015) fnv041.

[43] M.E. Pavan, N.I. López, M.J. Pettinari, Melanin biosynthesis in bacteria, regulation and production perspectives, Appl. Microbiol. Biotechnol. 104 (4) (2020) 1357–1370.

[44] H.A. Park, I. Yang, M. Choi, K.-S. Jang, J.C. Jung, K.-Y. Choi, Engineering of melanin biopolymer by co-expression of MelC tyrosinase with CYP102G4 monooxygenase: structural composition understanding by 15 tesla FT-ICR MS analysis, Biochem. Eng. J157 (2020) 107530.

[45] K.Y. Choi, Bioprocess of microbial melanin production and isolation, Front. Bioeng. Biotechnol. 9 (2021) 765110.

[46] X. Zhu, P.R. Williamson, Role of laccase in the biology and virulence of *Cryptococcus neoformans*, FEMS Yeast Res. 5 (1) (2004) 1–10.

[47] S. Singh, S.B. Nimse, D.E. Mathew, A. Dhimmar, H. Sahastrabudhe, A. Gajjar, et al., Microbial melanin: Recent advances in biosynthesis, extraction, characterization, and applications, Biotechnol. Adv. 53 (2021) 107773.

[48] M. Xiao, W. Li, Fundamentals and applications of optically active melanin-based materials, Melanins: Functions, Biotechnological Production, and Applications, Springer, 2023, pp. 127–146.

[49] M. Seiberg, C. Paine, E. Sharlow, P. Andrade-Gordon, M. Costanzo, M. Eisinger, et al., Inhibition of melanosome transfer results in skin lightening, J. Invest. Dermatol. 115 (2) (2000) 162–167.

[50] M. Seiberg, C. Paine, E. Sharlow, P. Andrade-Gordon, M. Costanzo, M. Eisinger, et al., The protease-activated receptor 2 regulates pigmentation via keratinocyte-melanocyte interactions, Exp. Cell Res. 254 (1) (2000) 25–32.

[51] M. Friedman, Food browning and its prevention: an overview, J. Agric. Food Chem. 44 (1996) 631–653.

[52] M.A. Route-Mayer, J. Ralambosa, J. Philippon, Roles of o-quinones and their polymers in the enzymic browning of apples, Phytochemistry 29 (1990) 435–440.

[53] M.J. Henley, A.N. Koehler, Advances in targeting 'undruggable' transcription factors with small molecules, Nat. Rev. Drug. Discov. 20 (9) (2021) 669–688.

[54] S.U. Rehman, N. Ullah, Z. Zhang, Y. Zhen, A.U. Din, H. Cui, et al., Recent insights into the functions and mechanisms of antisense RNA: emerging applications in cancer therapy and precision medicine, Front. Chem. 11 (2024) 1335330.

[55] S. Ghavami, A. Pandi, CRISPR interference and its applications, Prog. Mol. Biol. Transl. Sci. 180 (2021) 123–140.

[56] Y. Yuan, W. Jin, Y. Nazir, C. Fercher, M.A.T. Blaskovich, M.A. Cooper, et al., Tyrosinase inhibitors as potential antibacterial agents, Eur. J. Med. Chem. 187 (2020) 111892.

[57] Y.J. Kim, H. Uyama, Tyrosinase inhibitors from natural and synthetic sources: structure, inhibition mechanism and perspective for the future, Cell Mol. Life Sci. 62 (15) (2005) 1707–1723.

[58] S. Quni, Y. Zhang, L. Liu, M. Liu, L. Zhang, J. You, et al., NF-κB-signaling-targeted immunomodulatory nanoparticle with photothermal and quorum-sensing inhibition effects for efficient healing of biofilm-infected wounds, ACS Appl. Mater. Interfaces 16 (20) (2024) 25757–25772.

[59] A.L. Jackson, P.S. Linsley, Recognizing and avoiding siRNA off-target effects for target identification and therapeutic application, Nat. Rev. Drug. Discov. 9 (1) (2010) 57–67.

[60] C. Algieri, L. Donato, P. Bonacci, L. Giorno, Tyrosinase immobilized on polyamide tubular membrane for the L-DOPA production: total recycle and continuous reactor study, Biochem. Eng. J. 66 (2012) 14–19.

[61] T.S. Chang, An updated review of tyrosinase inhibitors, Int. J. Mol. Sci. 10 (6) (2009) 2440–2475.

[62] K.R. Alexander, P.E. Kilbride, G.A. Fishman, M. Fishman, Macular pigment and reduced foveal short-wavelength sensitivity in retinitis pigmentosa, Vis. Res. 27 (7) (1987) 1077–1083.

[63] K. Kalinderi, L. Fidani, Z. Katsarou, S. Bostantjopoulou, Pharmacological treatment and the prospect of pharmacogenetics in Parkinson's disease, Int. J. Clin. Pract. 65 (12) (2011) 1289–1294.

[64] J.M. Gillbro, M.J. Olsson, The melanogenesis and mechanisms of skin-lightening agents—existing and new approaches, Int. J. Cosmet. Sci. 33 (3) (2011) 210–221.

[65] N. Smit, J. Vicanova, S. Pavel, The hunt for natural skin whitening agents, Int. J. Mol. Sci. 10 (12) (2009) 5326–5349.

[66] M.R. Loizzo, R. Tundis, F. Menichini, Natural and synthetic tyrosinase inhibitors as antibrowning agents: an update, Compr. Rev. Food Sci. Food Saf. 11 (2012) 378–398.

[67] T.F. Kuijpers, H. Gruppen, S. Sforza, W.J. van Berkel, J.P. Vincken, The anti-browning agent sulfite inactivates Agaricus bisporus tyrosinase through covalent modification of the copper-B site, FEBS J. 280 (23) (2013) 6184–6195.

[68] R. Haudecoeur, A. Boumendjel, Recent advances in the medicinal chemistry of aurones, Curr. Med. Chem. 19 (18) (2012) 2861–2875.

[69] R. Haudecoeur, M. Carotti, A. Gouron, M. Maresca, E. Buitrago, R. Hardré, et al., 2-Hydroxypyridine-N-oxide-embedded aurones as potent human tyrosinase inhibitors, ACS Med. Chem. Lett. 8 (1) (2016) 55 17.

[70] S. Okombi, D. Rival, S. Bonnet, A.M. Mariotte, E. Perrier, A. Boumendjel, Discovery of benzylidenebenzofuran-3(2H)-one (aurones) as inhibitors of tyrosinase derived from human melanocytes, J. Med. Chem. 49 (1) (2006) 329–333.

[71] R. Haudecoeur, A. Ahmed-Belkacem, W. Yi, A. Fortuné, R. Brillet, C. Belle, et al., Discovery of naturally occurring aurones that are potent allosteric inhibitors of hepatitis C virus RNA-dependent RNA polymerase, J. Med. Chem. 54 (15) (2011) 5395–5402. Aug 11.

[72] R. Haudecoeur, A. Gouron, C. Dubois, H. Jamet, M. Lightbody, R. Hardré, et al., Investigation of binding-site homology between mushroom and bacterial tyrosinases by using aurones as effectors, Chembiochem 15 (9) (2014) 1325–1333.

[73] D. Nokinsee, L. Shank, V.S. Lee, P. Nimmanpipug, Estimation of inhibitory effect against tyrosinase activity through homology modeling and molecular docking, Enzyme Res. 2015 (2015) 262364.

[74] J.S. Chen, C. Wei, M.R. Marshall, Inhibition mechanism of kojic acid on polyphenol oxidase 1, J. Agric. Food Chem. 39 (11) (1991) 1897–1901.

[75] R.M. Halder, G.M. Richards, Topical agents used in the management of hyperpigmentation, Skin. Ther. Lett. 9 (6) (2004) 1–3.

[76] D.A. Case, T.A. Darden, T.E. Cheatham IIIet al., AMBER 12, University of California, San Francisco, CA, 2012.

[77] A.F. Zahoor, A. Hafeez, A. Mansha, S. Kamal, M.N. Anjum, Z. Raza, et al., Bacterial tyrosinase inhibition, hemolytic and thrombolytic screening, and in silico modeling of rationally designed tosyl piperazine-engrafted dithiocarbamate derivatives, Biomedicines 11 (2023) 2739.

[78] S. Saeed, M.J. Saif, A.F. Zahoor, H. Tabassum, S. Kamal, S. Faisal, et al., Discovery of novel 1,2,4-triazole tethered β-hydroxy sulfides as bacterial tyrosinase inhibitors: synthesis and biophysical evaluation through in vitro and in silico approaches, RSC Adv. 14 (22) (2024) 15419–15430.

[79] J.A. Imlay, The molecular mechanisms and physiological consequences of oxidative stress: lessons from a model bacterium, Nat. Rev. Microbiol. 11 (7) (2013) 443–454.

[80] E. Kurpejović, V.F. Wendisch, B. Sariyar Akbulut, Tyrosinase-based production of L-DOPA by Corynebacterium glutamicum, Appl. Microbiol. Biotechnol. 105 (24) (2021) 9103–9111.

[81] S. Faiz, A.F. Zahoor, M. Ajmal, S. Kamal, S. Ahmad, A.M. Abdelgawad, et al., Design, synthesis, antimicrobial evaluation, and laccase catalysis effect of novel benzofuran–oxadiazole and benzofuran–triazolehybrids, J. Heterocycl. Chem. 56 (2019) 2839–2852.

[82] Z. Shi, Z. Zhao, M. Huang, X. Fu, Ultrasound-assisted, one-pot, three-component synthesis and antibacterial activities of novel indole derivatives containing 1,3,4-oxadiazole and 1,2,4-triazole moieties, C. R. Chim. 18 (2015) 1320–1327.

[83] A. Irfan, S. Faiz, A. Rasul, R. Zafar, A.F. Zahoor, K. Kotwica-Mojzych, et al., Exploring the synergistic anticancer potential of benzofuran–oxadiazoles and triazoles: improved ultrasound- and microwave-assisted synthesis, molecular docking, hemolytic, thrombolytic and anticancer evaluation of furan-based molecules, Molecules 27 (2022) 1023.

[84] M. Kowalewska, H. Kwiecień, M. Śmist, A. Wrześniewska, Synthesis of new benzofuran-2-carboxylic acid derivatives, J. Chem. 2013 (2013) 183717.

[85] S. Parekh, D. Bhavsar, M. Savant, S. Thakrar, A. Bavishi, M. Parmar, et al., Synthesis of some novel benzofuran-2-yl(4,5-dihyro-3,5-substituted diphenylpyrazol-1-yl) methanones and studies on the antiproliferative effects and reversal of multidrug resistance of human MDR1-gene transfected mouse lymphoma cells in vitro, Eur. J. Med. Chem. 46 (2011) 1942–1948.

[86] A. Almasirad, S.A. Tabatabai, M. Faizi, A. Kebriaeezadeh, N. Mehrabi, A. Dalvandi, Shafiee A. Synthesis and anticonvulsant activity of new 2-substituted-5-[2-(2-fluorophenoxy)phenyl]-1,3,4-oxadiazoles and 1,2,4-triazoles, Bioorg Med. Chem. Lett. 14 (2004) 6057–6059.

[87] M.R. Barros, T.M. Menezes, L.P. Da Silva, D.S. Pires, J.L. Princival, G. Seabra, et al., Furan inhibitory activity against tyrosinase and impact on B16F10 cell toxicity, Int. J. Biol. Macromol. 136 (2019) 1034–1041.

[88] V. Shuster, A. Fishman, Isolation, cloning and characterization of a tyrosinase with improved activity in organic solvents from Bacillus megaterium, J. Mol. Microbiol. Biotechnol. 17 (4) (2009) 188–200.

[89] R. Kausar, A.F. Zahoor, H. Tabassum, S. Kamal, M. Ahmad Bhat, Synergistic biomedical potential and molecular docking analyses of coumarin–triazole hybrids as tyrosinase inhibitors: design, synthesis, in vitro profiling, and in silico studies, Pharmaceuticals 17 (2024) 532.

[90] F. Hafeez, A.F. Zahoor, A. Rasul, A. Mansha, R. Noreen, Z. Raza, et al., Ultrasound-assisted synthesis and in silico modeling of methanesulfonyl-piperazine-based dithiocarbamates as potential anticancer, thrombolytic, and hemolytic structural motifs, Molecules 27 (2022) 4776.

CHAPTER NINE

Biomedical applications of tyrosinases and tyrosinase inhibitors

Luigi Pisano[a], Martina Turco[b], and Claudiu T. Supuran[c,*]
[a]Section of Dermatology, Health Sciences Department, University of Florence, Florence, Italy
[b]Health Sciences Department (DSS), University of Florence, Florence, Italy
[c]Neurofarba Department, Pharmaceutical and Nutraceutical Section, University of Florence, Sesto Fiorentino, Florence, Italy
*Corresponding author. e-mail address: claudiu.supuran@unifi.it

Contents

Abstract

Tyrosinase is involved in several human diseases, among which hypopigmentation and depigmentation conditions (vitiligo, idiopathic guttate hypomelanosis, pityriasis versicolor, pityriasis alba) and hyperpigmentations (melasma, lentigines, post-inflammatory and periorbital hyperpigmentation, cervical idiopathic poikiloderma and acanthosis nigricans). There are increasing evidences that tyrosinase plays a relevant role in the formation and progression of melanoma, a difficult to treat skin tumor. Hydroquinone, azelaic acid and tretinoin (all-*trans*-retinoic acid) are clinically used in

The Enzymes, Volume 56
ISSN 1874–6047, https://doi.org/10.1016/bs.enz.2024.05.005

the management of some hyperpigmentations, whereas many novel chemotypes acting as tyrosinase inhibitors with potential antimelanoma action are being investigated. Kojic acid, hydroquinone, its glycosylated derivative arbutin, or the resorcinol derivative rucinol are used in cosmesis in creams as skin whitening agents, whereas no antimelanoma tyrosinase inhibitor reached clinical trials so far, although thiamidol is a recently approved new tyrosinase inhibitor for the treatment of melasma. Kojic acid and vitamin C are used for avoiding vegetable/food oxidative browning due to the tyrosinase-catalyzed reactions, whereas bacterial enzymes show potential in biotechnological applications, for the production of mixed melanins, for protein cross-linking reactions, for producing phenol(s) biosensors, of for the production of L-DOPA, an anti-Parkinson's disease drug.

1. Introduction

Melanin is a naturally occurring pigment responsible for the colouration of human skin, hair, and eyes. Its synthesis is governed by a complex enzymatic process, prominently involving the enzyme tyrosinase, as shown in several chapters of the book. This interplay between melanin and tyrosinase is essential for various biological functions and has significant implications for human health and well-being [1,2].

Melanin, produced by specialized dendritic cells located at the dermal-epidermal junction (called melanocytes), exists in different forms, including eumelanin and pheomelanin, each contributing distinct hues to the skin and hair [3]. After maturation, melanin is transferred into the surrounding keratinocytes. Beyond its role in colouration, melanin serves as a natural defence mechanism against the harmful effects of ultraviolet (UV) radiation from the sun. Eumelanin, with its dark color, absorbs and dissipates UV radiation more effectively than pheomelanin, offering protection against sunburn, photoaging, and potentially carcinogenic DNA damage. Thus, individuals with higher eumelanin content tend to exhibit greater resistance to UV-induced skin damage [3–6].

Tyrosinase (TYR), a key enzyme in melanogenesis, catalyzes several crucial steps in the biosynthetic pathway, beginning with converting the amino acid tyrosine into dopaquinone, a melanin precursor. This enzymatic cascade is finely regulated, and influenced by genetic, hormonal, and environmental factors, ultimately determining the quantity and quality of melanin produced [7,8].

Melanogenesis disruption can lead to a variety of pigment disorders, which are broadly classified as hypopigmentation or hyperpigmentation, and may occur with or without a change in the number of melanocytes.

Hypopigmentation conditions such as vitiligo result from decreased melanin production, while hyperpigmentation disorders like melasma are characterized by excessive melanin synthesis. These conditions have a significant impact on psychological well-being, prompting great interest in developing new effective treatments [9].

Due to the exponential increase in hyperpigmentation disorders linked to sun exposure, there is a growing demand for the discovery of potent inhibitors of melanogenesis. These inhibitors hold promise for pharmaceutical and cosmetic industries in developing products aimed at regulating melanin production and addressing various pigmentary concerns. Through continued research into the intricate mechanisms underlying melanin synthesis and its regulation, novel therapeutic interventions may emerge, offering hope for individuals affected by pigmentary disorders [10,11].

2. Melanin depigmentation disturbances

Disorders of hypopigmentation and depigmentation comprise a large group of conditions characterized by altered melanocyte density or melanin concentration. Conditions may be inherited, such as albinism, piebaldism, or Waardenburg syndrome; whereas others, like pityriasis versicolor, pityriasis alba, vitiligo, lichen sclerosus, and intralesional steroid injection-related hypopigmentation, are acquired [12].

The hypopigmented or depigmented lesions may also appear in a localized form as in pityriasis versicolor, pityriasis alba, exposure to chemical agents, morphea, lichen sclerosus, vitiligo and halo nevus, or in a diffuse pattern, as in idiopathic guttate hypomelanosis, punctate leukoderma and progressive macular hypomelanosis of the trunk [12]. Establishing the correct diagnosis for hypomelanotic skin disorders requires a detailed medical history, a careful physical examination, the use of special lighting techniques, such as Wood's light, and sometimes a biopsy of both the abnormally pigmented skin and the normally pigmented skin [13].

2.1 Vitiligo

Vitiligo is a skin disorder characterized by the gradual loss of melanocytes from the epidermis and associated structures, resulting in milk- or chalk-white macules and patches. The lesions have well-demarcated borders and may vary in number and size. They typically appear on certain areas of the body, such as the extremities, hands, feet and periorificial areas (around the eyes, nose, lips, genitalia). Vitiligo can manifest in various types, including focal, segmental, acrofacial, and vulgaris, each displaying distinct distribution patterns. It affects approximately 0.5% to 1% of the global population without significant gender or ethnic predilection, usually starting in childhood or young adulthood and displaying an unpredictable course marked by periods of stability and lesion spreading [14,15].

The etiopathogenesis of vitiligo remains unclear. Immune-mediated destruction of melanocytes, possibly involving circulating autoantibodies and cytotoxic T cells, is one of the proposed theories, while neural and biochemical hypotheses suggest direct or indirect neural or metabolic influences on melanocyte survival. Additionally, factors such as defective melanocyte growth factors and genetic predispositions may contribute to the depigmentation process. The convergence theory suggests that multiple factors may interact to trigger melanocyte destruction, with different pathways possibly involved in various clinical types of vitiligo. Recent research has also highlighted the role of melanocyte detachment and migration in lesion development, particularly in response to friction or trauma [16,17].

2.2 Idiopathic guttate hypomelanosis

Idiopathic guttate hypomelanosis (IGH) is characterized by small, asymptomatic, white patches, typically 1–3 mm in diameter, found on sun-exposed areas of the skin. It mostly affects those with fair skin and is associated with sun exposure and ageing. Genetic factors may contribute to its development [18].

Unlike vitiligo, the patches of IGH do not tend to expand or merge. Histologically, IGH lesions show reduced melanin and fewer melanocytes, along with features like hyperkeratosis and thinning of the skin's upper layers [19].

Distinguishing IGH from other skin conditions like pityriasis alba, tinea versicolor, and vitiligo relies on factors such as disease progression, age of onset, family history, and patch characteristics. Prognosis for IGH is generally benign, with lesions typically remaining stable and new ones possibly appearing over time [18,19].

2.3 Pityriasis versicolor

Pityriasis versicolor (PV) is a widespread benign superficial skin infection caused by the saprophytic yeast *Malassezia globosa* and sometimes other Malassezia species. Malassezia is considered part of the normal skin microbiota in humans and animals. The development of clinical disease occurs when Malassezia transforms from the yeast form to the mycelial form [20].

PV is characterized by the appearance of round to oval, slightly scaling macules that are most commonly found on the trunk, chest, back, and shoulders. PV is usually asymptomatic, although pruritus does occur in some cases. These lesions may vary in color. The depigmented form of PV, also called *pityriasis versicolor alba*, presents with white macules, whereas the hyperpigmented form is associated with pink, tan, brown, or black lesions. Recently, a variant with red maculae (*pityriasis versicolor rubra*) and another with black ones (*pityriasis versicolor nigra*) were described [21].

Wood's lamp and microscopy can confirm the clinical diagnosis of PV. Wood's lamp displays a yellow-orange fluorescence due to the porphyrin pityrialactone produced by Malassezia. Microscopy reveals the typical "spaghetti and meatballs" pattern, which reflects the presence of hyphae and blastospores [12].

2.4 Pityriasis alba

Pityriasis alba (PA) is considered a minor manifestation of atopic dermatitis and is typically associated with a history of atopy. It is often observed in preadolescent children, manifesting as white or pink irregular plaques with slight scaling and well-defined borders, typically ranging from 0.5 to 6 cm in diameter. It commonly affects the head and neck, trunk, and limbs. Males and individuals with darker skin tones are more prone to this condition [22]. The exact causes and mechanisms behind pityriasis alba remain poorly understood. However, excessive sun exposure without protection, frequent bathing and hot baths, are believed to contribute to its development [23].

3. Increased melanin biosynthesis diseases and their management

Hyperpigmentation disorders are a varied group of diseases that comprise both congenital forms, with different patterns of inheritance, and acquired forms secondary to cutaneous or systemic problems. Most of them are linked to alterations in the melanin pigment and can be classified as epidermal, due to an increase in the number of melanocytes or in the production of melanin, or dermal, either melanocytic or not [24].

3.1 Melasma

Melasma is a common chronic refractory disorder of pigmentation most commonly affecting people with darker skin types, mainly women. Clinically, it manifests as symmetric reticulated hypermelanosis, with stains in shades of brown to bluish-gray. It usually affects the face and neck, and less commonly arms and sternal region. It can appear in three predominant facial patterns: centrofacial, malar, and mandibular. Wood's lamp examination has been used to classify melasma based on the depth of melanin in the skin: epidermal, typified by a light brown colouration, dermal, exhibiting a bluish-gray color or mixed, seen in a dark brown shade. Multiple factors have been associated with the development of melasma. Notably, UV exposure and genetic predisposition stand out as significant contributors. Ultraviolet radiation stimulates melanocortin production in both melanocytes and keratinocytes, implicating this hormone in melasma's pathogenesis. Recent findings also suggest that high-intensity visible light may contribute to melasma's pigmentation escalation, particularly among individuals with skin phototypes IV-VI [25,26].

Additionally, various conditions contribute to melasma's pathogenesis, including pregnancy, oral contraceptive use, endocrine disorders, and hormonal therapies. Moreover, the application of certain cosmetics and medications, such as anticonvulsants and photosensitizing agents, has been implicated as potential causative or exacerbating factors for hyperpigmentation [27]. The objective of treating melasma is to reduce melanocyte activity, suppress melanosome production, and facilitate their breakdown. Depigmenting agents are classically regarded as the gold standard treatment for melasma. Hydroquinone (HQ), a tyrosinase inhibitor, is one of the most frequently used and well-studied lightening products in the world, with mild to moderate risk of local adverse effects. HQ in a 4% cream can be used safely twice daily for up to 4–6 months. Improvement in melasma was seen as early as 3 weeks. Another depigmenting agent, azelaic acid (AA), is a competitive inhibitor of tyrosinase. It has a similar efficacy to that of 4% HQ but may have a greater risk of irritant adverse effects. HQ has been used in combination with other topical agents such as retinoic acid (RA) and glycolic acid (GA). One of the first combination topical therapies developed for the treatment of hyperpigmentation was the Kligman–Willis formula, consisting of 5% HQ, 0.1% tretinoin (all-*trans* retinoic acid), and 0.1% dexamethasone. Tretinoin prevents the oxidation of HQ and improves epidermal penetration while the topical corticosteroid reduces irritation due to the other two ingredients and decreases cellular metabolism, further inhibiting melanin synthesis. A recent investigation illustrated that employing the triple combination proves to be more economical compared to utilizing double or single therapy options. Furthermore, the blend of 0.05% RA + 4% HQ along with 0.01% fluocinolone acetonide exhibits superior evidence quality and effectiveness as indicated by efficacy studies in melasma treatment [28,29]. Furthermore, thiamidol (isobutylamido thiazolyl resorcinol) was recently identified as the most potent strictly competitive inhibitor of human tyrosinase out of 50,000 screened molecules. With this mode of action, it has the advantage of lacking the potential side effects of compounds that are acting via transformation to quinones [30].

The persistent nature of melasma and its tendency to recur often deter patients from adhering to the prescribed treatment regimen, particularly concerning sunscreen usage. For cases resistant to topical medications, the consideration of adjunctive outpatient procedures such as peels, dermabrasion or Nd:YAG laser may be warranted, with particular attention to the risk of post-inflammatory hyperpigmentation [28].

3.2 Lentigines

Actinic lentigines, also known as solar lentigines or lentigo senilis, are brown spots or patches that appear predominantly in sun-exposed areas such as the hands, forearms, trunk, and face. They can be single or multiple and are associated with prolonged sun exposure, indicating a risk for skin cancers. They are more common in fair to medium skin tones and prevalent in India and other Asian countries, particularly among older individuals. While UV exposure is a significant factor in their development, the exact molecular mechanisms remain unclear, though keratinocyte growth factor and other dermal factors are implicated [31].

Treatments for solar lentigines can be categorized into two main groups: physical treatments and topical treatments. Physical treatment options include cryotherapy, laser therapy, pulsed light, and chemical peels. These methods often yield high success rates but should be balanced with potential side effects and recurrence risks. On the other hand, various topical therapies are available, such as hydroquinone, tretinoin, adapalene, and the recent stable combination of mequinol and tretinoin. While topical treatments generally require more time to show results compared to physical treatments, they offer patients greater control over their treatment and may have reduced side effects. Additionally, patients are typically advised to use sunscreens regularly to prevent further sun damage, thus maintaining the success of their treatment [32,33].

3.3 Post-inflammatory hyperpigmentation

Post-inflammatory hyperpigmentation (PIH) is characterized by increased skin pigmentation following cutaneous inflammation including acne, atopic dermatitis, allergic contact dermatitis, trauma, psoriasis, lichen planus, drug eruptions, and cosmetic procedures. Inflammation of the epidermis triggers the generation and release of various cytokines, prostaglandins, and leuko-trienes, which activate the epidermal melanocytes, causing heightened synthesis of total melanin [34].

Treatment outcomes for PIH are unpredictable, with frequent relapses. Topical hydroquinone and tretinoin are effective but require prolonged use. Other bleaching agents like kojic acid and azelaic acid are used with varying success. Chemical peels with glycolic or salicylic acid are effective, particularly in higher skin phototypes. Laser treatments, such as Q-swit-ched lasers, are used but carry the risk of hyperpigmentation in darker skin tones. Fractional photothermolysis shows promise but requires further study for efficacy assessment [35,36].

3.4 Periorbital hyperpigmentation

Periorbital hyperpigmentation (POH) is a common dermatological concern affecting individuals of all genders, ages, and ethnicities. POH typically presents as bilateral, homogenous hyperchromic areas in the infraorbital region, varying in intensity with factors like fatigue, lack of sleep, and ageing-related changes such as skin thinning and sagging. Atopic individuals or those allergic to airborne substances may be pre-disposed to POH due to frequent skin rubbing, triggering post-inflam-matory pigmentation [24].

Treatment for dark circles remains challenging due to multifactorial causes. Corrective makeup offers temporary relief, while therapeutic options often necessitate combination approaches. Topical hydroquinone is a commonly used depigmenting agent; topical retinoic acid also reduces pigmentation by inhibiting tyrosinase transcription. Superficial peels with trichloroacetic acid or glycolic acid are effective but carry risks of adverse events. Laser treatments such as Q-switched ruby laser or Nd:YAG can provide significant improvements, while safety precautions are essential due to the delicate nature of periorbital tissues [37].

3.5 Cervical idiopathic poikiloderma

Cervical idiopathic poikiloderma, also known as poikiloderma of Civatte (PC), is a benign dermatosis characterized by irregular pigmentary alterations, including hypo- and hyperpigmentation, superficial atrophies and telangiectasias. Typically asymptomatic, PC may rarely cause discreet burning or pruritus. The pigmentation is characterized by reticulated, reddish-to-brown patches with irregular and symmetrical distribution, affecting the hemifaces, neck, and upper chest while sparing the shaded chin area [24]. Although its exact pathophysiology is not fully understood, cumulative sun exposure exacerbated by photoallergic or phototoxic reactions from fragrances and cosmetics in susceptible individuals is believed to play a significant role. PC follows a slow, progressive, and irreversible course if aggravating factors persist [24,38].

Treatment aims to simultaneously address pigmented and vascular components through a combination of topical medications, including depigmenting agents, procedures targeting the vascular component, and broad-spectrum photoprotection. Chemical peels and topical retinoids may serve as adjuvants to counteract photoaging. Laser treatments such as argon lasers have been recommended, although they require caution due to potential complications. Intense pulsed light (IPL) is an alternative option that works across a wide range of wavelengths, effectively reducing pigment and telangiectasias with fewer risks of complications [24].

3.6 Acanthosis nigricans

Acanthosis nigricans (AN) is described as a cutaneous disorder of multiple etiologies, characterized by symmetric, dark, coarse, thickened, velvety appearing plaques commonly distributed on the neck, axillae, antecubital and popliteal fossae, inframammary, and groin areas. Recently, it has been categorized into eight subtypes: benign, obesity-related, syndromic, malignant, acral, unilateral, secondary to medication, and multifactorial. AN is often associated with endocrine disorders such as obesity, insulin resistance, type II diabetes mellitus, and polycystic ovary syndrome. The underlying mechanism involves hyperinsulinemia due to insulin resistance, leading to the stimulation of insulin-related growth factors and subsequent keratinocyte proliferation. In cases associated with malignancy, AN may precede or occur simultaneously with the neoplasm, commonly gastric adenocarcinoma, although it has been reported in various other cancers as well [39,40]. Improvement in AN lesions is dependent on resolving the underlying disease. Oral and topical retinoids have shown promising results, as have keratolytic agents like 12% ammonium lactate or triple therapy combining hydroquinone, retinoic acid, and fluocinolone acetonide. Topical agents containing urea and salicylic acid have produced varying outcomes. Oral treatments such as isotretinoin and metformin may be prescribed, especially in cases associated with hyperandrogenism. Dermabrasion, Alexandrite laser, and trichloroacetic acid application are among the procedural options available for AN treatment [39–41].

4. Melanoma and tyrosinase

TYR and tyrosinase-related proteins 1/2 (TYRP1, TYRP2) are also involved in cancerogenesis, more precisely in melanoma, an aggressive skin tumor originating from melanocytes, which is also characterized by a highly metastatic phenotype [42–44]. These proteins have different functions in such processes, many of which only recently investigated and in some cases still poorly understood. For example, not all melanoma cell lines express TYR (e.g., cell line A375 [43]), but the lines expressing the enzyme such as the mouse melanoma line B16F10 or the human melanoma line WM266–4 showed a significant overexpression of TYR, which seems to be involved in enhanced cell migration, by regulating cell motility through modulation of actin polymerization [43]. TYRP1 on the other hand, is also involved in melanin biosynthesis (together with the main player enzyme, which is TYR), stabilizing TYR and modulating its catalytic activity [45]. However, the gene encoding for TYRP1 possesses a non-coding function too, which promotes melanoma tumor cell proliferation when highly expressed in a cell line, by a poorly understood mechanisms to date [43]. It is thus clear that the catalytic activity of TYR/TYRP1/TYRP2 as well as their interactions, are necessary for melanoma cells survival and disease progression, which explains why the inhibition of these proteins with specific inhibitors has been proposed as an anticancer mechanism for the treatment, imaging and photodymanic therapy of these tumors [46–63]. Melanogenesis was in fact shown not only to affect melanoma behavior/ response to therapy, but also to inhibit immune responses, which implies that TYR inhibitors might be synergistic with other antitumor agents, such as immune checkpoint inhibitors, hypothesis recently confirmed in several interesting studies [55,64,65].

Most TYR inhibitors reported to date belong to the natural products (NPs) classes of compounds (see Chapters 5 and 6 of this book and Refs. [7,66]) but drug design studies of synthetic inhibitors were also reported, leading thus to various classes, belonging to many chemotypes, among which: sulfur analogs of tyrosine [46], phenyl-thioureas [48], arylpiper-azines incorporating cinnamic acid [49], nanocomposites derivatized with classical TYR inhibitors [53], substituted pyrazoles [54], PROTACs [57], 4-amino-2′,4′-dihydroxyindanones [58], bipiperidines [60], 1,3,5-triazines [61], calix[4]azacrowns [62], etc. The most investigated compounds were however the NPs, due to their facile availability, low toxicity and relatively simple structures, which allowed for a good understanding of their inhi-bition mechanism, at least in most cases investigated so far [7,66]. Many NP phenols [7,66], flavonoids [47,51,63], maltol [56], kojic acid and its deri-vatives [7,66], coumarins and other such compounds [7,50,55] have been extensively studied as TYR inhibitors [7,66] and in some cases also for their antimelanoma activity [56–62]. Although some of these derivative seem to have a significant antimelanoma activity in cell cultures, none of them progressed so far to clinical studies [55].

It should be mentioned that although there are many methods for TYR activity/inhibition assay [67–72], based on spectrophotometric, fluores-cence spectroscopy, oxygen consumption, light scattering or NIR tech-niques, erroneous reports/interpretation of such data also emerged [67]. In some reports available in the literature [67] some authors employed the same substrate (DOPA) and phenolic inhibitor concentrations during the TYR inhibition assay, which is conceptually incorrect [67]. Substrate (DOPA) concentrations employed in correct enzyme activity/inhibition assays should be chosen in such a way that saturation of the enzyme with the substrate is achieved [63], case in which competition for binding to the di-copper center between the enzyme and the inhibitor(s) does not occur, but this basic requirement for enzyme inhibition studies is not understood by some researchers [67]. Thus, poor interpretations [67] of TYR inhi-bition data abound in the literature [68].

A recent study [73] highlighted the connection between melanoma and neurodegenerative diseases, such as Parkinson's disease (PD). Indeed, co-occurrence of melanoma and PD seem to be 3-times higher than expected, presumably due to dysregulated TYR activity, which leads to loss of dopamine and neuromelanin in *substantia nigra* neurons (causing PD), as well as triggering melanoma formation, due to phenomena discussed above in the chapter. These authors proposed that detection and monitoring of TYR activity might represent a means of studying the early progression of melanoma and PD, which might open interesting opportunities for the diagnosis and treatment of both diseases [73].

5. Tyrosinase inhibitors in cosmesis and preclinical/ clinical settings

Some of the simple TYR inhibitors such as kojic acid **1**, hydroquinone **2**, its glycosylated derivative arbutin **3** or the resorcinol derivative rucinol **4** (Fig. 1) are used in cosmesis in creams (with 0–1–2% concentrations of the active ingredients) as skin whitening agents due to their inhibition of TYR enzymatic activity [10,66,74], or for the management of various skin pigmentation diseases (see above).

All of them are based on natural product (kojic acid) or diphenols (hydroquinone and resorcinol). Indeed, kojic acid is biosynthesized in various fungal species, such as for example *Aspergillus orizae*, and is found in various food/drink products, among which miso, shoyu and sakè [66].

Thiamidol **5** (Fig. 1) is a synthetic TYR inhibitor incorporating the resorcinol fragment, which was developed by Beiersdorf and recently approved for clinical use for the management of melasma [30,75,76]. Thiamidol is a potent inhibitor of human TYR, it is well tolerated and significantly improved melasma. In a 24-week, randomized, double-blind, vehicle-controlled, cosmetic clinical study to assess the efficacy and tolerability of **5** in melasma, thiamidol was well tolerated and superior to baseline vehicle and hydroqionone over the treatment period [30]. Other subsequent studies confirmed these data, with 4-times daily administration of the drug being well tolerated (and more effective than twice daily) for improving facial hyperpigmentation [77]. The melasma treatment achieved using 0.2% thiamidol **5** was also comparable to that of 4% hydroquinone **2** cream, but the new drug was better tollerated compared to hydroquinone [78].

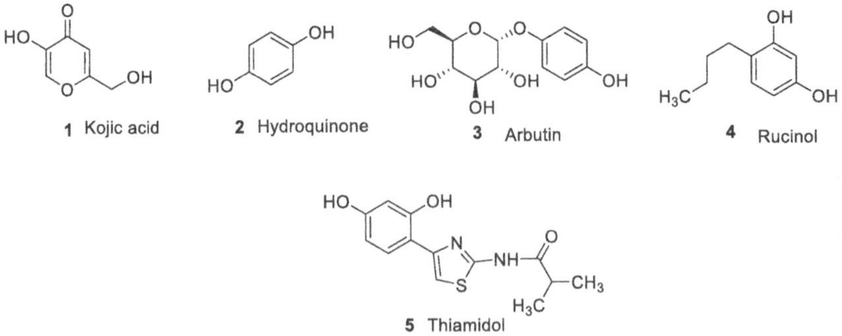

1 Kojic acid	**2** Hydroquinone	**3** Arbutin	**4** Rucinol

5 Thiamidol

Fig. 1 Kojic acid **1**, phenols **2–4** used in cosmetics as skin whitening agents and the recently approved agent thiamidol **5**.

6. Food applications of tyrosinase inhibitors

Kojic acid **1** and other compounds (e.g., vitamin C, 6'-O-caffeoy-larbutin **6** [79]—Fig. 2) are used for avoiding vegetable (and other food) oxidative browning as a consequence of TYR inhibition (of course, the enzyme present in the plants/animals used as food source should be inhibited) [79,80]. 6'-O-caffeoylarbutin, a natural flavonoid present in green tea, showed a low toxicity at the employed dosages (in mice) and preservation trials on apple juice demonstrated its good potential in reducing oxidative browning [79]. Cyrene **7** (Fig. 2) and its dihydrate (at the ketone moiety), were on the other hand shown to act as effective TYR inhibitors and to preserve sliced potatoes from browning [80].

Many other synthetic and NP derivatives acting as TYR inhibitors, among which chlorogenic acid **8** (Fig. 2) were also investigated in detail for their TYR inhibition mechanism and potential to be employed as anti-browning agents [81–87], but at the moment the only industrially used derivative seem to be kojic acid **1**, which is mainly employed for avoiding the browning of sea food [66].

7. Bacterial tyrosinases

TYRs are present in many bacteria (see also Chapter 1 of the book) where they play relevant functions connected with pigmentation, being also important factors of bacterial colonization and primary immune response from the host during wound healing after a bacterial infection [88]. Being dicopper enzymes with a strong preference for phenolic and diphenolic substrates [89], their catalytic activity on various substrates produces an activated quinone as the main product, thus also showing potential in biotechnological applications, such as the production of mixed melanins, for protein cross-linking reactions, for producing phenol(s)

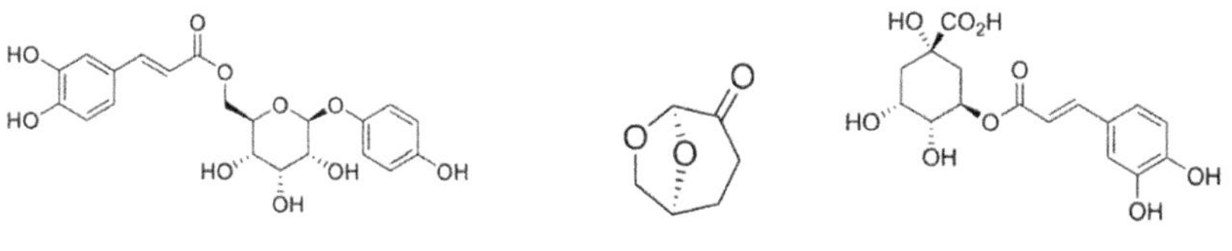

6   **7**   **8**

Fig. 2 Structures of TYR inhibitors 6'-O-caffeoylarbutin **6**, cyrene **7** and chlorogenic acid **8**.

biosensors, for the production of L-DOPA, an anti-Parkinson's disease medication, for obtaining structurally complex phenols or for dye removal and other biocatalytic processes involving phenol/diphenol oxidation reactions [89–93]. The *Streptomyces* sp. enzymes were the most investigated bacterial TYRs [88,89], but recently the TYR from other species, such as *Pseudomonas* sp. EG22 [90], *Verrucomicrobium spinosum* [91] or *Bacillus subtilis* [94] have also been cloned and characterized in detail. The possibility to consider bacterial TYR as an antibacterial drug target has also been recently recognized [94], but very few drug design such studies are available to date.

8. Conclusions

Being involved in several human diseases, including hypopigmentation/depigmentation conditions (vitiligo, idiopathic guttate hypomelanosis, pityriasis versicolor, pityriasis alba) as well as hyperpigmentations (melasma, lentigines, post-inflammatory and periorbital hyperpigmentation, cervical idiopathic poikiloderma and acanthosis nigricans) TYR is being intensely investigated as a drug target, in the search of agents useful for the management of these diverse and in several cases, difficult to treat conditions. Although creams incorporating simple TYR inhibitors such as hydroquinone, azelaic acid and tretinoin (all-*trans*-retinoic acid) are widely used for the management of some of these hyperpigmentation diseases, few novel compounds arrived to clinical trials, with the exception of thiamidol, a recently discovered and approved novel agents for the treatment of melasma. Furthermore, there are increasing evidences that TYR plays a crucial role in the formation and progression of melanoma, a difficult to treat skin cancer. Thus, a multitude of drug design studies were reported for the design of TYR inhibitors for the management of melanoma, with many novel chemotypes being reported and showing interesting action in the in vitro inhibition of melanoma growth in various cell lines, but not in animal models, for the moment. Kojic acid and vitamin C, well known natural product TYR inhibitors, are also used for avoiding vegetable/food oxidative browning due to the tyrosinase-catalyzed reactions. Bacterial TYRs are widespread in many species and started to be investigated in detail only in the last decades. They show potential in biotechnological applications, for the production of mixed melanins, for protein cross-linking reactions, for producing biosensors for the analysis of various phenols, or for the production of the Parkinson's disease drug L-DOPA. The intense research in the field may soon bring novel relevant discoveries which may be translated into innovative drugs for the management of skin disorders as those mentioned here or for different biotechnological applications.

References

[1] S.A. D'Mello, G.J. Finlay, B.C. Baguley, M.E. Askarian-Amiri, Signaling pathways in melanogenesis, Int. J. Mol. Sci. 17 (7) (2016) 1144, https://doi.org/10.3390/ijms17071144 PMID: 27428965; PMCID: PMC4964517.

[2] I.F. Videira, D.F. Moura, S. Magina, Mechanisms regulating melanogenesis, An. Bras. Dermatol. 88 (1) (2013) 76–83, https://doi.org/10.1590/s0365-05962013000100009 PMID: 23539007; PMCID: PMC3699939.

[3] Z. Rzepka, E. Buszman, A. Beberok, D. Wrześniok, From tyrosine to melanin: signaling pathways and factors regulating melanogenesis, Postepy Hig. Med. Dosw. (Online) 70 (0)) (2016) 695–708, https://doi.org/10.5604/17322693.1208033 PMID: 27356601.

[4] S. Ito, K. Wakamatsu, Quantitative analysis of eumelanin and pheomelanin in humans, mice, and other animals: a comparative review, Pigment. Cell Res. 16 (5) (2003) 523–531, https://doi.org/10.1034/j.1600-0749.2003.00072.x PMID: 12950732.

[5] A. Hennessy, C. Oh, B. Diffey, K. Wakamatsu, S. Ito, J. Rees, Eumelanin and pheomelanin concentrations in human epidermis before and after UVB irradiation, Pigment. Cell Res. 18 (3) (2005) 220–223, https://doi.org/10.1111/j.1600-0749.2005.00233.x PMID: 15892719.

[6] M. Snyman, R.E. Walsdorf, S.N. Wix, J.G. Gill, The metabolism of melanin synthesis—from melanocytes to melanoma, Pigment. Cell Melanoma Res. (2024), https://doi.org/10.1111/pcmr.13165 Epub ahead of print. PMID: 38445351.

[7] S. Zolghadri, A. Bahrami, M.T. Hassan Khan, J. Munoz-Munoz, F. Garcia-Molina, F. Garcia-Canovas, et al., A comprehensive review on tyrosinase inhibitors, J. Enzyme Inhib. Med. Chem. 34 (1) (2019) 279–309, https://doi.org/10.1080/14756366.2018.1545767 PMID: 30734608; PMCID: PMC6327992.

[8] C.A. Ramsden, P.A. Riley, Tyrosinase: the four oxidation states of the active site and their relevance to enzymatic activation, oxidation and inactivation, Bioorg Med. Chem. 22 (8) (2014) 2388–2395, https://doi.org/10.1016/j.bmc.2014.02.048 Epub 2014 Mar 4. PMID: 24656803.

[9] T. Pillaiyar, M. Manickam, S.H. Jung, Downregulation of melanogenesis: drug discovery and therapeutic options, Drug. Discov. Today 22 (2) (2017) 282–298, https://doi.org/10.1016/j.drudis.2016.09.016 Epub 2016 Sep 28. PMID: 27693716.

[10] A.M. Thawabteh, A. Jibreen, D. Karaman, A. Thawabteh, R. Karaman, Skin pigmentation types, causes and treatment—a review, Molecules 28 (12) (2023) 4839, https://doi.org/10.3390/molecules28124839 PMID: 37375394; PMCID: PMC10304091.

[11] A. Nautiyal, S. Wairkar, Management of hyperpigmentation: current treatments and emerging therapies, Pigment. Cell Melanoma Res. 34 (6) (2021) 1000–1014, https://doi.org/10.1111/pcmr.12986 Epub 2021 Jun 3. PMID: 33998768.

[12] I. Mollet, K. Ongenae, J.M. Naeyaert, Origin, clinical presentation, and diagnosis of hypomelanotic skin disorders, Dermatol. Clin. 25 (3) (2007) 363–371, https://doi.org/10.1016/j.det.2007.04.013 PMID: 17662902.

[13] M.S. Ceresnie, S. Gonzalez, I.H. Hamzavi, Diagnosing disorders of hypopigmentation and depigmentation in patients with skin of color, Dermatol. Clin. 41 (3) (2023) 407–416, https://doi.org/10.1016/j.det.2023.02.006 Epub 2023 Mar 27. PMID: 37236710.

[14] Y. Wang, S. Li, C. Li, Clinical features, immunopathogenesis, and therapeutic strategies in vitiligo, Clin. Rev. Allergy Immunol. 61 (3) (2021) 299–323, https://doi.org/10.1007/s12016-021-08868-z Epub 2021 Jul 20. Erratum in: PMID: 34283349.

[15] G. Iannella, A. Greco, D. Didona, B. Didona, G. Granata, A. Manno, et al., Vitiligo: pathogenesis, clinical variants and treatment approaches, Autoimmun. Rev. 15 (4) (2016) 335–343, https://doi.org/10.1016/j.autrev.2015.12.006 Epub 2015 Dec 23. PMID: 26724277.

[16] M.S. Al Abadie, C. Chaiyabutr, K.X. Patel, D.J. Gawkrodger, Vitiligo and psychological stress: a hypothesis integrating the neuroendocrine and immune systems in melanocyte destruction, Int. J. Dermatol. (2024), https://doi.org/10.1111/ijd.17148 Epub ahead of print. PMID: 38570937.

[17] F. Diotallevi, H. Gioacchini, E. De Simoni, A. Marani, M. Candelora, M. Paolinelli, et al., Vitiligo, from pathogenesis to therapeutic advances: State of the art, Int. J. Mol. Sci. 24 (5) (2023) 4910, https://doi.org/10.3390/ijms24054910 PMID: 36902341; PMCID: PMC10003418.

[18] F. Brown, J.S. Crane, Idiopathic guttate hypomelanosis, StatPearls [Internet], StatPearls Publishing, Treasure Island (FL), 2024 PMID: 29489254.

[19] J. Buch, A. Patil, G. Kroumpouzos, M. Kassir, H. Galadari, M.H. Gold, et al., Idiopathic guttate hypomelanosis: presentation and management, J. Cosmet. Laser Ther. 23 (1-2) (2021) 8–15, https://doi.org/10.1080/14764172.2021.1957116 Epub 2021 Jul 25. PMID: 34304679.

[20] A.K. Gupta, R. Batra, R. Bluhm, T. Boekhout, T.L. Dawson Jr., Skin diseases associated with Malassezia species, J. Am. Acad. Dermatol. 51 (5) (2004) 785–798, https://doi.org/10.1016/j.jaad.2003.12.034 PMID: 15523360.

[21] V. Crespo-Erchiga, V.D. Florencio, Malassezia yeasts and pityriasis versicolor, Curr. Opin. Infect. Dis. 19 (2) (2006) 139–147, https://doi.org/10.1097/01.qco.0000216624. 21069.61 PMID: 16514338.

[22] D.N. Givler, H.M. Saleh, A. Givler, Pityriasis alba, StatPearls [Internet], StatPearls Publishing, Treasure Island, FL, 2024 PMID: 28613715.

[23] M. Rao, K. Young, L. Jackson-Cowan, A. Kourosh, N. Theodosakis, Post-inflammatory hypopigmentation: Review of the etiology, clinical manifestations, and treatment options, J. Clin. Med. 12 (3) (2023) 1243, https://doi.org/10.3390/jcm12031243 PMID: 36769891; PMCID: PMC9917556.

[24] T.F. Cestari, L.P. Dantas, J.C. Boza, Acquired hyperpigmentations, An. Bras. Dermatol. 89 (1) (2014) 11–25, https://doi.org/10.1590/abd1806-4841.20142353 PMID: 24626644; PMCID: PMC3938350.

[25] S. Nouveau, D. Agrawal, M. Kohli, F. Bernerd, N. Misra, C.S. Nayak, Skin hyperpigmentation in Indian population: Insights and best practice, Indian J. Dermatol. 61 (5) (2016) 487–495, https://doi.org/10.4103/0019-5154.190103 PMID: 27688436; PMCID: PMC5029232.

[26] T. Passeron, M. Picardo, Melasma, a photoaging disorder, Pigment. Cell Melanoma Res. 31 (4) (2018) 461–465, https://doi.org/10.1111/pcmr.12684 Epub 2018 Jan 12. PMID: 29285880.

[27] S. Rajanala, M.B.C. Maymone, N.A. Vashi, Melasma pathogenesis: a review of the latest research, pathological findings, and investigational therapies, Dermatol. Online J. 25 (10) (2019) 13030/qt47b7r28c. PMID: 31735001.

[28] J. McKesey, A. Tovar-Garza, A.G. Pandya, Melasma treatment: an evidence-based review, Am. J. Clin. Dermatol. 21 (2) (2020) 173–225, https://doi.org/10.1007/s40257-019-00488-w PMID: 31802394.

[29] N. Neagu, C. Conforti, M. Agozzino, G.F. Marangi, S.H. Morariu, G. Pellacani, et al., Melasma treatment: a systematic review, J. Dermatol. Treat. 33 (4) (2022) 1816–1837, https://doi.org/10.1080/09546634.2021.1914313 Epub 2022 Mar 23. PMID: 33849384.

[30] D. Roggenkamp, A. Sammain, M. Fürstenau, M. Kausch, T. Passeron, L. Kolbe, Thiamidol® in moderate-to-severe melasma: 24-week, randomized, double-blind, vehicle-controlled clinical study with subsequent regression phase, J. Dermatol. 48 (12) (2021) 1871–1876, https://doi.org/10.1111/1346-8138.16080 Epub 2021 Oct 21. PMID: 34676600.

[31] A. Sevilla, Solar lentigines: sun-damaged skin on the hand, Indian J. Dermatol. Venereol. Leprol. 89 (5) (2023) 787, https://doi.org/10.25259/IJDVL_413_2023 PMID: 37436017.

[32] I. Mukovozov, J. Roesler, N. Kashetsky, A. Gregory, Treatment of lentigines: a systematic review, Dermatol. Surg. 49 (1) (2023) 17–24, https://doi.org/10.1097/DSS.0000000000003630 PMID: 36533790.

[33] J.P. Ortonne, A.G. Pandya, H. Lui, D. Hexsel, Treatment of solar lentigines, J. Am. Acad. Dermatol. 54 (5 Suppl 2) (2006) S262–S271, https://doi.org/10.1016/j.jaad.2005.12.043 PMID: 16631967.

[34] N. Elbuluk, P. Grimes, A. Chien, I. Hamzavi, A. Alexis, S. Taylor, et al., The pathogenesis and management of acne-induced post-inflammatory hyperpigmentation, Am. J. Clin. Dermatol. 22 (6) (2021) 829–836, https://doi.org/10.1007/s40257-021-00633-4 Epub 2021 Sep 1. PMID: 34468934.

[35] A. Shenoy, R. Madan, Post-inflammatory hyperpigmentation: a review of treatment strategies, J. Drugs Dermatol. 19 (8) (2020) 763–768, https://doi.org/10.36849/JDD.2020.4887 PMID: 32845587.

[36] B. Sofen, G. Prado, J. Emer, Melasma and post inflammatory hyperpigmentation: management update and expert opinion, Skin. Ther. Lett. 21 (1) (2016) 1–7 PMID: 27224897.

[37] L. Michelle, D. Pouldar Foulad, C. Ekelem, N. Saedi, N.A. Mesinkovska, Treatments of periorbital hyperpigmentation: a systematic review, Dermatol. Surg. 47 (1) (2021) 70–74, https://doi.org/10.1097/DSS.0000000000002484 PMID: 32740208.

[38] A.C. Katoulis, N.G. Stavrianeas, S. Georgala, E. Bozi, D. Kalogeromitros, E. Koumantaki, et al., Poikiloderma of Civatte: a clinical and epidemiological study, J. Eur. Acad. Dermatol. Venereol. 19 (4) (2005) 444–448, https://doi.org/10.1111/j.1468-3083.2005.01213.x PMID: 15987290.

[39] A.K.C. Leung, J.M. Lam, B. Barankin, K.F. Leong, K.L. Hon, Acanthosis Nigricans: an updated review, Curr. Pediatr. Rev. 19 (1) (2022) 68–82, https://doi.org/10.2174/1573396318666220429085231 PMID: 36698243.

[40] A. Das, D. Datta, M. Kassir, U. Wollina, H. Galadari, T. Lotti, et al., Acanthosis nigricans: a review, J. Cosmet. Dermatol. 19 (8) (2020) 1857–1865, https://doi.org/10.1111/jocd.13544 PMID: 32516476.

[41] S. Kapoor, Diagnosis and treatment of Acanthosis nigricans, Skinmed 8 (3) (2010) 161–164 quiz 165. PMID: 21137622.

[42] S. Honda, T. Matsuda, M. Fujimuro, Y. Sekine, Tyrosinase regulates the motility of human melanoma cell line A375 through its hydroxylase enzymatic activity, Biochem. Biophys. Res. Commun. 707 (2024) 149785.

[43] D. Gilot, M. Migault, L. Bachelot, F. Journé, A. Rogiers, E. Donnou-Fournet, et al., A non-coding function of TYRP1 mRNA promotes melanoma growth, Nat. Cell Biol. 19 (11) (2017) 1348–1357.

[44] H. Kamo, R. Kawahara, S. Simizu, Tyrosinase suppresses vasculogenic mimicry in human melanoma cells, Oncol. Lett. 23 (5) (2022) 169.

[45] T. Kobayashi, G. Imokawa, D.C. Bennett, V.J. Hearing, Tyrosinase stabilization by Tyrp1 (the brown locus protein), J. Biol. Chem. 273 (48) (1998) 31801–31805.

[46] M. Granada, E. Mendes, M.J. Perry, M.J. Penetra, M.M. Gaspar, J.O. Pinho, et al., Sulfur analogues of tyrosine in the development of triazene hybrid compounds: a new strategy against melanoma, ACS Med. Chem. Lett. 12 (11) (2021) 1669–1677.

[47] H.A.S. El-Nashar, M.I.G. El-Din, L. Hritcu, O.A. Eldahshan, Insights on the inhibitory power of flavonoids on tyrosinase activity: a survey from 2016 to 2021, Molecules 26 (24) (2021) 7546.

[48] E. Jung, I. Shim, J. An, M.S. Ji, P. Jangili, S.G. Chi, et al., Phenylthiourea-conjugated BODIPY as an efficient photosensitizer for tyrosinase-positive melanoma-targeted photodynamic therapy, ACS Appl. Bio Mater. 4 (3) (2021) 2120–2127.

[49] R. Romagnoli, P. Oliva, F. Prencipe, S. Manfredini, M.P. Germanò, L. De Luca, et al., Cinnamic acid derivatives linked to arylpiperazines as novel potent inhibitors of tyrosinase activity and melanin synthesis, Eur. J. Med. Chem. 231 (2022) 114147.

[50] Y. He, T.L. Suyama, H. Kim, E. Glukhov, W.H. Gerwick, Discovery of novel tyrosinase inhibitors from marine cyanobacteria, Front. Microbiol. 13 (2022) 912621.

[51] T. Hwang, H.J. Lee, W.S. Park, D.M. Kang, M.J. Ahn, H. Yoon, et al., Catechin-7-O-α-L-rhamnopyranoside can reduce α-MSH-induced melanogenesis in B16F10 melanoma cells through competitive inhibition of tyrosinase, Int. J. Med. Sci. 19 (7) (2022) 1131–1137.

[52] Y. Tamura, A. Ito, K. Wakamatsu, T. Torigoe, H. Honda, S. Ito, et al., A sulfur containing melanogenesis substrate, N-Pr-4-S-CAP as a potential source for selective chemoimmunotherapy of malignant melanoma, Int. J. Mol. Sci. 24 (6) (2023) 5235.

[53] S. Li, G. Zhang, Y. Peng, P. Chen, J. Li, X. Wang, et al., Tyrosinase-activated nanocomposites for double-modals imaging guided photodynamic and photothermal synergistic therapy, Adv. Healthc. Mater. 12 (23) (2023) e2300327.

[54] S.T. Boateng, T. Roy, K. Torrey, U. Owunna, S. Banang-Mbeumi, D. Basnet, et al., Synthesis, in silico modelling, and in vitro biological evaluation of substituted pyrazole derivatives as potential anti-skin cancer, anti-tyrosinase, and antioxidant agents, J. Enzyme Inhib. Med. Chem. 38 (1) (2023) 2205042.

[55] R. Logesh, S.R. Prasad, S. Chipurupalli, N. Robinson, S.K. Mohankumar, Natural tyrosinase enzyme inhibitors: a path from melanin to melanoma and its reported pharmacological activities, Biochim. Biophys. Acta Rev. Cancer 1878 (6) (2023) 188968.

[56] N.R. Han, H.J. Park, S.G. Ko, P.D. Moon, Maltol has anti-cancer effects via modulating PD-L1 signaling pathway in B16F10 cells, Front. Pharmacol. 14 (2023) 1255586.

[57] M. Xu, Z. Zhang, P. Zhang, Q. Wang, Y. Xia, C. Lian, et al., Beyond traditional methods: unveiling the skin whitening properties of rhein-embedded PROTACs, Bioorg Med. Chem. 96 (2023) 117537.

[58] L.M. Lazinski, M. Beaumet, B. Roulier, R. Gay, G. Royal, M. Maresca, et al., Design and synthesis of 4-amino-2',4'-dihydroxyindanone derivatives as potent inhibitors of tyrosinase and melanin biosynthesis in human melanoma cells, Eur. J. Med. Chem. 266 (2024) 116165.

[59] S. Jilani, J.D. Saco, E. Mugarza, A. Pujol-Morcillo, J. Chokry, C. Ng, et al., CAR-T cell therapy targeting surface expression of TYRP1 to treat cutaneous and rare melanoma subtypes, Nat. Commun. 15 (1) (2024) 1244.

[60] K.M. Khan, Z.S. Saify, M.T. Khan, N. Butt, G.M. Maharvi, S. Perveen, et al., Tyrosinase inhibition: conformational analysis based studies on molecular dynamics calculations of bipiperidine based inhibitors, J. Enzyme Inhib. Med. Chem. 20 (4) (2005) 401–407.

[61] N. Lolak, M. Boga, M. Tuneg, G. Karakoc, S. Akocak, C.T. Supuran, Sulphonamides incorporating 1,3,5-triazine structural motifs show antioxidant, acetylcholinesterase, butyrylcholinesterase, and tyrosinase inhibitory profile, J. Enzyme Inhib. Med. Chem. 35 (1) (2020) 424–431.

[62] M. Oguz, E. Kalay, S. Akocak, A. Nocentini, N. Lolak, M. Boga, et al., Synthesis of calix[4]azacrown substituted sulphonamides with antioxidant, acetylcholinesterase, butyrylcholinesterase, tyrosinase and carbonic anhydrase inhibitory action, J. Enzyme Inhib. Med. Chem. 35 (1) (2020) 1215–1223.

[63] K. Jakimiuk, S. Sari, R. Milewski, C.T. Supuran, D. Şöhretoğlu, M. Tomczyk, Flavonoids as tyrosinase inhibitors in in silico and in vitro models: basic framework of SAR using a statistical modelling approach, J. Enzyme Inhib. Med. Chem. 37 (1) (2022) 421–430.

[64] T.C. Albershardt, A.J. Parsons, R.S. Reeves, P.A. Flynn, D.J. Campbell, J. Ter Meulen, et al., Therapeutic efficacy of PD1/PDL1 blockade in B16 melanoma is greatly enhanced by immunization with dendritic cell-targeting lentiviral vector and protein vaccine, Vaccine 38 (17) (2020) 3369–3377.

[65] A.S.H. Chan, T.O. Kangas, X. Qiu, M.T. Uhlik, R.B. Fulton, N.R. Ottoson, et al., Imprime PGG enhances anti-tumor effects of tumor-targeting, anti-angiogenic, and immune checkpoint inhibitor antibodies, Front. Oncol. 12 (2022) 869078.

[66] S. Carradori, F. Melfi, J. Resetar, R. Simsek, Tyrosinase enzyme and its inhibitors: an update of the litearture, in: C.T. Supuran, W.A. Donald (Eds.), Metalloenzymes—From Bench to Bedside, Elsevier – Academic Press, London, UK, 2024, pp. 533–546.

[67] H. Wojtasek, Oxidation of flavonoids by tyrosinase and by o-quinones-comment on "Flavonoids as tyrosinase inhibitors in in silico and in vitro models: basic framework of SAR using a statistical modelling approach" published by K. Jakimiuk, S. Sari, R. Milewski, C.T. Supuran, D. Şöhretoğlu, and M. Tomczyk (J Enzyme Inhib Med Chem 2022;37:427-436), J. Enzyme Inhib. Med. Chem. 38 (1) (2023) 2269611.

[68] M. Tomczyk, Reply letter to Dr. Wojtasek regarding: oxidation of flavonoids by tyrosinase and by o-quinones-comment on "Flavonoids as tyrosinase inhibitors in in silico and in vitro models: basic framework of SAR using a statistical modelling approach" published by K. Jakimiuk, S. Sari, R. Milewski, C.T. Supuran, D. Söhretoglu, and M. Tomczyk (J Enzyme Inhib Med Chem 2022;37:427-436) (all co-authors), J. Enzyme Inhib. Med. Chem. 38 (1) (2023) 2269613.

[69] Q. Chen, L. Zheng, X. Deng, M. Zhang, W. Han, Z. Huang, et al., A fluorescence biosensor for tyrosinase activity analysis based on silicon-doped carbon quantum dots, Chem. Pharm. Bull. (Tokyo) 71 (11) (2023) 812–818.

[70] M.X. Li, K.W. Kang, M. Huang, R. Cheng, W. Wang, J. Gao, et al., Simple and rapid detection of tyrosinase activity with the adjustable light scattering properties of CoOOH nanoflakes, Anal. Bioanal. Chem. 415 (18) (2023) 4569–4578.

[71] Q. Dai, Z. Qi, Z. Yan, B. Yu, J. Li, B. Ge, et al., A blue/NIR ratiometric fluorescent probe for intracellular detection of Tyrosinase and the inhibitor screening, Talanta 254 (2023) 124175.

[72] Y.F. Fan, S.X. Zhu, F.B. Hou, D.F. Zhao, Q.S. Pan, Y.W. Xiang, et al., Spectrophotometric assays for sensing tyrosinase activity and their applications, Biosens (Basel) 11 (8) (2021) 290.

[73] W. Jin, S.J. Stehbens, R.T. Barnard, M.A.T. Blaskovich, Z.M. Ziora, Dysregulation of tyrosinase activity: a potential link between skin disorders and neurodegeneration, J. Pharm. Pharmacol. 76 (1) (2024) 13–22.

[74] J.C. Zilles, F.L. Dos Santos, I.C. Kulkamp-Guerreiro, R.V. Contri, Biological activities and safety data of kojic acid and its derivatives: a review, Exp. Dermatol. 31 (10) (2022) 1500–1521.

[75] T. Mann, W. Gerwat, J. Batzer, K. Eggers, C. Scherner, H. Wenck, et al., Inhibition of human tyrosinase requires molecular motifs distinctively different from mushroom tyrosinase, J. Invest. Dermatol. 138 (7) (2018) 1601–1608.

[76] C. Arrowitz, A.M. Schoelermann, T. Mann, L.I. Jiang, T. Weber, L. Kolbe, Effective tyrosinase inhibition by thiamidol results in significant improvement of mild to moderate melasma, J. Invest. Dermatol. 139 (8) (2019) 1691–1698.

[77] W.G. Philipp-Dormston, A. Vila Echagüe, S.H. Pérez Damonte, J. Riedel, A. Filbry, K. Warnke, et al., Thiamidol containing treatment regimens in facial hyperpigmentation: an international multi-centre approach consisting of a double-blind, controlled, split-face study and of an open-label, real-world study, Int. J. Cosmet. Sci. 42 (4) (2020) 377–387.

[78] P.B. Lima, J.A.F. Dias, D.P. Cassiano, A.C.C. Esposito, L.D.B. Miot, E. Bagatin, et al., Efficacy and safety of topical isobutylamido thiazolyl resorcinol (Thiamidol) vs. 4% hydroquinone cream for facial melasma: an evaluator-blinded, randomized controlled trial, J. Eur. Acad. Dermatol. Venereol. 35 (9) (2021) 1881–1887.

[79] D. Xie, W. Fu, T. Yuan, K. Han, Y. Lv, Q. Wang, et al., 6'-O-Caffeoylarbutin from Quezui tea: a highly effective and safe tyrosinase inhibitor, Int. J. Mol. Sci. 25 (2) (2024) 972.

[80] J. Cytarska, J. Szulc, D. Kołodziej-Sobczak, J.A. Nunes, E.F. da Silva-Júnior, K.Z. Łączkowski, Cyrene™ as a tyrosinase inhibitor and anti-browning agent, Food Chem. 442 (2024) 138430.

[81] N.A. Hassanuddin, E. Normaya, H. Ismail, A. Iqbal, M.B.M. Piah, S. Abd Hamid, et al., Methyl 4-pyridyl ketone thiosemicarbazone (4-PT) as an effective and safe inhibitor of mushroom tyrosinase and antibrowning agent, Int. J. Biol. Macromol. 255 (2024) 128229.

[82] J. Chen, Z. Zhang, H. Li, H. Tang, Exploring the effect of a series of flavonoids on tyrosinase using integrated enzyme kinetics, multispectroscopic, and molecular modelling analyses, Int. J. Biol. Macromol. 252 (2023) 126451.

[83] Z. Peng, G. Wang, J.J. Wang, Y. Zhao, Anti-browning and antibacterial dual functions of novel hydroxypyranone-thiosemicarbazone derivatives as shrimp preservative agents: Synthesis, bio-evaluation, mechanism, and application, Food Chem. 419 (2023) 136106.

[84] Z. Peng, G. Wang, Y. He, J.J. Wang, Y. Zhao, Tyrosinase inhibitory mechanism and anti-browning properties of novel kojic acid derivatives bearing aromatic aldehyde moiety, Curr. Res. Food Sci. 6 (2022) 100421.

[85] G. Wang, M. He, Y. Huang, Z. Peng, Synthesis and biological evaluation of new kojic acid-1,3,4-oxadiazole hybrids as tyrosinase inhibitors and their application in the anti-browning of fresh-cut mushrooms, Food Chem. 409 (2023) 135275.

[86] W. You, C. Wang, J. Zhang, X. Ru, F. Xu, Z. Wu, et al., Exogenous chlorogenic acid inhibits quality deterioration in fresh-cut potato slices, Food Chem. 446 (2024) 138866.

[87] M. He, J. Zhang, N. Li, L. Chen, Y. He, Z. Peng, et al., Synthesis, anti-browning effect and mechanism research of kojic acid-coumarin derivatives as anti-tyrosinase inhibitors, Food Chem. X 21 (2024) 101128.

[88] H. Claus, H. Decker, Bacterial tyrosinases, Syst. Appl. Microbiol. 29 (1) (2006) 3–14.

[89] M. Fairhead, L. Thöny-Meyer, Bacterial tyrosinases: old enzymes with new relevance to biotechnology, N. Biotechnol. 29 (2) (2012) 183–191.

[90] S.M.A. El-Aziz, A.H.I. Faraag, A.M. Ibrahim, A. Albrakati, M.R. Bakkar, Tyrosinase enzyme purification and immobilization from Pseudomonas sp. EG22 using cellulose coated magnetic nanoparticles: characterization and application in melanin production, World J. Microbiol. Biotechnol. 40 (1) (2023) 10.

[91] M. Fekry, K.K. Dave, D. Badgujar, E. Hamnevik, O. Aurelius, D. Dobritzsch, et al., The crystal structure of tyrosinase from Verrucomicrobium spinosum reveals it to be an atypical bacterial tyrosinase, Biomolecules 13 (9) (2023) 1360.

[92] O.F. Restaino, P. Manini, T. Kordjazi, M.L. Alfieri, M. Rippa, L. Mariniello, et al., Biotechnological production and characterization of extracellular melanin by Streptomyces nashvillensis, Microorganisms 12 (2) (2024) 297.

[93] M. Agunbiade, M. Le Roes-Hill, Application of bacterial tyrosinases in organic synthesis, World J. Microbiol. Biotechnol. 38 (1) (2021) 2.

[94] A.F. Zahoor, F. Hafeez, A. Mansha, S. Kamal, M.N. Anjum, Z. Raza, et al., Bacterial tyrosinase inhibition, hemolytic and thrombolytic screening, and in silico modeling of rationally designed tosyl piperazine-engrafted dithiocarbamate derivatives, Biomedicines 11 (10) (2023) 2739.

Printed and bound by CPI Group (UK) Ltd, Croydon, CR0 4YY

08/05/2025

01864966-0002